The Chemical Industry at the Millennium

The Chemical Industry at the Millennium

Maturity, Restructuring, and Globalization

Peter H. Spitz

Editor

Foreword by
William S. Stavropoulos

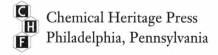

Chemical Heritage Press
Philadelphia, Pennsylvania

Printed in the United States of America.

For information about CHF publications write
Chemical Heritage Foundation
315 Chestnut Street
Philadelphia, PA 19106-2702, USA
Fax (215) 925-1954
www. chemheritage.org

Library of Congress Cataloging-in-Publication Data
The chemical industry at the millennium : maturity, restructuring, and globalization /
edited by Peter Spitz.
    p. cm.
    ISBN 0-941901-34-3 (hardcover : alk. paper)
1. Chemical industry. I. Spitz, Peter H. II. Title.
    HD9650.5.C524 2003
    338.4'766—dc21
                                                                      2003010798

Book interior designed by Patricia Wieland
Book jacket designed by Willie•Fetchko Graphic Design
Printed by Hamilton Printing Company

∞ The paper used in this publication meets the minimum requirements of the American National Standard for Information Sciences—Permanence of Paper for Printed Library Materials, ANSI Z39.48-1984.

Contents

Foreword

As chairman of Dow Chemical, I am gratified to write the foreword to Peter H. Spitz's informative and provocative new book. It is also an honor to follow in the footsteps of a distinguished former chairman of Dow, Robert W. Lundeen, who wrote the foreword to Spitz's previous book, *Petrochemicals: The Rise of an Industry*.

Peter Spitz brings impressive credentials to his work. A recognized expert on the history of the industry, he has served as a consultant for many chemical companies and governments around the world. In this volume he has also enlisted the help of numerous experts to write chapters on critical subjects related to the industry, including restructuring, specialty and performance chemicals, strategic planning, information technology, Responsible Care, quality and reengineering, and globalism.

In his earlier book Spitz outlined the remarkable rise of the chemical industry, focusing on the period from 1930 to 1980. In the new volume he highlights the industry's changes and strategic adjustments in the often-tumultuous last two decades of the twentieth century.

Consolidation and Cyclicality

He reviews the relentless consolidation and cyclicality that have characterized the recent era. Spitz notes that the industry's boom periods are becoming shorter in duration, while the "busts" have become longer. He identifies four other key developments:

- First, product demand in emerging and newly industrialized countries increased substantially, while it declined to gross domestic product levels in the industrialized nations;
- Second, countries with low-cost feedstocks, particularly in the Middle East, have economic advantages in overall production costs;
- Third, plants in new markets received tariff protection that made their products less expensive than imports; and
- Fourth, traditional customers in industrialized countries moved portions of their manufacturing capacity to nations with new chemical capacity.

This last development accelerated globalization in the industry. It drove the move toward global supply chains, including the trend for companies such as Dow to establish production facilities around the world, creating global organizations that are more efficient and that take full advantage of the talents of people of many nationalities.

The major developments of the past twenty years reflect Spitz's key insight: "What was once largely a regional industry with some global participation by multinationals has now become a truly global industry."

Spitz does not hide his occasional exasperation with the industry he loves. He says it has reached what he calls "a somewhat ambiguous maturity." He is referring to the industry's habit of reacting to periods of prosperity by building excess capacity. It is an industry sometimes too quick to rely on nostalgia about yesterday's robust demand—and too slow to grasp today's and tomorow's marketplace realities.

Buffeted by Change

In the past generation our industry has been buffeted by change. We have endured two major oil shocks, an unexpected natural gas price spike, and numerous recessions. We have seen major shifts in markets. We have encountered an almost ceaseless drumbeat of criticism from environmentalists—much of it well-founded—accompanied by a heavy degree of regulation by governments. We have also watched many old and distinguished companies basically disappear as independent entities, consumed by relentless consolidation. It was just a few short decades ago that Union Carbide, a company acquired by Dow just recently, was larger than Dow and widely considered to be the gold standard for the industry. Furthermore, we have seen the industry's capacity for innovation slow. Finally, we have watched the market value of our companies decline, relative to faster-growing businesses.

Given the tendency of many chemical companies to overbuild in the face of uncertainty, Spitz asks an old rhetorical question: "Will the industry ever learn?" The unsettling answer seems to be: perhaps—or perhaps not.

The central questions facing businesspeople in the chemical industry are these: What can we do to ensure that our industry not only survives in a new century but also flourishes? How can we serve our customers better, grow faster, and be more attractive to investors?

There will be a variety of answers to those questions. I firmly believe that our industry's success depends on dynamic change in four areas:

- *Culture:* Every employee of our companies must have his or her interests aligned with investors' interests. This means compensation should be directly tied to the total shareholder return. It also means a relentless effort to raise the talent and training of our employees. The last true source of competitive advantage in our industry is the quality of the workforce. Raw materials are relatively inexpensive. Technology can be rapidly duplicated or leapfrogged. And capital is available to enterprises around the globe.

- *Productivity:* Companies must continuously drive to lower costs for the simple reason that customers will always demand high-quality products at the lowest price. And because of customer consolidation, globalization, and technological advancement, they can and will get these things.
- *Portfolio redeployment:* Gone are the days when companies could diversify into areas beyond their expertise. Successful companies of the future must concentrate on the things they do best and must establish leading positions in their businesses in order to establish and maintain competitive advantage.
- *Growth:* It is axiomatic that no industry can save its way to growth. Our industry is no exception. Although in the short to medium term acquisitions can generate growth, in the long term research and development is the only sustainable way this industry can grow. And I am convinced that chemistry still has a rich potential for growth stemming from new discoveries.

No Substitute for Growth

As an industry we have to use our knowledge capital more effectively if we are to recapture higher growth rates. More vigorous innovation will enable us to create greater value for our customers and investors. Our job is not just to make products. Our job is to provide science-based solutions.

Of course, the answer is not simply to throw more money at R&D but, rather, to increase research productivity. That means, among other things, collaborating with academic and government researchers and adapting technologies from other industries, such as the combinatorial chemistry used so successfully in pharmaceuticals.

There are huge new areas of opportunity in chemical innovation—in biotechnology, nanotechnology, and synthetic materials. In these areas R&D can truly bring about "better living through chemistry."

Although we should never dismiss the value of incremental improvements, especially process advances and product upgrades, there is no substitute for breakthrough products. Truly innovative products generate nearly four out of every ten revenue dollars—and more than six out of every ten profit dollars.

We must rely more heavily on R&D if we are to recapture higher growth rates and create greater value for our customers and investors. By applying a financial meat-ax to R&D, we can save a few dollars—but only at the expense of mortgaging our future.

Strategic improvements, a reenergized workforce, and an emphasis on R&D are all vitally important to our industry, but alone they are not enough. If we do not as an industry change public perceptions of the chemistry business, we risk strangulation by excessive regulations and public distrust.

An Enabling Industry

There is an urgent need for the chemistry business to gain credibility—with all stakeholders. In large part credibility depends on our helping establish a real two-way dialogue with stakeholders, including not only our friends but also our critics. It means we must sharpen our listening skills. It also means we must report, along with our financial progress, our environmental and social results, good and not-so-good.

There is an urgent need for us to build upon—and communicate—the industry's considerable accomplishments in protecting the environment, health, and safety through the Responsible Care program. We must also lead the way in testing the compounds we produce, in doing fundamental research about the effects of chemicals on the environment and human health, and in communicating the findings to the public. In this effort, third-party verification of the findings is a *sine qua non*.

Another vital step involves cultivating allies through partnering—not only with such natural allies as customers and community residents but also, as I said, with our critics, including environmentalists.

Finally, in gaining credibility, our industry has to reject our traditional reticence, our tendency to simulate a punching bag. We have to move away from the position where we allow others to define the chemistry business—often in negative terms. Instead, we need to define ourselves as an industry that is dynamic, responsible, and essential.

We are an industry whose products are essential to the new age of information technology, and so we should project ourselves as the catalyst for the economy of tomorrow. To get that message across, we need to harness the power of information technology as effectively as we have harnessed the power of chemistry.

If our industry takes farsighted steps, it conceivably could sustain itself into the twenty-second century. As Peter Spitz points out, the important concept of sustainable development recognizes our obligation as businesspeople to leave the world a better place than we found it.

Sustainable Development

As we have learned, sustainable development does not focus on a single bottom line. Instead, it recognizes a triple bottom line: economic progress, environmental protection, and social responsibility. Sustainability is a mindset, a leadership philosophy, and probably the only way to do business successfully in our era. In fact, companies that are weaving economic performance; environmental, health, and safety concerns; and social responsibility are increasingly the norm.

Social responsibility goes beyond philanthropic contributions. It also involves the way we treat employees by setting examples of respect and ethical behavior; providing training, education, and opportunities for advancement; putting safety first; and creating a truly diverse workplace. The latter point involves having employees who reflect—in race, gender, nationalities, background, perspective, and experience—the diverse world in which we do business.

It is becoming clear to many of us that in the twenty-first century companies can be successful only if they achieve success in all of the bottom lines—not just in one. Sustainable development in a global context contains no hidden agenda. It reflects only our commitment to good jobs, better standards of living, cleaner environments, safer working places, and healthier communities. Companies and even countries cannot sustain themselves as islands of prosperity in a sea of economic misery, environmental degradation, and social decline. Our success is intertwined with the success of our customers, our employees, and our communities.

In my case I take great pride in the nearly four decades I have spent in this vital industry. The challenges we face, as Peter Spitz points out, are daunting. But those very challenges make this an exciting time to be in the chemistry business—one whose opportunities far exceed its perils. That is because, as in the past, we have so much to contribute to economic development and a better quality of life throughout the world.

William S. Stavropoulos
Chairman, Dow Chemical Company

Acknowledgments

While my first book, *Petrochemicals: The Rise of an Industry*, largely chronicled a period starting in the mid-1930s, this book is mostly about much more recent history, the problems of the chemical industry in the 1980s and 1990s as it entered a mature state, similar to other industries that no longer exhibit rapid growth. When writing about events that took place only a decade or two ago, authors inevitably leave themselves open to the fact that many readers, and particularly coworkers in the industry, will be very familiar with the subject matter. Unquestionably, some of them will have different and sometimes more accurate recollections, they will disagree with some of the facts and with some of the conclusions that have been drawn.

To help in this respect, all the chapter drafts were read by friends either still working in or recently retired from the industry or with recent knowledge gained in relevant occupations. Their input has been extremely valuable and has made this book immeasurably better. I would especially like to thank Fernand Kaufmann, who made many helpful suggestions on the chapters dealing with industry restructuring and strategy. Many others read at least one chapter and made a number of useful comments. Roger Longley and John Philpot, both retired from Chem Systems' London office and Darrell Aubrey from the Houston office reviewed chapters and provided important material. Chem Systems, now part of Nexant, Inc., provided a number of historical exhibits, including graphs and tables, as did Chemical Market Associates, Inc. (CMAI). Some of these were used for background; some are included as figures or tables. Bruce Burke, Andrew Swanson, and Bob Bauman of Chem Systems collaborated on petrochemical material. Ron Cascone of Chem Systems reviewed the environmental chapter and, along with Tony Holloway of Nexant/Chem Systems' London office, provided the material on European and Japanese attitudes and actions in this area. Kevin Swift and others at the American Chemistry Council also provided much information on environmental regulations and provided various statistics on the U.S. chemical industry. Individuals who reviewed chapters or were generally helpful in making information available include John Aalbregtse, Gary Adams, Arnold Allemang, Gene Allspach, Dave d'Antoni, Keith Belden, Sally Boyer, Dave Buzzelli, Ernie Deavenport, Duane Dickson, Richard Doyle, Liz Eck, Michael Eckstut, Michel Hartfeld, Karen Heller, Roger Hirl, Alan Hirsig, William Joyce, Stephen Kemp, David Kepler, Yaarit Silverstone, Kenneth Smith, Frederick Webber, Edward Wolynic, Pedro Wongtschowski, Terry Yosie, and Jeffrey Zekauskas. In Japan, Ryota Hamamoto and Hiroyoshi Kamata provided data and views. Andrew Stratos at Chem Systems helped with some of the statistical information, while Denise Phillips in Nexant/ Chem Systems' library was extremely helpful in providing references and re- searching articles and databases.

Since many of the figures and tables came from Chem Systems' presenta- tions, I want to express my appreciation and thanks to Jeffrey Plotkin for his help in assembling and transforming them into a format that could be used by the publisher and to Donald Bari at Chem Systems for getting permission from Nexant to publish this material.

Much of the research for the book was done at the Chemical Heritage Foun- dation in Philadelphia, where I work part-time as a senior research fellow. I therefore appreciate my appointment to this position from Arnold Thackray, who heads the foundation and who also agreed to have CHF publish the book. Several of the librarians there were extremely helpful with my research, in- cluding Elsa Atson, associate director of the Othmer Library; Kristen Graff, document delivery coordinator; and Laura Wukovitz and Christopher Stan- wood, assistant librarians for access services. Shelley Geehr, at Chemical Heri- tage Press, was excellent to work with and greatly improved the book through her efforts and dedication. Patricia Wieland, who edited the book, though a severe taskmaster, was a pleasure to work with and a great professional.

I particularly want to thank my wife, Hilda, for her patience in letting me spend so much time on writing, which obviously takes time away from family pursuits.

Introduction

Peter H. Spitz

The global chemical industry is huge, one of the largest industries in the world. In 2000, in the United States alone shipments amounted to $460 billion, up from $309 billion in 1990—a 4 percent compounded annual growth. Capital expenditures in 2000 were $31.2 billion, almost the same as research and development (R&D) expenses at $31.1 billion. The industry contributes greatly to the nation's exports, with $79.6 billion in chemicals shipped abroad in 2000; however, imports totaled $73.6 billion.[1] As discussed in Chapter 8, the U.S. net trade in chemicals kept rising for many years until 1995, when it reached a positive balance of about $20 billion. Within six years, by 2001, the chemical trade balance had actually turned negative, with imports exceeding exports. The reasons for this are important to understand, since they are a manifestation of the problems that have beset this industry.

Despite the fact that the chemical industry is as large as it is and still growing, investors appear to have less and less interest in it. In 1968 the Standard and Poor 500 Index (S&P 500) category called "basic materials," which includes chemicals, made up 15 percent of the market capitalization for all of the companies in the S&P 500. Basic materials was then the largest segment in terms of combined market capitalization, followed by energy (14 percent), capital goods (14 percent), consumer cyclicals (13 percent), and communication services (10 percent). In 1998 basic materials was down to 3 percent and energy to 6 percent, while technology was leading at 19 percent, consumer staples at 15 percent, and finance at 14 percent. In March 2001, chemicals (today about 47 percent of basic materials) was down to one percent of the S&P 500 market cap! Something very serious has been going on, since the investing public and institutional investors increasingly prefer to put their money into industries that promise greater earnings growth, which usually translates to higher company valuations and correspondingly higher market caps.

What has happened is that much of the chemical industry has matured, like the steel, aluminum, and cement industries; so it is becoming another "smokestack" industry in many people's eyes. Nevertheless, it is a fascinating industry that keeps improving our standard of living and provides many of the essentials that our society needs.

Chemicals cover a broad range of products, and it is important to recognize the industry's different sectors. The American Chemistry Council (ACC) divides the chemical industry into four categories: basic chemicals, specialties, life sciences, and consumer products. This book is largely about basic chemicals and specialties. Much of the book focuses on petrochemicals (part of basic chemicals), the products made from hydrocarbon feedstocks, which are based on natural gas and refinery feedstocks. This focus has been adopted because many of the current woes of the industry stem from the fact that the industrial countries no longer have a competitive advantage in the manufacture of many of these products. Given this slant, it can be argued that the book is not really about the entire chemical industry. Readers should understand, however, that we are focusing on the industry segments that are troubled and whose future is unclear. The ACC's *Guide to the Business of Chemistry* is quite instructive. It shows that growth prospects in all but the life sciences category (which covers pharmaceuticals and agricultural chemicals) are fairly poor. It also indicates that all categories except life sciences had low returns on capital over the last ten years, which shows that we are dealing with a mature industry.[2]

Maturity is, of course, a relative condition. The firm of Arthur D. Little has often been associated with the "life cycle" concept, where the life of a product or industry is shaped like a rising S-curve, divided into four periods: embryonic, or developing; rapidly growing; mature; and declining. Much of the chemical industry clearly fits into the third category. A mature industry suffers from lack of innovation, as evidenced by the longtime conclusion of industry observers and participants that linear low-density polyethylene (LLDPE), commercialized in the early 1980s and now having an installed U.S. capacity of about ten billion pounds, is probably the last new multibillion-pound polymer—and it was invented more than twenty years ago. Similarly, polyester and nylon, still the two synthetic fibers used most for textiles and carpeting, were commercialized forty or fifty years ago. While it is conceivable that a polymer based on starch or another biodegradable material might become as big a seller as polyester or nylon, most industry people would say that the probability of that is slight. It would mean that the polymer industry would have jumped to another S-curve for renewed growth, a concept that was described in Richard Foster's book published in 1986.[3] But basic materials industries can seldom jump to another S-curve for renewed innovation, even though the steel industry's mini-mills, which eliminated the use of blast and open-hearth furnaces, might fall into this category. In any case the concept of jumping to a new S-curve holds more for individual processes and products, which cannot transform an entire industry. Some relatively new polymers, such as polyethylene bottle resins and several specialty fibers, have come along more recently, but these are not breakthrough products that will transform an industry. The rate at which new processes and products are developed and commercialized is therefore another indication of where the industry finds itself in its life cycle and at what degree of maturity.

In my first book, *Petrochemicals: The Rise of an Industry*, I described the origins and explosive growth of this relatively young industry, which was able to produce a vast array of synthetic materials from hydrocarbons, a much more abundant feedstock than the coal gases and coal tars used earlier. In the late 1980s when the book was completed, the entire chemical industry was beginning to enter its mature phase in the industrial countries, although petrochemicals were just becoming a growth industry in the developing world, presaging a major diversification and shift of suppliers and customers. The period between 1980 and 2000 then became a turbulent time for the global chemical industry, as traditional companies sought to find their proper role in the new environment. Only a few decades earlier new synthetic polymers coming on the market were considered the exciting materials of the future, as memorialized in the 1960s film *The Graduate*, in which Dustin Hoffman is advised to go into plastics. Now most plastics have become commodity materials sold at low prices with little or no profit by global producers locked in intractable competition. Synthetic rubber, still the basic material for automobile tires, is a product shunned by most larger petrochemical firms owing to its commodity status with low or nonexistent profits and low growth prospects. Synthetic fibers, such as nylon and polyester, are no longer of great interest to the multinational firms, which have sold or will probably be selling these businesses either to companies based in developing countries or to financial buyers. More and more of the textile industry is moving to Asia and Latin America, where it can operate with much lower labor costs.

Companies operating in North America, Western Europe, and Japan in the 1980s and 1990s have been coping with a series of extremely difficult problems. Slowing markets for their traditional products, correspondingly lower business growth rates, and heavy competition in many of their product lines have resulted in slowdowns in earnings growth. Companies have thus been subjected to strong pressure from institutional investors, which have seen other sectors of the market, principally technology-oriented companies, exhibit stronger growth and more excitement. The managements of chemical firms became very much aware that their most important mission was to increase shareholder value. How to do that has been the principal question occupying the minds of chief executive officers for the last two decades.

In their attempts to create value, the managers of chemical companies have taken many different paths, none of which has yet proved generically successful. While some of these companies had strong top executives who provided vision and strategic direction, they also availed themselves of the best consulting talent money could buy to look at different growth and earnings possibilities for their firms, to reshape their portfolios, or to completely reinvent their companies. But many of the industries' problems have seemed intractable, and the usefulness of some of the consulting advice they paid for is in question. Loss of business from customers or entire customer industries moving abroad was one issue. Pervasive cyclicality of commodity businesses, in both petrochemicals and inorganics (e.g., chloralkali and titanium dioxide), was another. The U.S. Gulf Coast area no longer provided its original comparative advantage from a feedstock cost-and-availability standpoint, as new producers from the Middle East and elsewhere took over large chunks of the global petrochemical business. Finally, many producers could not resist playing the market-share game, which has had disastrous consequences owing to furious price cutting and periodic overbuilding to maintain market share. This has resulted in years of relatively low operating rates and little or no profit.

As a managing director for a firm that specialized in consulting for the global chemical industry, I have been privileged to work closely with a number of businesses facing these problems. Because the last twenty years of this maturing industry's history was an interesting time, I decided to begin a second book to chronicle it. I asked several friends and coworkers, all experts in their field, to write parts of the book. Their contributions are stand-alone chapters that cover the period in question from their particular vantage points.

This leads to the organization and purpose of this book. A number of important themes are covered, such as technology, strategy, globalization, information technology, and environmental issues. The chapters develop the individual stories or themes chronologically. They usually discuss some of the same issues (e.g., cyclical industry behavior, the highly competitive nature of the industry, and the evolving maturity of processes and products) but from the different authors' perspectives. My aim was to provide a book that could be read either from cover to cover or in pieces, according to the reader's interests.

My first chapter on early industry restructuring is followed by a chapter on technology by Jeffrey Plotkin, who heads up a successful multiclient program in this area for Chem Systems and who for a time wrote monthly articles for *Innovation* magazine, published by the American Chemical Society. Plotkin laments the fact that innovation in the chemical industry has slowed substantially from earlier decades, partly because the efficiency of producing many commodity chemicals is now very high, which gives less incentive to develop new technology. Moreover, there has been much less process research and more emphasis on application research and process improvement. Recent initiatives are aimed at developing better catalysts for existing plants and trying to use cheaper but less reactive (that is, more difficult to convert) feedstocks.

The third chapter, on specialty chemicals, was written by Andrew Boccone, who for many years was the president of Kline and Company, considered by many to be the premier consulting firm in this area. Boccone gives a great deal of anecdotal evidence about the slowdown in growth rates for this industry and the reasons why participating firms are consolidating.

Michael Eckstut, who headed up strategy consulting in chemicals at Booz Allen and Company and later at A. T. Kearney, cooperated with me on the chapter on strategy. Beginning in the 1970s, companies along with several well-known consultants strove mightily to find tools and methodologies that executives could use to analyze their businesses to determine which to keep and expand and which to sell or shut down. Chemical firms were among the largest users of strategy consulting firms. If it were possible to graph the financial success of firms relative to the amount of strategy consulting they used, there would be some surprises at the probable lack of such a correlation. However, the executives of these firms, which often had too many businesses in their portfolios, clearly wanted outside help to figure out their options when business operations became increasingly complex, while shedding much of their internal planning staff to cut overhead expenses.

As business results and prospects declined, companies felt they had no choice but to improve their businesses, a process probably long overdue. In the 1980s many thought that Japan was about to take over the commercial world because of the efficiency of its companies' operations and the high quality of their products. This fact had much to do with the quality-control programs that were subsequently installed by most firms in most industries, including the chemical industry. But quality-control programs were mainly about continuous improvement, and so a number of strategy practitioners, such as Michael Hammer (see Chapter 5), came up with "reengineering" as a discipline for breaking down the so-called silos, or "fortresses," represented by various corporate functions and installing more efficient business processes that would get the product to the customer more rapidly and at a lower cost.

With business processes changing and cheap computing power becoming available, companies turned to information technology to replace many tasks that had been carried out previously by clerks and engineers using spreadsheets and other now-antiquated methods of planning, inventory control, and invoicing. Such consulting firms as Accenture and IBM helped companies install the new computers and software, which became increasingly complex, with so-called enterprise resource planning (ERP) systems becoming the vogue in the 1990s. These systems provided a series of software programs to monitor and control the entire transaction, from ordering the product to receipt of cash for the sale. David Crow, the head of the global chemicals division at Accenture, wrote the chapter on this subject.

Perhaps the most significant issues the chemical industry faced in the closing decades of the twentieth century were the many regulations placed on the industry and the various effects of environmental issues. The regulations were partly a consequence of previous "sins" and partly of greater public and corporate awareness of air, water, and land pollution problems as well as worker and public exposure to certain toxic chemicals. Industry leaders, who were at first somewhat recalcitrant, became convinced of the need to take appropriate actions and later became dedicated stewards for ensuring compliance. Such industry associations as the American Chemical Council developed broad programs, including Responsible Care, to deal with the many issues involved. The chapter describing how the industry dealt with environmental issues is one of the longest in the book and yet hardly scratches the surface of an immensely important subject. These issues continue to give the chemical industry one of the lowest ratings in the public mind, despite the fact that everybody agrees that chemicals are essential to our way of life.

During the last twenty years the chemical industry became a truly global industry. The multinational companies already operating in more than one country recognized that their customers were moving many of their operations around the world, particularly into developing countries with lower labor costs, and therefore began to stress globalization even more. They adopted strategies to serve customers in every location around the world. This philosophy went well beyond normal exports, as companies recognized the need to create global supply chains to serve customers from several plants located at strategic sites. How this evolved is covered in Chapter 8.

I have already mentioned the growing influence of the financial community on public companies as investors and strategic planners forced managements to trim and reshape portfolios and to regain earnings momentum. John Roberts, formerly the chief chemical equity analyst at Merrill Lynch, discusses this in Chapter 9, which also gives background on how valuations were carried out and how the managers of companies were given incentives through stock options to align their personal interests with that of the firm they were managing.

The concluding chapter describes how the industry eventually reshaped itself and how a number of companies tried, some more successfully than others, to reinvent themselves into firms that would thrive in the new environment. Some ideas on how technology might yet allow a rebirth of the industry and how companies should strive to adapt themselves to the new environment are also included.

In many respects this book features the interplay between companies, regions, and governments and their agencies, while emphasizing the relatively recent creation of a global industry that has not been particularly kind to most of its participants. Sectors of an industry can be quite profitable when competitive advantage is established through proprietary technology, low-cost raw materials, or high entry barriers or when favorable supply-and-demand conditions exist for a time. But in the last twenty years there has been a steady erosion of the earlier, more benign industry climate in the industrial countries. This is one of the key themes of the book.

In a book published in 1986, Joseph Bower, a professor at the Harvard Business School, describes the plight of the petrochemical industry in the early 1980s as a consequence of the economic recession that followed the second oil shock.[4] This period was arguably the worst in the history of the young industry in terms of operating losses, particularly in Europe and Japan but also in North America. Among Bower's important themes were national interest, state-owned enterprise, company advantage, competition, and government policy. Some current problems with the petrochemical industry date back even to the 1980s, notably irrational competition in businesses in which there are many more global players now than there were two decades ago, in spite of extensive regional rationalization. One favorable development since the 1980s, however, has been the withdrawal of governments from ownership of chemical enterprises. The last two decades have seen extensive privatization of chemical businesses in Western Europe, Latin America, and Asia. Most governments in the industrial countries have left the chemical industry to suffer through its own economic problems, except for periodic governmental help on trade issues. But governments have also strongly increased regulatory activities, which have caused companies to incur vast expenditures for various types of compliance.

The landscape of the traditional chemical industry in the United States, Western Europe, and Japan has changed fairly rapidly. Some of the old-line companies, such as Allied Chemical, Union Carbide, American Cyanamid, Rhône-Poulenc, and Hoechst, have disappeared as a result of acquisitions or mergers. Other companies, such as Monsanto and Imperial Chemical Industries (ICI), have completely transformed into specialty companies. Consolidation in the petroleum industry has combined the petrochemical activities of such firms as Exxon and Mobil, Phillips and Chevron, British Petroleum (BP) and Amoco, as well as those of Total, Atochem, and Petrofina. The chemical divisions of other oil companies, such as Texaco and Arco, were sold or taken public. In Japan, where rationalization is just beginning, the chemical companies under the Mitsubishi name and those under the Mitsui name were combined some time ago. Now Sumitomo Chemical is thinking about merging with the Mitsui chemical group, and other mergers are highly likely for this industry, which is still less competitive than its global peers. From a global standpoint it is becoming clearer every year that the only long-term survivors in commodity petrochemicals will be companies that either have substantial back-integration into attractively priced hydrocarbon feedstocks or are large enough to withstand the economic hardships companies face at the bottom of the so-called petrochemical cycle, whose up-and-down nature seems slated to continue into the future.

In the United States and in Western Europe the structure of the chemical industry has profoundly changed and will continue to change. While consolidation from mergers and acquisitions has somewhat reduced the number of players, other changes are more significant. The era of the large multinational companies concentrating mostly on basic chemicals is almost at an end. Arguably, only Dow and BASF remain in this category, and even these firms consider such specialties as crop sciences and performance chemicals to be the most important areas for future growth and value creation. The industry recognizes that the manufacture of basic chemical commodities is a low-profit business that does not attract investors, although it does provide good cash flow from high depreciation charges and earnings. The relatively high cash flow is probably the reason so many competitors continue in these businesses and why merchant banks, investment funds, and highly leveraged private firms like Huntsman and Ineos have bought a number of these businesses. The most logical owners of many of the commodity petrochemical businesses, however, are the oil companies' chemical divisions and such firms in oil-rich nations as Saudi Basic Industries, or SABIC, since they will be low-cost producers at the end of the day and will have the required staying power. This has been evident for some time to the many multinationals, which started to withdraw from these businesses. After suffering through the dismal financial results of the early 1980s, a number of companies that made petrochemicals (e.g., DuPont, Monsanto, Allied, PPG Industries, and Hercules) started to divest these businesses. With each successive down cycle more chemical firms left the petrochemical industry, significant events perhaps being the divestments of Celanese and Rhodia by, respectively, Hoechst and Rhône Poulenc when they formed Aventis, a life sciences company, and the sale of Bayer's half of the Erdölchemie joint venture to BP Chemicals. Other firms have divested their petrochemical activities into joint ventures, such as Equistar in the United States, which combined the petrochemical activities of Millennium Chemicals, Oxy Chemical, and Lyondell. Even BASF and Shell Chemical, two certain survivors in petrochemicals, decided to place their polyolefin activities into a joint venture, Basell, rather than keep their parts as a core business. These joint ventures are currently struggling because polyolefins are historically among the hardest hit of the cyclical businesses.

The industry's evolution is not yet over, and some important questions must be resolved. Can highly leveraged firms, particularly in commodities, survive and continue to compete against their better-positioned and better-capitalized peers? Will there be more mergers between smaller specialty chemical firms to achieve sufficient size and growth prospects to maintain investor interest? Can the Japanese chemical industry sufficiently improve its operations through mergers and cost cutting to become a viable competitor in the burgeoning Asian market? Will the new European Union Chemicals Policy Review (called the "White Paper") require massive new toxicity testing of chemicals at huge additional cost to the industry?

As longtime participants in the chemical industry, the authors of this book remain interested and dedicated spectators as the chemical industry keeps evolving.

Endnotes

1. *Guide to the Business of Chemistry* (Arlington, Va.: American Chemistry Council, 2001).
2. Ibid.
3. Richard N. Foster, *Innovation: The Attacker's Advantage* (New York: Summit Books, 1986).
4. Joseph L. Bower, *When Markets Quake* (Boston: Harvard Business School Press, 1986).

Chapter 1

Restructuring

The First Wave

Peter H. Spitz

U nderstanding how the chemical industry has been transformed over the last twenty years—roughly the period covered in this book—is not an easy task. Why focus on these two decades? This period became in a sense a "rite of passage" for the chemical industry, a time when it was no longer considered a growth industry but had in many observers' opinion entered maturity. In the 1970s the world had been subjected to an unprecedented upheaval: the forceful actions of the once-docile Organization of Petroleum Exporting Countries (OPEC) cartel, which imposed as a result of political chaos in the Middle East fourfold and eventually tenfold price hikes on crude oil and a short-term embargo on exports. These events were the so-called oil shocks of 1973 and 1978–79. Long gasoline lines at the pumps, surges in heating oil prices, and frantic moves by utilities and industries to switch from oil to natural gas or coal fuels were but some of the broader effects. For chemical companies it was an extraordinary time, as they attempted to cope with limited oil supplies and wildly swinging prices. All was not bad, of course, since the value of preshock inventories surged and product pricing soared, producing short-term windfall profits for petrochemical firms. The resulting inflation, caused largely by high energy prices, was soon followed by a severe recession that tested the mettle of the management of firms that had only recently seen unprecedented profits. These events set in motion, or at least greatly accelerated, a trend for petrochemical companies to either leave this sector (that is, through sale, merger, or shutdown) or try to strengthen their position in the industry. They would then be better able to cope with uncertainty, the cyclical behavior of the industry, and the emergence of new producers in the developing world, now much surer of their future after having seen OPEC use its oil weapon to obtain the multinational oil companies' tacit agreement to help them "go down-stream" into refining and petrochemicals production.

The emergence of new competitors, coupled with a slowdown in demand growth and increasing maturity in manufacturing technology, set in motion forces that caused what is arguably the most dramatic realignment of assets seen in any industry to date. The unprecedented industry downturn and ac-companying uncertainties following the second oil shock (1978–79) were the main factors that led to the first wave of chemical industry restructuring, as commodity chemical producers suffered through recession and overcapacity in the first part of the 1980s. In the United States the so-called Reagan recession of the early 1980s followed the highly inflationary effects of the oil shock, pro-ducing a double-digit jobless rate. Poor demand for durables and consumer goods hit the chemical industry hard, resulting in what was the worst period the industry had seen for many decades. The new era of high prices for crude oil and feedstock unsettled many of the companies, even though prices moder-ated substantially and then peaked again as the decade progressed. More im-portant, the cyclical characteristics of the petrochemical industry became a primary factor in convincing a number of firms with conservative manage-ments that the industry was too unpredictable. Some company leaders simply felt that it was high time to diversify portfolios heavily weighted toward com-modities. How a number of companies in the United States, Western Europe, and Japan coped with this challenge will be explored in this chapter, which examines the industry through the end of the 1980s.

In many respects the chemical industry's restructuring initiatives were simi-lar to but more dramatic than those in many other commodity-oriented indus-tries, including steel, automobiles, and appliances. Globalization, slower demand growth, growing competition, and proliferation of technology were similarly affecting these other industries, driving companies to reduce costs and over-heads, to combine with or acquire competitors, and to look for new growth models.

Chemical industry restructuring took two forms. The first was an attempt at consolidation of specific industry segments (for example, ethylene, polyethyl-ene, polyvinyl chloride [PVC], and ethylene glycol), which would reduce the number of players and shut down uncompetitive plants, both steps intended to improve the "quality of the industry" and to increase operating rates. The sec-ond form involved more broadly scoped change, where some long-term par-ticipants in the industry would decide to get rid of some or most of their petrochemical operations and look for more attractive businesses.

The 1980s were therefore a wrenching time for the global petrochemical industry. The business-as-usual approach was becoming a thing of the past. Feedstock costs were now going to be unpredictable, and supplies would be subject to disruptions. New plants were being built in some regions where natural gas was previously flared (that is, burned rather than used as a fuel or feedstock) and, when now recovered as feedstock, was therefore extremely low priced, creating new, serious competition. With oil for a while at $3 or even $12 per barrel, recovering ethane to make ethylene from natural gas in remote locations had previously made little economic sense. However, with oil prices at $20 or more per barrel the picture changed dramatically. The second oil shock brought into the world of chemicals not only the Middle East but also western Canada, whose inexpensive gas had up to then been unable to compete as a feedstock in a low-cost oil environment, owing to the remote location and the need to build world-scale plants in the small Canadian market. Demand growth for petrochemicals had slowed markedly from the double-digit growth rates of the 1950s and 1960s, largely because substitution of natural with synthetic materials—previously the main engine for growth—had more or less run its course. In fact, the high oil prices were raising concern that economics might no longer favor plastics and synthetic fibers over the traditional materials—for example, wood, paper, glass, natural rubber, and aluminum—that these petrochemicals were replacing. There was also a growing problem with finding sites for new plants, as the environmental movement gained popular support. Federal and state regulatory agencies had recently become more heavily involved in the siting of plants and in plant expansion permits as a result of the Clean Air Act and accompanying amendments (see Chapter 7) and other legislation intended to lower levels of pollution caused by plant emissions.

The feedstock issue was probably of greatest concern. The U.S. oil-producing industry had traditionally been protected from "low-cost" foreign oil supplies through the long-standing Mandatory Oil Import Program. Without this program lower-cost foreign oil would have flooded into the United States, and crude-based feedstocks would have been even cheaper. Now the situation had turned on its head, and foreign oil was setting the global price of crude oil so that domestic petrochemical companies were no longer able to predict their future raw materials costs, which in the case of petrochemicals represents a large part of total cash operating costs. Competition from the newly built plants in the Middle East that were based on cheap flare gas was beginning to target traditional U.S. and European export markets with ethylene derivatives, threatening existing producers with higher feedstock costs. As Saudi Arabian plants came on stream, however, Saudi Basic Industries (SABIC), the government-owned petrochemical monopoly, became a responsible marketer, establishing joint marketing programs with Exxon Chemical and other partners for products made at its Al Jubail complex in eastern Saudi Arabia.

Western Canada was also establishing a highly competitive petrochemical industry, adding to a growing supply crunch. Further, as the 1980s began, the industry realized that it would henceforth probably be susceptible to endemic cyclicality, continuing the trend established in the 1970s. Industry participants in the United States, Europe, and Japan were just beginning to understand the negative consequences of overbuilding with excess cash generated during "boom" periods, but they could not seem to resist the temptation, even though the heady times would inexorably be followed by a period of low operating rates, resulting in low or nonexistent profits. In the late 1960s little new capacity had been built owing to a recessionary environment. This turned the first oil shock into a profitable period for producers, as customers, concerned about even higher prices, went on a buying spree, forcing operating rates into the 90 percent to 100 percent range, with consequent high pricing. Subsequent capacity additions in the mid-1970s were higher than prudent, but the second oil shock nevertheless created huge profits, which always occurs when there are shortages and prices fly up to abnormal levels. Between January and May 1979 benzene prices rose from $420 to $680 per ton, paraxylene from $500 to $900 per ton, and styrene from $590 to as high as $1,050 per ton.[1] Inventories of these products had been produced from cheap naphtha, which until January had cost $200 per ton but had risen to as high as $360 per ton by May. Inventories based on much cheaper crude were now sold at amazing prices. As a result of overinvestment decisions in 1978–79, much more capacity came on stream in 1980 and 1981, which produced disastrous results as an economic recession hit the world. The 1980s began to hear the mantra of "will the industry finally have learned its lessons?" This became an even louder chorus a few years later after another price fly-up period (that is, a steep price increase) in 1988 was followed by another dramatic downturn. The presumed "lesson" was that companies should refrain from overinvesting during periods of high cash flow, since this would inevitably lead to overcapacity and resulting poor margins when the new plants came on stream. Writing now, early in the twenty-first century, it is hard to say whether the lesson has been learned, though there has been no price fly-up since 1995 and therefore no period of "lumpy" reinvestment. (A lumpy period is said to occur when large increases in supply are made by several producers in the same timeframe.)

During the 1980s the major players had to make wrenching decisions regarding their future. Staying in the petrochemical industry meant believing that the production costs of key plants would be low enough to match or beat those of competitors, including those of international competitors on a "landed" basis (i.e., including freight and duty added to the cost for material produced in regions with low feedstock cost and shipped to traditional markets). The term *competitive advantage* would be heard more and more, to the increasing disadvantage of U.S. commodity players. Another problem for companies in this now-cyclical industry was the fact that severely fluctuating earnings would make it difficult to maintain, to say nothing of even increasing, annual dividend payments. It was therefore a time that forced management to make a fundamental review of its firm's strengths and weaknesses and to determine what alternatives the firm might have if it was decided that petrochemicals were not going to be an important part of its future. But implementing such a decision did not necessarily happen quickly, as firms found various types of barriers to exiting.

For example, some uneconomical plants were part of a much larger complex so that shuttering these plants would hurt the economics of some of the other plants, and in Europe many plants were owned by government entities, making shutdowns more difficult because of unemployment issues. When another price fly-up came in 1988, a number of the firms did wonder briefly whether petrochemicals were such a bad business after all. But in the 1990s it became clear that restructuring of the petrochemical industry was essential and that only certain types of companies could hope to succeed in this field, which would continue to be subject to major cycles of boom and bust proportions.

The companies most likely to stay were the major oil companies, such as Exxon, Shell, and British Petroleum (BP), which had greater access to cheaper feedstocks and in many respects had more reasons to stay in this field, linked as it was to their refinery operations, compared with chemical companies that were not back-integrated into hydrocarbons and also had more options to diversify. The oil companies in any case had deeper pockets and published papers telling the other players that this was a business to get out of if they could not stand the heat.

Companies that did decide to stay in the industry tended to cite steady longer-term demand growth as their dominant theme for reinvestment. When BP Chemicals decided to build its large cracker extension at Grangemouth, Scotland, in the 1980s, it included in its public relations material a chart that showed that ethylene demand went up steadily over time and that it was therefore a safe bet to build more capacity. An up-to-date version of that concept developed by Chem Systems—a chemically oriented consulting firm specializing in technology, markets, and business analysis—had shown this confidence in demand growth to be correct.[2] Many industries in fact would be pleased to experience a smooth growth in demand such as this. While adding capacity in large, "lumpy" amounts by a combination of firms is not desirable, steady long-term growth, as discussed above, does support companies deciding to reinvest and also supports those who in the 1980s entered the industry for investment purposes at low cost, such as Gordon Cain and Jon Huntsman, as will be discussed later. In any case, while some chemical companies left the field in the 1980s and early 1990s, a number of others—for example, Dow Chemical, Solvay, BASF, and Union Carbide—were not intimidated by the major oil companies' advantages owing to their back-integration into hydrocarbon feedstocks and their size. Petrochemical restructuring during the 1980s was, however, extensive and continued through the next decade and beyond.

Bad Times for the Industry

As discussed, chemical companies were exceptionally hard hit by the downturn in the early 1980s. Now the companies were facing a devastating combination of recession-induced low demand along with, for a time, still quite high feedstock pricing. Natural gas and gas liquids pricing had risen sharply in tandem

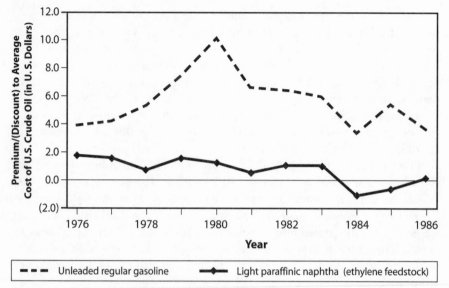

Figure 1. Comparative pricing trends: gasoline versus petrochemical feedstocks. *Source:* Nexant, Inc./Chem Systems, by permission.

with crude oil, as energy and feedstock costs equilibrated. Ethane prices had doubled from 14 to 28 cents per gallon and propane from 20 to 45 cents per gallon.

Recognizing the uncertainties occasioned by developments in the 1970s, some chemical firms were taking steps to strengthen their position in feedstocks, which they hoped would make them more independent relative to their hydrocarbon suppliers. This approach took three forms: back-integration into hydrocarbons, so-called chemical refineries, and development of technologies to produce feedstocks from sources other than crude oil. Two notable examples of back-integration were DuPont's acquisition of Continental Oil Company (Conoco) and Dow's construction of a "feedstock refinery" in Freeport, Texas. Both companies were initially lauded for these steps. Ultimately they did not really help either company in their petrochemical operations. Dow never operated the refinery, since the world oil situation had again normalized by the time it came on stream in the early 1980s. The project was eventually written off, as Dow learned to its dismay that a refinery that does not produce gasoline will probably not be profitable. Figure 1 shows how the production of gasoline adds much more value to crude oil than converting crude to petrochemical feedstock naphtha.

DuPont pursued a different strategy, on the one hand acquiring an oil company and on the other withdrawing from some of its petrochemical operations. As time went on, the second initiative became dominant, as the firm recognized that it was unable to obtain real synergy from its investment in Conoco and that its future lay in the manufacture of more highly valued, differentiated chemicals rather than competing with such firms as Shell and Exxon. More than a decade later DuPont decided to sell Conoco to the public.

Chemical refineries were a favorite topic for a short period. The idea was to build a refinery that would produce feedstocks ideally suited for a cracker that was close-coupled to crude oil distillation, which would feed everything from ethane to heavy gas oil, and a catalytic reformer to produce benzene-toluene-xylene (BTX) aromatics. Such a complex would produce little or no gasoline, as was the case for the Dow refinery. The economics of chemical refineries turned out to be not particularly attractive, given the large offsite investment required for crude oil docks, receiving and product shipping, and extensive loading and unloading facilities, piping, and storage; the poor price realization (netback) for heavy fuel oil, which was not suitable as cracker feed; and again the fact that gasoline is usually the most lucrative refinery product.

Figure 2 tracks the calculated profitability of conventional refinery operation, a theoretical petrochemical refinery, and naphtha cracking for ethylene from 1976 to the mid-1980s. Over a period like the second oil shock (1978–79), the economics of a petrochemical refinery looked good on paper, with petrochemical prices in a fly-up mode. But when the petrochemical industry went into decline, the economics for a petrochemical refinery also tanked. Only one such refinery was built, by Petrosar in Corunna, Ontario. Its originally conceived economics were fairly good, since it was to include a plant for producing needle coke, an expensive material used for constructing graphite electrodes for aluminum plants. Unfortunately for the venture, Union Carbide

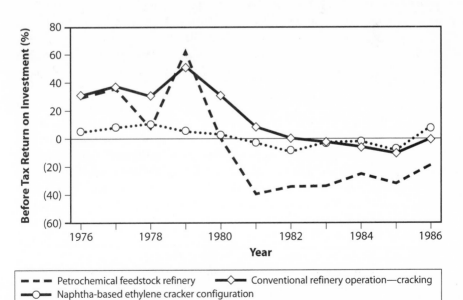

Figure 2. Comparative refining profitability: gasoline versus petrochemical feedstocks. *Source:* Nexant, Inc./Chem Systems, by permission.

Canada, which was a 16.3 percent partner in Petrosar (other partners were Polysar, DuPont Canada, and Koch Canada Fuels) and had initially spearheaded the project, decided it did not need more needle raw material for its global graphite electrode business and canceled plans for the Canadian plant. The chemical refinery went ahead anyway but seldom made any money. It is now operated by Nova Chemicals, the last remaining sizable Canadian petrochemical company.

Several companies decided to step up their research efforts to develop processes for making hydrocarbons, using synthesis gas produced through gasification of natural gas or coal. It was already clear that methanol could be transformed into hydrocarbon liquids, as evidenced by Mobil's natural gas–to–gasoline project in New Zealand. Other technologies could produce fuels in the diesel range. Coal gasification yielded "petrochemicals," although at high cost, when all the offsites, such as coal storage and pollution control, were added in. These technologies would be competitive if crude oil prices were to rise and stay much higher, which many observers in 1979 were predicting would happen. Among the most active companies carrying out research on synthesis gas to give petrochemicals were Shell, Hoechst, Union Carbide, and Mobil. The most promising technologies involved methanol cracking to ethylene and a number of possible routes to other traditional petrochemicals, such as ethylene glycol, vinyl acetate, and styrene.[3] However, economics developed during that period showed that crude oil would have to be in the $75-per-barrel range or higher for olefins and other petrochemicals to be produced competitively from coal-derived synthesis gas. BASF's Mathias Seefelder stated that ethylene could be produced economically from methanol if the latter were available at $150 per ton, about two-thirds of the then-current price. Union Carbide was working on a process to make ethylene glycol from carbon monoxide and hydrogen using a rhodium carbonyl cluster. The company's chairman, William Sneath, had earlier implied that Carbide would probably use this process in the firm's next glycol plant.[4] Sasol, in South Africa, had for a long time been producing fuels and chemicals via the Fischer-Tropsch synthesis, using cheap domestic coal, although the economics of this operation had depended on government tax relief. A series of economics cases created by Chem Systems indicated that ethylene from natural gas–derived synthesis gas using the Fischer-Tropsch synthesis was probably the most likely route if crude oil prices kept escalating.[5] Oil and chemical firms were in any case carrying out extensive research in these areas while hoping that "rational" oil pricing would soon resume. And in the mid-1980s crude oil prices did again drop below $20 per barrel, causing most of these firms to put their research into alternate feedstock sources on the back burner. (In 2000, with crude oil prices briefly rising again into the $30- to $35-per-barrel range, many of these ideas were revived, only to be shelved again in 2001.)

In the meantime much uncertainty prevailed, and the economic outlook

Table 1. Energy Consumption for Plastics versus Natural Materials
(Metric Tons of Petroleum, or Equivalent)

Product (Unit)	Quantity
Packaging film (1 million square meters)	
Polypropylene	112
Cellulose	153
Fertilizer sacks (1 million units)	
Polyethylene	470
Paper	700
One-inch diameter service pipe (100 kilometers)	
Polyethylene	52
Copper	65
Galvanized steel	228
One-liter containers (1 million units)	
Polyvinyl chloride	90
Glass	220

Source: Imperial Chemical Industries.

was poor. The inflation-adjusted increase in gross national product (GNP) from 1980 to 1981 was just 0.2%. Demand growth for chemicals was mostly flat or even negative. With relatively high product prices—even though these did not even closely reflect raw materials costs—there was concern that one of the important drivers for synthetic fibers and plastics growth rates—continuing replacement of natural materials—would slow down the historical inexorable rise in demand for synthetics. In 1979, when oil and petrochemical prices had risen dramatically, Imperial Chemical Industries (ICI) and other firms had published papers to the effect that the energy input for producing steel, glass, paper, and other traditional materials was higher than that for plastics, thus trying to assure the industry that it was not in danger of losing its major growth driver (Table 1).[6] As it turned out, there was little switching back to natural material–based products because their manufacture was mostly more energy intensive than that of petrochemicals.

In 1982, with demand flat, industry participants were not so sure that healthy growth would resume. A paper by Exxon Chemical presented a chart showing that U.S. primary petrochemical production, which had grown at a compounded annual growth rate of 17 percent between 1950 and 1960 and 13 percent between 1960 and 1970, had slowed to 4.6 percent between 1970 and 1980 and was projected to stay at or below that figure in the 1980s.[7]

As mentioned earlier, the plight of chemical producers operating in a severe recessionary environment was made worse because these firms had decided to expand capacity substantially in the late 1970s. In 1976 total U.S. ethylene capacity was 28.4 billion pounds, but it rose to 36.5 billion pounds in 1979, 39.3 billion in 1980, and 40.7 billion in 1981 as additional plants came on stream. Meanwhile, ethylene demand stayed relatively constant at around 29 billion pounds from 1979 to 1981, resulting in operating rates falling from 89 percent in 1979 to 71 percent in 1980. Since feedstock prices had doubled, producers found it difficult to raise ethylene prices in response and therefore operated with slim or nonexistent margins.

Vinyl chloride monomer (VCM) was another product in trouble because of overcapacity. Between 1977 and 1982, U.S. VCM capacity had increased from 6.4 billion pounds per year to 9 billion pounds, an increase of 41 percent, while production over the same period had risen by only 8 percent. This resulted in abysmally low operating rates in the early 1980s. The situation had been exacerbated by the construction of over a billion pounds of U.S. capacity by Shintech (Japan) and Formosa Plastics (Taiwan), firms that had concluded after the second oil shock that the United States still represented a good region for manufacture, much of the driving force being exports to their homeland and to their traditional Asian export markets.

Variable and cash cost margins had already dropped substantially by 1981, and the situation worsened in 1982 as production of ethylene and many other petrochemicals receded from 1981 levels. Chemical company profits in 1982 dropped sharply. Thus Dow's income from the first quarter of 1982 was off 44 percent, while Union Carbide's was down 49 percent, American Cyanamid's was down 41 percent, and Uniroyal's chemical and plastics operations were down 55 percent.[8] Celanese reported in early 1982 that while composite raw materials prices for chemicals, fibers, plastics, and specialties had risen by 180 percent since 1977, selling prices had not kept up, having risen only 144 percent.[9]

Oil companies' chemical operations were no better off. Mobil Chemical lost $12 million in the last quarter of 1982 compared with a $23 million profit a year earlier, while Shell's chemical operations widened its 1981 last-quarter loss of $11 million to $26 million in the comparable period in 1982.[10] The recession continued into 1983, and chemicals producers stayed in a retrenching mode. For the first time since 1968, capital spending would be less than in a previous year, down 5.1 percent from 1982.[11] And companies were also cutting back on R&D expenditures, as they sharply curbed their appetite for new plant construction based on new technologies.

During this period Chem Systems approached Dick Carlson, head of process research at Dow, to obtain financial support for developing and commercializing a technology (which would then eventually be owned by Dow) to make vinyl chloride directly from ethane, using as a basis some work originally done by ICI and licensed by Chem Systems' laboratory.[12] Dow research people reviewed the background work and Chem Systems' proposal and liked what they saw. But Dow turned down the proposal. "We would have no problem in

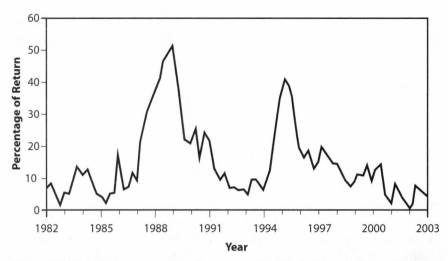

Figure 3. U.S. industry return on replacement capital (before taxes). *Source:* Nexant, Inc./Chem Systems, by permission.

getting authorization for spending the $20 million for completing the R&D and funding the pilot plant work," Carlson said. "But given the current 70 percent operating rate in the vinyl chloride industry and the almost nonexistent cash margins we are experiencing, we could never convince our board to build a large new vinyl chloride plant based on new technology. So we're not going to go to the board to ask for research funds."[13] (Interestingly, a form of this technology was piloted fifteen years later by European Vinyls Corporation [created in the 1980s through a merger of ICI's and Ente Nazionale Idrocarburi's vinyl chloride and PVC businesses], and will apparently be commercialized in the near future.)

The profitability problems of the petrochemical industry during the first half of the decade are clearly evident in Figure 3, which calculates a composite pretax return on investment (ROI) for a basket of thirty major petrochemicals on a quarterly basis. It shows that pretax returns in the early 1980s were extremely poor, the worst period in the history of the petrochemical industry to that time, with operating rates below 70 percent in 1982. For some petrochemicals the returns were actually negative.

While high feedstock costs and recession-induced low pricing were important reasons for poor performance, extreme competition in many product lines was certainly another factor. Any notion of "price discipline" breaks down in an industry containing a large number of producers for a specific chemical. The desire of individual companies to maintain or gain market share and the ability of customers to play competitors against each other makes for a chaotic market, particularly during periods of low operating rates. In certain businesses, where there are few producers and where one firm has a relatively large market share, price discipline can at times be maintained if the leader takes some capacity offline to improve the supply-and-demand balance. This occurs, however, fairly rarely. More often, industry conduct is chaotic, and pricing breaks down to levels at which even the most efficient producers do not realize much more than their cash costs. At such times the industry "laggards" barely cover their variable costs and either try to hang on for better times or in a few cases decide to shut down.

In the United States, in 1982, there were nineteen producers of ethylene, many of which were in the merchant market, where discounting became extreme. Among the largest were Shell, with 14 percent of the total U.S. capacity, and Dow, with 10 percent. There were eighteen producers of high-density polyethylene (HDPE), with Celanese, Allied, and Soltex Polymer each having over 10 percent of the total U.S. capacity.[14] The following year an ethylene producer with a relatively small plant of 500 million pounds per year—designated as a "laggard" using Chem Systems' methodology—had a cash cost margin of 1.63 cents per pound and a pretax ROI of –6.45 percent. A "leader" ethylene plant with a capacity of 1.1 billion pounds per year had a cash cost margin of 3.73 cents per pound and a pretax ROI of 1.64 percent. A styrene laggard operating a 600-million-pounds-per-year plant had a cash cost margin of 2.18 cents per pound and a pretax ROI of –1.34 percent.[15] With conditions like these, it is no wonder the U.S. petrochemical industry was experiencing profitability problems.

In Western Europe the situation was even worse. Chem Systems' West European Profitability Index (1978 = 100), which had risen to over 300 in 1979, dipped to zero in the second quarter of 1981, rose slightly later in 1981, and then almost dropped to zero again in mid-1982. The disastrous industry performance in the 1980–82 period is shown in a 1990 paper published by Shell that detailed the ups and downs of the industry's performance from 1967 to 1987.[16]

Polymer businesses were particularly hard hit, as indicated by the Chem Systems Polymer Index (1978 = 100), which had risen to 400 in 1979, only to drop below zero in early 1981. The calculated average return on new capital for a group of petrochemicals serving as an index dropped below zero in 1980 and only recovered somewhat by early 1984. Chem Systems' Petroleum and Petrochemical Economics (PPE) program, a quantitative analysis of industry performance headed by David Glass, which was subscribed to by almost every European petrochemical producer, was at times used as a benchmark for producers to look at typical industry economics covering both leader and laggard plants, allowing them to share in the misery. Possibly the hardest hit company at that time was BP Chemicals, which lost $257 million in 1980. Undoubtedly, this loss was aggravated by its acquisition of Union Carbide's European polyethylene and ethylene oxide and glycol assets and Monsanto's European styrene business. The U.S. firms had decided that their lack of back-integration in Europe had made these businesses fatally vulnerable. Other U.S. firms withdrawing from Europe included National Distillers and Chemicals, which sold its Benelux polyethylene operations to Exxon Chemical; Hercules, which sold its European pigments business to Ciba Geigy; Gulf Oil Chemical, which exited all its production, including ethylene and cumene; and Monsanto, which sold several fiber and fiber-related plants.

In Japan the situation was also grim. The industry's problems were outlined in a paper presented to the European Chemical Marketing Research Association in 1993.[17] The figures showed a total collapse of profitability in the 1980–82 period, which H. Kamata and M. Hongoh from Idemitsu Petrochemical Company blamed on overcapacity, excessive competition, poor international competitiveness of Japanese plants, and the emergence of new, strongly cost-competitive producers worldwide. In the authors' views the solution could only come from scrapping a number of small, uncompetitive plants, which was difficult in Japan, where a culture of domestic price protection had always kept small producers in business. However, in Japan steps were being considered to form so-called "joint sales companies" that would comprise groups of producers and thereby reduce the number of competitors contending for business, which actually happened soon after the paper was published.

This then was the situation in the early 1980s, a time for serious reflection on the future. The calls for industry restructuring became more and more urgent, but the way ahead was not clear. As described in the subsequent section, Western Europe's producers were even more concerned than their American counterparts and more inclined to take action as a result of discussions among the players. Chem Systems played a role in allowing these producers to talk to each other effectively by making an industrywide cost and business analysis of European producers available by means of a multiclient study, as discussed below. But exit barriers, including government ownership of plants, and cultural problems stood in the way.

In the United States, producers could not communicate in the same way used in Europe, partly because Chem Systems did not get sufficient industry support at that time to launch a study comparable to its PPE program in Europe. Producers in North America took unilateral actions to help bring capacity closer to demand, but these actions were insufficient to raise operating rates to a comfortable level. Only a few years of steady demand growth without capacity additions could really solve the problem.

Restructuring in Western Europe

If petrochemical producers in North America were suffering through what they considered the most challenging period they had ever experienced, European producers were even worse off. Europe's petrochemicals production was essentially based on naphtha, derived from relatively high-priced crude oil, while U.S. ethylene crackers were still largely based on ethane and liquefied petroleum gas (LPG), which is normally a considerably less expensive natural gas–based feedstock. Furthermore, the average size of petrochemical plants in Europe was substantially smaller than that of comparable plants in North America, resulting in higher fixed costs per pound. Europe's petrochemical industry was dispersed all over the continent, resulting in high distribution costs and inefficient logistics, while in the United States most of the plants were in Texas and Louisiana, with a substantial interconnecting pipeline grid and extensive opportunities for barge shipments. In contrast, much of Europe's petrochemical output was moved by rail and truck. As if that were not bad enough, companies wanting to move product both ways across the Alps also had a logistical problem owing to poor transportation facilities through Switzerland and Austria. Thus, excess capacity in northern Europe could only be moved at relatively high cost to the southern part of the Continent, and vice versa, effectively separating Western Europe into northern and southern halves.

Every European country had at one time wanted to participate in the exciting petrochemical industry (right after establishing a domestic airline). From an early logistical standpoint, when plants were small and could be located close to equally small markets, there had been some rationale for building these plants. Oil refineries in all these countries could theoretically produce petrochemical feedstocks. Countries with small markets but large refineries thus saw no particular obstruction to building petrochemical plants, a good example being Austria, which built plants near Vienna and Linz, the two largest cities. In Sweden, there was a different rationale, when Mo Och Domsjo, a paper company, decided to diversify into petrochemicals at a new site near Stenungsund because it saw that wood-based products were being replaced with petrochemical polymers. But because of small domestic markets, such plants needed to export much of their product to other European countries that unfortunately often had similar plants. A particularly egregious case, some people felt, was the construction of petrochemical plants in Portugal at Sines, which came on stream in 1981 and operated unprofitably for many years. In another example the Italian government in the 1960s decided to promote and finance petrochemical and associated fiber-producing plants in the poorer, so-called *Mezzogiorno* region, which includes Sardinia, southern Italy, and Sicily. Many plants had been located there for political reasons, simply to provide work in a region of notoriously low employment but strong political power. These plants were often in the worst possible locations, with awkward interconnections to upstream and downstream plants.

All these factors combined to bring about by the early 1980s a European petrochemical landscape consisting of too many small petrochemical and polymer production plants, many of which were struggling to survive. This was the situation faced by the European petrochemical community when it began to think about bringing capacity in line with weak demand growth, as almost every executive making speeches at industry meetings was then recommending. But there were two problems that would stand in the way of a relatively quick fix: government ownership of companies and plants (for example, in France, Italy, and the Netherlands), which would make shutdowns politically difficult as workers were thrown out of jobs, and the existence of isolated cracker complexes in France, Germany, and Italy, where the shutdown of, for example, an uncompetitive derivatives plant would reduce the operating rate and therefore the economics of the cracker and adversely affect the economics of the other derivatives plants. Such a snowball effect might eventually mean shutting down the entire complex, even if there were several other relatively viable plants. At another site the cracker itself might be small, and it therefore transferred ethylene at high cash costs so that the associated derivatives plants would all be in a poor position relative to similar units in other complexes that were receiving ethylene from a world-sized, or large-scale, cracker. This problem occurred less frequently in the United States, where pipelines link most complexes and plants. There, inefficient crackers could be shut down, with the remaining derivatives plants fed with ethylene via pipeline. Alternatively, if derivatives plants were shuttered, excess ethylene from a relatively efficient cracker could be fed into the pipeline grid and sold into the merchant market.

Shutdown of plants with multiple ownership presented another almost insoluble problem in Europe. In a paper presented during this period, John Philpot, a managing director at Chem Systems, said that "trying to get agreement to close plants which are multi-owner joint ventures or to transfer various plants between several producers can be almost impossibly complex. The notion of complexity being proportional to the square of the number of participants has some appeal."[18]

European producers agreed in principle to promoting the shutdown of uncompetitive plants. But what this would mean and particularly who would do it was unclear. Discussions were sponsored by the Conseil Européen des Fédérations de l'Industrie Chimique (CEFIC; Council of European Industrial Chemical Federations), a private industry group, as well as by the European Commission, including the European Union Industry Directorate. But there was almost always an attitude of distrust among participants: the weaker players believed they were at a disadvantage to large multinational firms like Shell, Dow, Exxon, ICI, and BASF, which were pushing for weaker players to shut down and, it was hoped, even leave the industry. Eventually, the European Commission and CEFIC agreed to sponsor the preparation of a report that would address the subject of how the industry should reduce overcapacity. The so-called Gatti-Grenier report, issued in the early 1980s, generally recommended cutbacks in proportion to volume produced rather than to efficiency of production. It also suggested that companies withdrawing from a business compensate the companies remaining. The report was strongly criticized by British and German firms, and its recommendations were generally ignored.[19]

Shortly before this report was issued, Chem Systems had done a multiclient

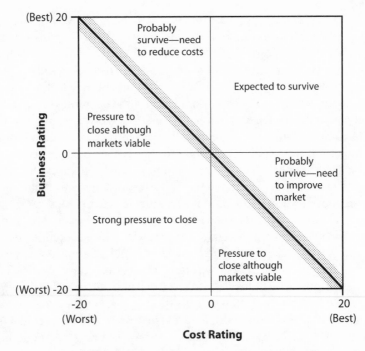

Figure 4. Plant survival matrix. *Source:* Nexant, Inc./Chem Systems, by permission.

study aimed at allowing producers to understand their position in the industry and give them the information needed to make difficult decisions. The project was conceived and managed by the head of Chem Systems' strategy group, Roger C. M. Longley.[20] The firm had approached European producers of ethylene, polyethylene, and the vinyls chain (chloralkali, VCM, and PVC) with the idea that representatives of Chem Systems would meet with all the firms and receive data in confidence about the production economics and "business position" of all their plants for these products. The consultants would then take these data back to their office and eventually characterize every plant and business on a business- and cost-rating matrix (Figure 4).[21] The study covered 150 separate plants owned by 30 or so companies. The "cost rating" corresponded directly to production economics, that is, the cash cost of production for product coming from the individual plants. The "business rating" reflected commercial strength as a function of product mix, geographic location, and additional judgment factors. When the study ended, participants would be given the entire matrix with unidentified points for all industry participants, but with the subscriber's own plants identified on the matrix. In this way subscribers would see how their plants rated within the entire spectrum of all plants operated in Western Europe. The place to be, of course, was toward the upper right-hand quadrant, where those companies that were expected to survive were placed. Diagonal lines were drawn to indicate how much cumulative capacity could be shut down as more and more plants fell off the lower-left quadrant. One matrix for sixty-six crackers covered a total ethylene capacity of 16 million tons (in 1983) against a forecast for ethylene demand of 12 million tons in 1990. Assuming a desirable utilization rate of 90 percent, this suggested that 3 million tons of cracker capacity should be shut down to reach a reasonable supply-and-demand balance (Figure 5).[22] The report went so far as to identify the geographical distribution of capacity recommended for shutdown: Benelux (Belgium, the Netherlands, and Luxembourg), 0.6 million tons; France, 0.4 million tons; Germany, 1.2 million tons; Italy, 0.4 million tons; Spain, 0.2 million tons; and the United Kingdom, 0.5 million tons. The total recommended for shutdown was 3.3 million tons. For low-density polyethylene (LDPE), the Chem Systems report reached the detailed conclusions shown in Table 2.

Chem Systems' report was generally acclaimed and undoubtedly helped some

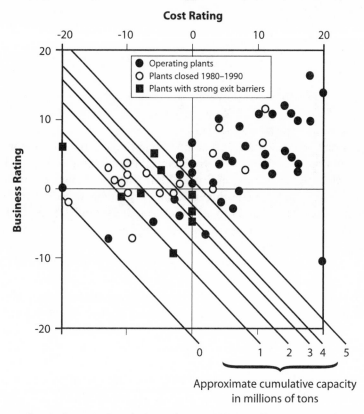

Figure 5. Survival matrix for ethylene plants in Western Europe (1983 forecast of industry restructuring through 1990). *Source:* Nexant, Inc./Chem Systems, by permission.

Table 2. LDPE Capacity Ranking of the Least Competitive Plants

Plant Location	Noncompetitive*	Grey Area†	Held in by Exit Barriers‡
	(In Metric Tons per Year)		
Benelux (Belgium, Netherlands, Luxembourg)	—	150	—
France	—	60	—
Germany/Austria	120	100	—
Italy	250	320	—
Scandinavia	—	100	160
Spain/Portugal	140	150	200
United Kingdom	80	60	—
Total (= 1,890)	**590**	**940**	**360**

Source: Nexant, Inc./Chem Systems.
* Candidate for shutdown.
† Possible additional candidate for shutdown.
‡ Shutdown unlikely owing to specific site issues.

companies to make the hard decisions. It did not help, however, that some firms suggested that the Chem Systems study had been promoted and supported by several of the strongest firms.

An important step toward eventual industry rationalization was the formation in 1985 of the Association of Petrochemical Producers in Europe (APPE) by thirty-four large petrochemical firms. While this association had no power or even a formal role in promoting agreement on industry restructuring, it proved useful in collecting and providing information on industry capacity expansions and reductions.[23] The Chem Systems report was said to have been particularly helpful to APPE in this respect.

By 1985 a substantial amount of European petrochemical capacity had been shut down. A paper by C. M. Doscher, a former employee of Dow but then working for Enichem, disclosed that over the previous four years the net reduction in capacity for ethylene was 25 percent; for LDPE, 24 percent; for HDPE, 17 percent; and for PVC, 12 percent.[24] During the same period European capacity utilization had increased markedly, although operating rates were still in a relatively unprofitable mode (Table 3).

During 1981–82 Italy had shut down 49 percent of its ethylene capacity and 42 percent of its LDPE capacity. The United Kingdom had shut down 33 percent of its ethylene capacity, while West Germany shut down 24 percent.[25] The start-up of new, highly efficient units was often accompanied by shutdowns of older, less efficient plants operated by the same firms. Thus, when the new Shell-Exxon cracker was started up in Mossmoran, Scotland, an Exxon Chemical ethylene plant in Cologne and a Shell cracker in Carrington, England, were shut down.

These shutdowns surprisingly resulted from unilateral action, as no cooperative effort was supported by CEFIC. Joint decisions on shutdowns, which might be seen as a form of industry collusion, had never been strongly supported by the larger players, particularly not by U.S. firms, which were more concerned with possible antitrust problems involving the Directoire Generale IV (DGIV), the antitrust arm of the European Commission.

In Italy, the government, which controlled the country's energy arm, Ente Nazionale Idrocarburi (ENI), took it upon itself to restructure the petrochemical industry. It prompted the consolidation of the smaller petrochemical firms, so that, for example, Rumianca and Societa Italiana Resine (SIR) disappeared into ENI. The firm then caused Montedison and ANIC Petrochemical (ENI's chemical arm, later called Enichem) to swap a number of businesses in an effort to strengthen both firms. Enichem would have all the crackers and most of the petrochemical plants, while Montedison would have many of the specialty and fine chemicals but would keep its strong position in polypropylene.

In France the situation was grim, with important firms in 1981 all sustaining losses (CDF Chimie, 1.2 billion French francs [FF]; Pechiney Ugine Kuhlmann [PCUK], FF 800 million; Rhône Poulenc, FF 330 million; Chloe Chimie, FF 370 million; Atochimie, FF 130 million; and EMC, FF 100 million) during a year when Hoechst and BASF were still making profits.[26] Among the effects of the government-conducted restructuring was the splitting up of PCUK, with most of its parts going to Rhône Poulenc, and the emergence of Elf Aquitaine's Atochem subsidiary as the dominant petrochemicals producer in the country. CDF Chimie (later called Orchem), always a weaker firm, was eventually split between Elf Atochem and Total, with Atochem receiving most of the petrochemicals and Total some specialty chemicals businesses.

Table 3. Improvement in European Operating Rates (1981–1985)

Product	Operating Rate (%)	
	1981	1985
Ethylene	71	84
Low-density polyethylene	62	89
High-density polyethylene	64	84
Polyvinyl chloride	64	82
Polystyrene	60	83

Source: C. M. Doscher, "Restructuring: What Still Needs to Be Done," European Chemical Marketing Research Association (ECMRA) Conference, Antwerp, 1986.

Companies at that time were also starting to consider portfolio exchanges, which became known as "portfolio swaps." The most important of these involved a decision by ICI and BP Chemicals to exchange their respective polyethylene and vinyls chain portfolios in 1982. While this was at first challenged by DGIV, it was later approved by the agency. It must have been difficult for ICI, the inventor of polyethylene, to exit this business, but it was only an example of things to come. Two years later ICI and Enichem placed all their (largely unprofitable) vinyls assets into a joint venture, European Vinyls Corporation, which was spun off to the public several years after that.

When industry restructuring reduces the number of players and the remaining producers try harder to maintain or even increase prices, collusion is always a possibility. Europe's chemical industry historically has had cartels, and this mentality may have been at play as agreements on shutdowns and information-sharing arrangements aimed at coordinated reduction of capacity took place. A number of chemical companies in the polypropylene, polyethylene, and PVC industries were found to have violated antitrust rules and were fined by the European Commission in 1986 and 1988.[27] Ironically, their "cooperation" was only related to pricing and not to constructive ideas about rationalization.

Western Europe's restructuring in the 1980s focused almost entirely on shutting down uncompetitive capacity in an effort to improve the supply-and-demand balance. It left a number of isolated crackers and associated derivatives plants in poor competitive and logistical positions. Significantly, the restructuring seldom led to companies completely exiting a line of chemicals or polymers, and, similar to the United States, it still left too many producers and a highly competitive environment, with too many players chasing market share.

In North America—More Urgency

Dismal financial results in petrochemicals prompted a somewhat different response in North America. There were a number of reasons for the difference, including greater pressure on U.S. company managers to improve financial results, a better logistical situation for U.S. firms to shut down uneconomical plants, fewer exit barriers than in Europe, and the emergence of "petrochemical entrepreneurism," which brought about a number of management buyouts, thus facilitating the implementation of exit strategies by some of the traditional petrochemical producers.

Similar to producers in Western Europe, North American firms were shutting down uneconomical ethylene capacity but doing it a little faster. By the end of 1982, 1.7 billion pounds had been permanently shut down, and another 2.9 billion pounds had been shuttered for possible restart later. About 25 percent of North American capacity had been taken out of service.[28] However, for the major ethylene derivatives, there was much less net capacity reduction, and the situation was not improving as fast as in Europe. The lack of improvement was caused in part by the commissioning of new Canadian plants in Alberta, which produced low-cost ethylene, polyethylene, and ethylene glycol and which were largely built to supply West Coast and Asian markets but were also targeting the U.S. Midwest. With Albertan natural gas apparently in almost unlimited supply, the provincial government had decided to promote vigorously a petrochemical center, making gas available for as low as 35 cents per million BTUs. Plants were also built for producing methanol, much of which was targeting the U.S. market, eventually triggering a successful antidumping action by DuPont and Celanese.[29]

In the polyethylene industry 1.7 billion pounds of conventional capacity were shut down in North America, which was more than balanced by the construction of plants to produce 3 billion pounds of linear low-density polyethylene (LLDPE), a new form of the product that was commercialized and subsequently heavily licensed by Union Carbide. In the vinyl chloride industry Stauffer and Shell shut down plants, but overcapacity remained. In the styrene industry U.S. capacity had climbed from 3.2 billion pounds in 1979 to 4.2 billion pounds in 1984, with only minor shutdowns by Dow and Union Carbide.

One of the first public indications that some companies were becoming disenchanted with petrochemicals came in April 1982, when Cities Service abruptly changed its mind about this business. Over several years the firm had conducted a diversification program in chemicals, tied in part to its mining operations and in part to its refineries. The high debt incurred for these moves and the dismal situation in petrochemicals led to a large writeoff for its plastics operations in 1981 and a determination to get back to profitability and to the reduction of corporate debt. The company, which had hired investment advisers, announced the shutdown of two ethylene plants at Lake Charles, Louisiana; the shutdown of its 700-million-pounds-per-year polyethylene plant; and the discontinuation of the construction of a new HDPE plant in Bay City, Texas. These plants were now put up for sale, while the company's new chief executive, Charles J. Waidelich, tried to assure investors that "redeployment of resources into an expanded search for oil and gas reserves worldwide will provide the company and its shareholders the growth they are seeking." These steps were taken against a background of rumors that Cities Service would soon be taken over by another firm. Management therefore needed to tell institutional investors "that the company has savvy managers who know which operations belong and which don't."[30] Cities Service's petrochemical assets were eventually bought by Westlake Polymers, a Taiwanese-owned private firm, and restarted later when the timing was more propitious.

In late 1982 a new kind of transaction was being formulated by Gordon Cain, a former general manager of Conoco Chemicals (and now a private consultant), and John Burns, then managing the petrochemical operations of DuPont's new subsidiary Conoco. The two old friends concluded that DuPont might be willing to sell these operations, which did not fit particularly well with any of DuPont's other businesses. After hooking up with financial advisers, the team met with senior DuPont officials and found that the company would entertain an offer for the facilities, which included a world-scale ethylene plant, a detergent business, and plants for the production of vinyl chloride and PVC.[31] As this was the industry's first large leveraged buyout (LBO), financing of the transaction by the Sterling Group, headed by Gordon Cain, was complicated, took almost two years, and was not as financially favorable to the buyers as later transactions of this kind. But it was successfully concluded in the summer of 1984, leading to the formation of Vista Chemical, and was followed by a number of other buyouts over the next several years.

Both parties felt that the transaction accomplished an important goal. Gordon Cain and his team were convinced that the petrochemical business would rebound after a few years, given the inexorable rise in demand that would undoubtedly push operating rates back into the 90 percent range or even higher. And DuPont sold assets it did not really want to own, whose operations had done little for DuPont's bottom line, and that would require further investment to stay competitive.

This LBO was followed over the next three years by several others, including Sterling Chemical by the Sterling Group (Monsanto's Texas City complex, which included styrene, acrylonitrile, acetic acid, and plasticizers); Cain Chemical (two large crackers in Corpus Christi and Chocolate Bayou, Texas) and polyethylene and ethylene glycol plants from DuPont, ICI, and the Pittsburgh Plate Glass Company (PPG) by the Sterling Group; Georgia Gulf by the management of Georgia Pacific's petrochemical operations (plants for the production of cumene, phenol, chloralkali, vinyl chloride, PVC, ammonia, and methanol); Rexene by a financial group (plants included ethylene, polyethylene, polypropylene, and specialty films); and Borg Warner Chemicals by an investor group headed by Merrill Lynch Capital Partners (ABS [acrylonitrile-butadiene-styrene] resins, additives, and compounds).

These transactions had a common thread. In all these cases the selling companies had decided to exit petrochemical operations owing to a lack of faith in the future of these businesses. The fact that the Cain Chemical deal could be done in mid-1987, when operating rates for petrochemicals were already beginning to rise sharply, could only mean that DuPont, PPG, and ICI were no longer committed to the commodity parts of this business and were ready to get out, even if it meant walking away from what was likely to be a very attractive but probably short period of tight supply and demand and high operating rates.

The Vista Chemical buyout had involved a complicated scheme for covering the subordinated debt, greatly favoring the banks putting up the money. In the case of Sterling Chemical at Texas City, the buyout had been facilitated via take-or-pay contracts[32] with BP Chemicals (for acetic acid) and Monsanto (for styrene), but these deals limited considerably the potential upside for the buyers. After that, LBOs became easier to finance.

Chem Systems was heavily involved in several of these buyouts, since its projections for the businesses being bought were used by the banks for their financing models. Two transactions in which the firm was the lead consultant provide further details on how they were carried out.

In the case of Georgia Gulf, Chem Systems was asked by General Electric Credit Corporation, which financed the deal, to attend the first key meeting with the principals, who were headed by Jim Kuse. Kuse, along with his management team at Georgia Pacific Corporation, had built a series of world-class petrochemical plants when this pulp and paper manufacturer decided in the early 1970s that plastics would soon supplant wood and paper in many applications. But the firm's view of petrochemicals had soured in the early 1980s, and it was therefore receptive to a buyout bid. It was later understood that Kuse and his group had scraped together only one or two million dollars via new mortgages and small loans to put up a little "earnest money" for equity in the LBO. This stake had given them enough ownership to become multimillionaires several years later, when their company went public.

To assist with the Cain Chemical transaction, Chem Systems was hired by Chase and Morgan Stanley, which would end up with large equity stakes, along with Gordon Cain and the Sterling Group. This transaction had been germinating for some time, but when the pieces began to fit into place, it proceeded with lightning speed. There were so many plants involved that the Chem Systems due-diligence team barely had time to rush from one plant to the other. Fortunately, these plants were all clean, excellently maintained, and hooked into the extensive Gulf Coast pipeline grid. Financing was easy because it was evident from Chem Systems' projections that Cain Chemical soon stood to become very profitable, given the rapid tightening of supply and demand, particularly for ethylene. The timing was perfect, and Chem Systems thereby became a key participant in what has arguably been the biggest "home run" LBO in the petrochemical industry. Within a year Cain Chemical was acquired by Occidental Petroleum Corporation for a price that provided the equity holders of Cain Chemical a profit of close to a billion dollars.[33]

By the time Cain Chemical was formed in 1987, the investment community had begun to use "junk" bonds (the name recently coined to describe lower-than-investment-grade bonds) instead of subordinated debt, since a substantial market for such bonds had been created by Michael Milken of Drexel Burham Lambert and others. Availability of this type of money for financing highly leveraged transactions would mean that LBOs were here to stay.

The 1980s also saw the emergence of Huntsman Chemical as a private firm that used leveraged financing for a string of acquisitions that continued through the 1990s. Jon Huntsman had created Huntsman Container Corporation in 1970, and in 1983 he started Huntsman Chemical with the acquisition of Shell's polystyrene plant in Belpre, Ohio. This was followed by many other transactions, including polystyrene acquisitions from Shell in Carrington, United

Kingdom, and from Hoechst-Celanese in Virginia and Illinois; formation of the Polycom-Huntsman joint venture in plastics compounding; a joint venture with General Electric Plastics in polystyrene; acquisition of a Shell polypropylene plant in Woodbury, New Jersey; and acquisition of Hoechst-Celanese's styrene plant in Bayport, Texas. Huntsman's total revenues grew from $75 million in 1975 to $1.3 billion in 1990.[34]

Industry restructuring in the United States also involved a number of major acquisitions by foreign firms, largely European multinationals. These firms were becoming convinced that they needed to expand their sales in North America and that manufacturing economics and the climate for allowing new plant construction there would be better than in Europe. Until 1980 Europe's chemical firms had a limited presence in the United States, except for a few joint ventures (for example, Mobay [Monsanto-Bayer] and BASF Wyandotte) and minor manufacturing positions by ICI, Hoechst, and a few others. In eastern Canada, ICI and BASF had established manufacturing in a few areas of their strength, such as chloralkali and explosives for ICI. And, of course, Shell Chemical, although a U.S. company, had been set up by the Shell parent firms in Houston, Texas, decades before. But during the 1980s European chemical companies made a number of important acquisitions, both entire companies and individual businesses being divested. BASF bought Inmont Chemicals, a manufacturer of inks and coatings; Polysar's latex businesses in Canada and the United States; Celanese's carbon fibers business; and the part of Wyandotte it did not already own. Bayer bought out its partner in the Mobay polyurethane and polycarbonate resins joint venture. Hoechst acquired Celanese, and Akzo acquired the balance of Akzona, a fiber joint venture. Atochem acquired Pennwalt, Unilever acquired National Starch and Chemicals, and ICI acquired Glidden Paints and Beatrice Chemical.

The emergence of Shintech, Formosa Plastics, and Westlake Chemicals as important new players also indicated the mounting presence of foreign firms. Dai Nippon Ink acquired Sun Chemical and Reichhold Chemical. A group of Mitsubishi companies headed by Mitsubishi International, the trading firm, acquired what was once the chemical division of U.S. Steel, now named Aristech Chemical.

In most cases the foreign firms were looking for downstream businesses that complemented their positions in the home market. But companies' investment strategies differed: ICI, which had participated in a cracker at Corpus Christi, Texas, and had built an ethylene oxide plant there, sold these assets to Cain Chemical, thus exiting commodity chemicals in the United States. Earlier, Solvay had built polyolefin plants on the Gulf Coast, but with the sale of its share of the Corpus Christi cracker, it remained a nonintegrated producer. Solvay, one of the world's oldest and largest PVC producers, with substantial operations in Europe and elsewhere, had never entered the vinyls chain in the United States—probably a wise decision in retrospect.

In some respects the most significant acquisition was that of Celanese by

Hoechst, particularly because Celanese was largely a petrochemical company, whereas other European multinationals were acquiring downstream assets. Hoechst historically had fewer petrochemical assets in Europe than its two sisters, BASF and Bayer, so the move was a surprise. But there was a good fit in acetyls, polyester and acetate fibers, oxo alcohols, and engineering resins. Just as important, the rationale for the merger was to globalize Hoechst's operations by obtaining a U.S. base from which to export to traditional customers in Latin America and Asia.

Two other transactions from this period are noteworthy. First, after Unilever acquired Cheesebrough Ponds in 1987, it decided to sell Stauffer Chemical, which had earlier been acquired by Cheesebrough. In a complicated transaction in which ICI paid $1.69 billion but had prenegotiated simultaneous divestment of certain businesses to other firms, the three parts of Stauffer ended up as follows: the basic chemicals (for example, sulfuric acid and sulfur dioxide) were bought by the U.S. subsidiary of Rhône Poulenc; the catalyst business and the licensing division were bought by Akzo; and the balance remained with ICI, which was primarily interested in Stauffer's agricultural chemicals business.

The split-up of Stauffer Chemical ended the history of one of the best-known chemical firms in the United States, a victim of industry restructuring. (The firm had never really recovered from the retirement and death of its founder, Hans Stauffer.)

In the other transaction Montedison purchased Hercules's interest in Himont, the world's largest polypropylene producer. The consolidation of Himont by Montedison, the successor company to Montecatini, where Giulio Natta had invented polypropylene, was also noteworthy. It led to further advances in polymer chemistry by Paolo Galli, making Himont a dominant, worldwide player in polypropylene and advanced polymers derived from the same technology.

While chemical firms were engaging in a number of these dramatic restructuring moves, the chemical divisions of U.S. oil companies had different responses to the problems of the 1980s. Large firms like Exxon Chemical and Shell Chemical were determined to ride out the storm, although not without a serious review of their portfolios. When John Akitt of Exxon Chemical spoke at the Society of Chemical Industry meeting in Florence in 1983, he pointed to the shutdown of 15 percent of Exxon's total U.S. production capacity, which involved nine petrochemical products. These businesses were considered unable to produce an average 10 percent return on capital and represented $500 million of gross plant investment by the company.[35] Akitt also pointed to similar steps taken by Union Carbide and Hercules to restructure their portfolios, shedding businesses in which these firms felt they could not be leading competitors.

In the 1980s there was also a dramatic restructuring of the U.S. oil industry. In addition to DuPont's acquisition of Conoco, the decade saw Gulf become a part of Chevron, Getty acquired by Texaco, and Diamond Shamrock bought by Occidental Petroleum. All these transactions increased the chemical assets of the surviving firms, although this was not the primary rationale for these transactions. Some of the oil companies, in addition to Cities Service (see earlier in this section), were also becoming disenchanted with their chemical subsidiaries. Atlantic Richfield's subsidiary Arco Chemical was forced to divest its two large crackers and other commodity units into Lyondell Petrochemical, which was then spun off to the public. Arco Chemical also sold several specialty chemicals operations it had acquired over the previous few years. In a paper presented in 1986 before the Society of Chemical Industry in Madrid, Harold Sorgenti, CEO of Arco Chemical, said that

> no matter which route the restructuring takes, whether it be by government action as in Japan, by LBOs and assets sales as in the U.S. or by swaps and joint ventures as in Europe, the end result is [that] the basis of competition has changed. Fewer companies that are stronger in market position, integration, technology or product differentiation are competing more effectively in the world market place. This bodes well for improved profitability of our industry.[36]

The U.S. chemical industry saw little growth from 1980 to 1986. Data collected by the Bureau of the Census and the Chemical Manufacturers Association (CMA) showed that industry shipment of "chemicals and allied products," the broadest category for the industry, grew at a historically rather weak 3 percent compounded rate over this period, about 1 percent less than "all manufacturing." Significantly, "industrial chemicals, except pigments" grew at a compounded rate of only 1.6 percent.[37] During the same period ethylene had a compounded growth of 2.3 percent and ethylene oxide, 0.6 percent. Other petrochemicals showed somewhat better growth. This six-year period tried the determination of many companies and shaped the industry in ways that could not have been anticipated a decade earlier.

Restructuring in Japan

Because Japan had no domestic crude oil resources, its economy was plunged into a grave depression soon after the second oil shock began, and the depression wore on for several years. The skyrocketing costs of energy and feed-

Table 4. Feedstock and Energy Costs in Japan before and after the Second Oil Shock

Variable	FY 1978	FY 1981
Crude oil (U.S. $/barrel)	13.89	36.9
Naphtha (U.S. $/barrel)	22.8	56.6
Power cost (cents/kwh)	12.53	21.46

Source: Ministry of International Trade and Industry (MITI) and Japanese Petrochemical Association.

Table 5. Slump in Japanese Petrochemicals in the Early 1980s

Variable	FY 1979	FY 1980	FY 1981	FY 1982
Ethylene capacity (in million tons)	6.20	6.23	6.24	6.35
Ethylene production (in million tons)	4.84	3.86	3.62	3.56
(operating rate)	(78%)	(62%)	(58%)	(56%)
Polyolefin capacity (in million tons)	3.78	3.79	4.01	4.13
Polyolefin production (in million tons)	3.25	2.62	2.65	2.68
(operating rate)	(86%)	(69%)	(66%)	(65%)

Source: Ministry of International Trade and Industry (MITI) and the Japanese Petrochemical Association.

stocks, which continued into the next decade (Table 4), coupled with the decline in the price of chemical products because of a stagnant market, more imports from overseas, and fewer exports owing to loss of competitiveness, had a harsh effect on the domestic chemical industry (Table 5). All petrochemical companies large and small were forced to take drastic rationalization measures to survive this unprecedented situation.

The situation in which Japanese chemical and other companies found themselves had some relatively unique and, with respect to restructuring efforts, troublesome features. Management-labor contracts made it difficult for Japanese companies to lay off workers, thereby barring companies from withdrawing from uncompetitive lines of business. Furthermore, bank support for Japanese companies owing to interlocking holdings is strong. Thus banks will not easily restrict lending to first-rank companies as long as they are paying interest, even if they are in the red. This situation evidently limited the banks' role as possible agents for restructuring. The Japanese petrochemical industry was characterized as "the boxer who, even though hurt, can't get out of the ring."[38]

When the economic troubles of the Japanese chemical industry did not improve through 1982 and early 1983, the Japanese government itself was forced to take action. Recognizing the serious problems of the industry, the government asked the companies to implement a unique scheme to improve their competitive position. It enacted the "Temporary Measures Law for the Stabilization of Specific Depressed Industries" in May 1983. Commonly referred to as the "Industrial Structure Adjustment Law," it remained in effect until June 1988, when global petrochemical prices again rose on a strong upward trend. The law, which was applied to eleven basic industries, including fertilizers and synthetic fibers, allowed firms to curtail their excessive production capacities and set up alliances with competitors, and provided certain financial rescue measures as well as special treatment on taxation. In the case of the ethylene industry, all fourteen producers agreed to shut down their old crackers, for good or to "mothball" them (that is, to shut them down in anticipation of possibly restarting them later), sometimes for agreed periods. Japanese ethylene capacity was thereby reduced by 36 percent. The seventeen Japanese polyolefin producers also organized four joint sales companies—Dia Polymer, Ace Polymer, Mitsui-Nisseki Polyolefin, and Union Polymer—that integrated their production, marketing, and logistics operations. To allow this, the Fair Trade Commission of Japan moderated its interpretation of Japanese antimonopoly law. The joint sales companies mainly dealt with commodity grades of polyethylene and polypropylene and achieved a reported combined cost-savings target of 40 billion yen. In the PVC industry no joint sales companies were set up, but producers individually took steps to shut down small, uncompetitive plants, resulting in a 24 percent overall reduction.

Similar to the steps taken by their North American and Western European competitors, the Japanese chemical companies, beginning in 1981, took some actions to improve their cost structure, attacking both variable costs (feedstock and utility consumptions and pricing) as well as plant scale[39] and corporate fixed costs. Some unprofitable businesses were shut down, and product portfolios partly shifted from commodities to various specialties.

The Japanese economy fortunately perked up toward the end of 1983. As a result of a better economic climate and of actions to rationalize capacities, reduce costs, and emphasize noncommodity products, the cumulative profits of the twelve companies operating ethylene plants in Japan went back into the black marginally in 1983 and then continued to improve. This lessened the urgency for additional reforms, such as strategic mergers, which would have further improved the competitiveness of the Japanese industry. The price fly-up in 1988 provided considerable encouragement for maintaining the status quo. As a result the efforts to make Japan's petrochemical industry more globally competitive slowed down, and the industry entered a ten- to fifteen-year "hibernation" period.[40]

Many of the Japanese firms, however, did take steps to get into or emphasize more specialized high-tech products, such as electronic chemicals, optoelectronics, ceramics, biochemicals, and structural composites. This change in strategic direction helped the companies to diversify their portfolios enough to cope better with the next downturn.[41]

Restructuring in Brazil

Although other South American countries (significantly Argentina, Venezuela, Chile, and Peru) had some petrochemical assets, Brazil had the most developed industry. Petroquimica Uniao, with a naphtha cracker and typical derivatives plants, was started in 1972, with ownership partly by the government through Petroquisa, the Petrobras chemical subsidiary, and partly private. A tripartite ownership model was later established, involving 33 percent participation in projects by Petroquisa, private Brazilian groups, and foreign multinationals. This led to the planning and construction of a second so-called petrochemical pole in Bahia (Camacari), with a cracker (Copene) and a number of petrochemical and other chemical plants owned in accordance with that model. A third complex near Porto Allegre in the south (Copesul) was started up just as the global petrochemical industry went into its dramatic slump in the early 1980s. Important local groups participating in these poles were Odebrecht, Ultra, Suzano, Mariani, Banco Economico, and Iparanga.

Up to the early 1980s the local petrochemical market was highly protected by tariff and nontariff barriers and had built up a good export business. In the global crisis situation of the early 1980s, with export markets lower and extremely competitive, Brazilian petrochemical producers were faced with oversupply partly because of the new plants in Bahia and in the south. Petrobras came to the rescue, making cheap naphtha available to help the industry. Naphtha was inexpensive primarily owing to the Brazilian move to drive cars on sugar cane–based alcohol rather than gasoline, given Brazil's vast sugar industry and its lack of adequate crude oil resources. Thus, Brazilian petrochemicals were also inexpensive on the world market and could be exported favorably. Petrobras further helped by using its trading arm, Interbras, to help place exported material. Under this arrangement naphtha was priced low enough to allow producers a margin amounting to at least 10 percent over variable costs, this margin being shared along the chain.[42] Meanwhile, few foreign petrochemicals could come into the protected Brazilian market. Nevertheless, Brazil had an unfavorable balance of trade in chemicals, with 1979 imports valued at $2.15 billion and exports at $400 million. Much of the imported material consisted of agricultural chemicals, and pharmaceutical raw materials, with the balance commodity chemicals and petrochemicals not made in Brazil. These figures improved markedly by 1985, when imports were $1.3 billion and exports were $1.4 billion, the latter including substantial amounts of petrochemicals.[43]

Given the protected nature of the Brazilian market, no strong incentive existed to restructure the Brazilian petrochemical industry. In fact, Brazil's producers may have been the most profitable global producers in the 1980s, having been protected during the downturn and then enjoying all the benefits of the 1988–89 price fly-up. Restructuring did, however, occur in the 1990s as the government, with Fernando Collor at the helm, abolished price controls and dramatically slashed import duties, both steps to make local end-product industries more competitive. Given the somewhat smaller-than-world-scale size of most Brazilian plants and the lack of need to become efficient under a regime of a protected market, the industry became more vulnerable to imports. The government further mandated Petrobras to sell its stake (i.e., the Petroquisa holdings) in the industry as quickly as possible because the government strongly endorsed privatization. By the mid-1990s the Brazilian petrochemical industry was largely in private hands.

Good Times Return—Followed by Another Downturn

Industry cyclicality is generally a good thing for investment bankers and consultants. The bankers usually do well in bad times, when companies divest assets or are merged or acquired, and they do well in good times when companies engage in acquisition sprees, using their high-priced stock as currency as well as their large cash flows. Consultants particularly like cyclical industries, since companies are more unsure about the future and therefore need to develop flexible strategies to deal with their uncertain environment. Industries that are relatively more predictable and produce fairly steady results for participants (for example, the way utilities used to be) generally have less use for strategy consultants.

Bankers and consultants were busy with chemical clients during the early and mid-1980s, trying to help them get through this frustrating period. This was also true as the end of the decade approached, and prices flew up—clients again began to enjoy coming to work. Bankers would now present grandiose acquisition ideas, while consultants would find clients flush with money to pay for various types of studies that had been deferred during more difficult times. The period from mid-1987 to mid-1989 was a happy time for the global petrochemical industry and, to a considerable extent, for much of the rest of the chemical industry and the professionals serving it.

The circumstances that bring about a price fly-up period are familiar enough to industry participants. At some point demand begins to catch up to supply because in a cyclical industry, as well as in most others, demand continues to rise from year to year, usually even in poor times; conversely, supply stays relatively constant or even declines a little since companies are not reinvesting (because of lack of cash flow) and may shut down some uneconomical capacity. As time goes on, rising demand will inexorably raise operating rates—the percentage of effective operating capacity for the product in question. In some cases planned or unexpected shutdowns may further curtail supply, thus pushing up operating rates another notch. As these rates reach and then exceed 90 percent, customers become nervous. They recognize that the situation will probably only get worse (for them) and that prices, already on the rise, will probably rise even more, as producers become short of product and start to allocate material. The domestic spot market price[44] will then invariably rise above the contract price, clearly indicating a tight supply or a shortage. The export price will rise even faster, as producers tend to take care of their domestic contract customers first and begin to curtail exports, which in poorer times tend to take care of some of their excess production. Customers concerned about shortages or playing the export market begin to order more than their normal needs, thus pushing "demand" up to an even higher level. For some chemicals the calculated operating rate will now reach or even exceed 100 percent as companies push production above the capacity for which the plant was

Table 6. The Price Fly-Up in 1988–89

Product	Leader Cash-Cost Margins (Cents/Pound)		
	1985	1988	1989
Ethylene glycol	6.0	16.0	24.2
Styrene	2.7	22.8	16.3
High-density polyethylene	2.9	13.4	9.7
Polystyrene	1.1	6.9	5.4
Pipe-grade PVC	2.0	8.5	7.0

Source: Nexant, Inc./Chem Systems.

designed, sometimes overloading reactors, even though this may shorten cata-lyst life and lead to earlier shutdowns for catalyst replacement. But by this time prices are so high that it does not matter, with the export market booming. It is time to make up for the lack of profits during the lean years and to push out as much product as possible.

As operating rates in 1987 began to rise more rapidly, the price of most petrochemicals shot up, reaching a peak about mid-1988. Ethylene, which had been in the 15-cents-per-pound range from 1984 to 1986, increased to well over 30 cents by mid-1988. Polyethylene rose from 30 cents to over 50 cents per pound, in large part reflecting the higher price of its main raw material. Ethylene glycol, which had seen prices in the 20- to 25-cents-per-pound range through most of the 1980s shot up to 45 cents per pound. Amazingly, the price of hydrocarbon raw materials remained fairly steady, producing windfall prof-its for producers of olefins and other petrochemical building blocks.

In mid-1988 Chem Systems was asked by a client to determine how high ethylene glycol prices were likely to go, with operating rates at 100 percent, a starved export market, and prices already reaching 45 cents per pound, con-trasted with the 20- to 25-cent range through most of the 1980s. Would the price go to 60 or 80 cents per pound, and what was the limiting mechanism? Chem Systems dispatched a consultant to the Far East, where demand and prices for glycol were very high, to find an answer. After a couple of weeks of work, along with that of another consultant from the firm's Tokyo office, he came back with what seemed like a possible answer, which turned out to be provided by a "value-in-use" analysis, a curve that shows at what price users of a given product would resort to some form of substitution rather than pay an even higher price. In this case the marginal buyers for ethylene glycol turned out to be suppliers of polyester fiber to the Taiwanese textile companies. The Taiwanese firms would pay up to 50 cents per pound for ethylene glycol as used to produce polyethylene terephthalate for polyester yarn. Above that price their customers would switch their polyester-cotton blends to use more

Table 7. High Petrochemical Prices Help the Bottom Line

Company	Net Income		
	1986	1987	1988
Union Carbide (in millions of U.S.$)	496	232	662
Quantum Chemical (in millions of U.S.$)	42	131	360
ICI Petrochemicals & Plastics (in millions of pounds sterling)	230	320	416

Source: Company annual reports.

cotton, thus effectively reducing their demand for polyester fiber and thus for ethylene glycol. The Chem Systems consultant also developed a cost curve that showed that the marginal suppliers to the Taiwanese were small, previously shuttered Japanese ethylene glycol plants, which had been restarted because they could produce and ship ethylene glycol to Taiwan for a landed price of 45 to 50 cents per pound with a small profit (at a time when large glycol producers' netback was 20 to 25 cents per pound higher than their manufacturing cost). This analysis made sense, but no one was ever really sure whether this rather unusual set of factors was in fact the global price-setting mechanism. Glycol never did reach 50 cents per pound in the United States.

The dramatic change that took place in the fortunes of the petrochemical industry as a result of the almost universal price fly-up in 1988 is vividly illustrated in Table 6, which contrasts cash-cost margins—the difference between price and cash cost—for several chemicals in 1988 and 1989 with those for 1985, only three years earlier.[45]

These high margins translated into robust profits, particularly for companies heavily involved in petrochemicals and plastics production. Good examples are Union Carbide, Quantum Chemical, and ICI's Petrochemicals and Plastics Division, whose financial results for 1986 to 1988 are shown in Table 7. For many other companies, including Dow, 1988 produced record earnings.

Several records were set in 1988. For the entire U.S. chemical industry, net after-tax income reached $23.7 billion versus $16.6 billion in 1987, $12.9 billion in 1986, and an average of $10 billion to $11 billion in the early 1980s. For petrochemical producers the difference between those years was much more dramatic. The overall industry operating rate was 83.9 percent (versus 71 percent in 1982), the production index was at 105.4 (versus 81.4 in 1984), and chemical exports hit $32 billion (versus an average of $20 billion in the early 1980s).[46] It was a heady time for the chemical industry: cash flows for some of the companies were truly outstanding. Sterling Chemical's net income in 1988 was $213 million, which is more than the amount the company was acquired for a couple of years earlier. Owing to the strength of both the domestic and export markets, styrene monomer—Sterling's most important product—was then arguably the most profitable petrochemical in the world.

This period also saw a number of highly unusual financial transactions, spurred by the large cash flow being generated during the petrochemical price fly-up. John Stookey, chairman of Quantum Chemical (formerly National Distillers and Chemicals), had acquired a number of ethylene and polyethylene assets during the 1980s (including Chemplex, Northern Petrochemical, Arco Chemical's polyethylene plants, and Amoco Chemical's HDPE plant at Chocolate Bayou). Now Quantum was generating a large amount of cash as the country's largest producer of polyolefins, which were selling for prices not seen for many years. Quantum Chemical's stock had jumped from around the $25 range only a few years back to well over $100 per share. Stookey and his bankers decided to pay a $50 per share "dividend" to Quantum shareholders, partly to reward them and partly to use up a lot of the cash being generated, which had made Quantum an attractive acquisition target.

Stookey asked Chem Systems to work with the bankers on a plan they had devised, which involved a one-billion-dollar loan for financing what was termed the "Quantum recap" (recapitalization). The amount was that large because Quantum was also intent on adding more ethylene and polyethylene capacity. The consulting work had to be done before the end of the year, and Chem Systems was slated to do a great deal of price forecasting (covering the industry as well as Quantum) and financial modeling. The work would be done over a two-week period, with consulting fees no object. The banking fees on this transaction were on the order of $70 million!

In retrospect, this transaction and the consulting engagement now seem like a paradigm for the whole era that culminated with the leveraged takeover of RJR-Nabisco, later described in the popular book *Barbarians at the Gate*.[47] The excesses during the latter part of the decade were fed by overexuberance, a strong economy, and almost unlimited liquidity for transactions, largely owing to the advent of junk-bond financing.

Chem Systems' financial forecasting and modeling for the Quantum recap included several scenarios, representing different economic and industry outlooks. While the *base case* included a fairly favorable scenario for the next couple of years, one of the cases showed a downturn coming in the third quarter of 1989. In this case the cash-cost margins for polyethylene started to come down fairly rapidly after that time as a result of a drop in export demand, more capacity coming on stream earlier, and other factors. It was clear that the recap, which involved huge interest and principal payments, could not have been financed with this case, but little attention was paid to Chem Systems' downturn scenario by either the bankers or Quantum Chemical.

In the second quarter of 1989, China, which was the largest customer for U.S. polymer exports, stopped purchasing chemicals and polymers for a time,

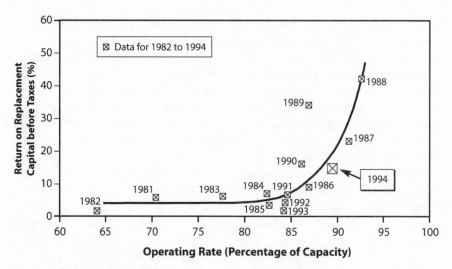

Figure 6. Industry profitability versus operating rate (1981–1994). *Source:* Marshall Frank, "Petrochemical Industry Profitability Cycle: Where Are We Now? Where Are We Going?" Chem Systems Annual Chemical Conference, Houston, Jan. 1995. Nexant, Inc./Chem Systems, by permission.

creating a sudden drop in demand. Export-oriented material began to be directed back into the domestic market. Customers who had stocked up with high inventories saw prices dropping sharply and therefore stopped buying as well, recognizing that they could replenish their inventories with lower prices later. The industry's condition now completely reversed. Chem Systems' West European profitability index (1979 = 100), which had risen above 600 in 1988, suddenly dropped to 350 in mid-1989 and to 275 by the first quarter of 1990. Similarly, Chem Systems' U.S. petrochemical industry pretax ROI for a basket of petrochemicals, which had reached 40 percent in 1988, dropped below 20 percent by 1990. This convincing evidence for the incredibly wide profitability swings of the industry is illustrated in Figure 6. The figure, which graphs the ROI against the industry operating rate from 1981 to 1994, illustrates the dramatic effect of high and low operating rates. Overinvestment during periods of high cash flow was identified as the main culprit for low operating rates.

By 1989 the industry had already decided to reinvest again, using some of the enormous cash flow generated during the twenty-month period from late 1987 to mid-1989. Of course, not all the cash went into new projects. Dow used its record-setting profits in 1988 ($2.398 billion), in part for a large-share repurchase program, to pay down long-term debt, and it spent $345 million on acquisitions, including Essex Chemical. But its capital expenditures were up 27 percent. Air Products planned to spend $550 million on capital expenditures and acquisitions in 1989. Union Carbide's capital expenditures rose from $502 million in 1987 to $671 million in 1988.[48]

Petrochemical reinvestment after the 1988–89 price fly-up was not as high as after the second oil shock, but it nevertheless helped set the stage for the next downturn. Table 8 gives a few examples of capacity increases.

U.S. production of chemicals fell in 1989, with ethylene down 5.3 percent, refinery propylene down 3.4 percent, and styrene down 6.2 percent.[49] Production picked up again after that, but cash-cost margins kept dropping. The Gulf War produced some pricing blips for a short period, particularly for aromatics derivatives, as crude oil prices rose sharply but then fell again. More North American and European capacity authorized during the 1988–89 fly-up period came on stream, while exports dropped, with more foreign capacity coming on line. In the United States spot VCM prices fell from 17 to 18 cents per pound at the start of the year to 12 cents per pound in April.[50] PVC suspension grade pricing dropped from 45 cents per pound in early 1989 to 30 cents by January 1990 and to 25 cents by early 1991.[51] In Europe nearly 1.4 million tons of total polyethylene capacity was going to come on stream during 1991–92. LDPE prices plunged in 1991, from DM 2.10 per kilogram in January to DM 1.25 per kilogram in May.[52] From 1987 to 1990 more than 1.5 million tons of worldwide p-xylene capacity had been added, and fierce competition in the Asia-Pacific region drove prices down to very low levels.[53]

As shown in Figure 7, the Chem Systems index of cash-cost margins, which had been over 300 in 1988 and still close to 200 in 1990, dropped to 100 in 1992, almost equivalent to the dismal margins characteristic of the early 1980s. The industry had gone full cycle, and prices did not reach another sustained peak until 1994–95.

In the early 1990s companies had ample opportunity to ponder the strategies they had adopted in the 1980s and in particular whether they had been wise in either divesting petrochemical assets or staying the course and enjoying the short, exciting period of high profits during the recent price fly-up. At that time Chem Systems communicated with firms like DuPont, Allied, and Monsanto, which had previously shed petrochemical assets, and was given to

Table 8. Reinvestment after 1988 Price Fly-up

Location and Product	Total Capacity at End of Year (Metric Tons × 1,000)				
	1988	1989	1990	1991	(1988–1991)
United States					
Ethylene	17,004	17,351	18,608	19,803	16% increase
Styrene	4,170	4,104	4,399	4,807	15% increase
Western Europe					
Ethylene	14,680	15,175	16,095	17,625	20% increase
High-density polyethylene	2,475	2,605	2,902	3,317	34% increase

Source: Nexant, Inc./Chem Systems.

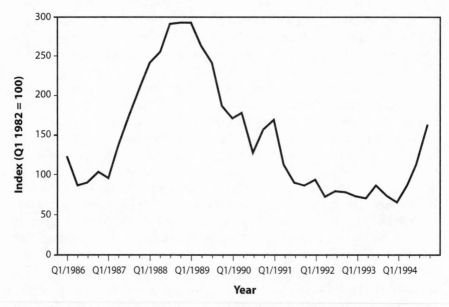

Figure 7. The petrochemical cycle (1986–1995). Chem Systems' profitability index calculated every quarter based on a weighted average of the cash margins of twenty petrochemicals indexed to first quarter 1982 = 100. *Source:* Nexant, Inc./Chem Systems, by permission.

understand that these firms were not upset about having missed a year or two of high profitability. They claimed that they were well along in implementing strategies that would emphasize higher-value specialty chemicals exhibiting less cyclicality and higher average margins. However, some of the larger firms that had chosen to remain in petrochemicals seemed resigned to the fact that the business was going to be unpredictable, with inevitable cyclicality. Their approach was to be or become a low-cost producer, to keep reinvesting, and to cut costs enough so as not to lose money at the bottom of cycles.

Around the same time Chem Systems was hired by an ethylene producer and asked to survey other producers to determine how these firms felt about reinvestment economics for olefin production. Before undertaking this survey, Chem Systems modeled a number of ethylene plants over the cycle and found that "leader" plants would just about return their firm's weight average cost of capital (WACC). The roughly two years of very high pricing and margins in the late 1980s were absolutely necessary for these businesses to earn the WACC, given the negligible profits over much of the rest of the cycle.

Chem Systems asked people from Dow and Exxon Chemical whether a return of not much more than WACC was considered satisfactory, and the answer seemed to be "yes." Both companies were heavily forward-integrated into downstream products and would expect to make better returns on their derivatives plants and, in the case of Exxon, on merchant sales of ethylene not needed for internal consumption. "What if the 1988–89 period had not been as good?" Chem Systems asked. This question elicited a curious response. These olefin producers said that the length and height of the fly-up period was in their judgment an "inevitable" outcome involving producers and customers interacting in a market that needed to provide enough return for suppliers to keep reinvesting. It was almost as if Adam Smith's invisible hand had caused the 1988 fly-up to happen and to last as long as it did. In any case these firms were evidently committed to the petrochemical industry and would not undergo the wrenching restructuring moves that took some of their competitors out of the industry to escape the inevitable cyclicality. But Dow and other companies staying the course would undertake extensive downsizing initiatives to be able to survive the bottom of the cycle.

A number of other petrochemicals producers had had enough. How the industry continued to restructure in the 1990s will be discussed in a later chapter.

The Chemical Landscape at the End of the 1980s

Chemical and Engineering News each year ranks the largest U.S. chemical producers by sales and profits. For many years it listed the top hundred largest producers, although by 2000 the wave of mergers and acquisitions shrank the magazine's list to the top fifty producers. All producers of chemicals were covered, including oil companies' chemical divisions and diversified producers that made chemicals. Only chemical sales were considered in making the ranking. Domestic and foreign sales were included. Table 9 shows how the industry lineup and revenues changed over the decade, from 1979 to 1989. The companies' positions provide an overview of how some rankings did or did not change.

The analysis found that DuPont, Dow, and Exxon Chemical were steadfast in pursuing their growth strategies. By 1990 the sales of the three largest firms had accelerated considerably. Several other chemical divisions of oil companies were now among the top fifteen producers, with Oxy and Chevron boosted by their respective acquisitions of Cain Chemical and Gulf Oil's chemical division. Shell Chemical had not grown much over the period, owing in part to restructuring divestments (vinyl chloride and polystyrene) and in part to diverting investments to western Canada and Saudi Arabia. Relatively large new players were Hoechst, which acquired Celanese, adding this firm to Hoechst's existing dyes, pigments, and resins operations; Oxy Chemical, which had acquired Cain Chemical; General Electric, which had acquired Marbon Chemical, an ABS resin producer, from Merrill Lynch Capital Partners; BASF, which had acquired Wyandotte, Inmont, Celanese's carbon fibers, Polysar Latex, and other businesses; and Quantum Chemical, which had acquired a number of olefin and polyolefin businesses, as discussed earlier.

Monsanto and Union Carbide were arguably the two firms whose size had

Table 9. Changes in Lineup of Top Companies through Restructuring

Top Ten Firms (1979)	Chemical Revenues (U.S.$ Millions)	Top Ten Firms (1989)	Chemical Revenues (U.S.$ Millions)
DuPont	9,700	DuPont	15,249
Dow Chemical	6,634	Dow Chemical	14,179
Exxon	5,807	Exxon	10,559
Union Carbide	5,300	Union Carbide	7,962
Monsanto	5,215	Monsanto	5,782
Celanese	3,010	Hoechst Celanese	5,658
W. R. Grace	2,619	Occidental Petroleum	5,203
Shell Oil	2,599	General Electric	4,929
Gulf Oil	2,437	BASF	4,461
Allied Chemical	2,150	Amoco	4,274

Source: Chemical & Engineering News (C&EN) 58 (9 June 1980), 44, and C&EN 68 (18 June 1990), 46.

been affected to the greatest extent by restructuring actions. The events that transformed these two firms in the 1980s are briefly described below.

Monsanto

The dismal performance of chemical companies in the early 1980s did not spare Monsanto. Accordingly, in 1983, Monsanto's CEO, John Hanley, undertook a massive restructuring of the firm.[54] The main objectives were to reduce the company's petrochemicals business, then accounting for 80 percent of sales, and to restructure the firm through reorganization and by taking out layers of management. DuPont's acquisition of Conoco, which had a joint-venture cracker with Monsanto at Chocolate Bayou, had already helped the divestiture program. Conoco bought out Monsanto's interest as a result of an antitrust action related to nylon fibers, which were being produced by both firms. Business restructuring now identified polymer products, industrial chemicals, fibers and intermediates, agricultural products, and high-technology products (silicon wafers and electron-beam equipment) as the surviving entities. Over the next several years Monsanto shifted much of its research into its successful agricultural chemicals program as well as into biotechnology. It acquired G. D. Searle as a vehicle to commercialize its prospective pharmaceutical products from biotech research, and it divested its entire Texas City petrochemical complex (styrene, acrylonitrile, acetic acid, and plasticizers) to Sterling Chemical, a leveraged buyout company created by Gordon Cain and Virgil Waggoner. It also divested a number of European operations, including acrylic fibers in the United Kingdom, Ireland, and Germany, and fiber intermediates in the United Kingdom.[55]

By the late 1980s the resulting restructured firm was focused strongly on agricultural chemicals, pharmaceuticals, and biotechnology, while it depended on its chemically related businesses to fund its further transformation. (In the late 1990s these businesses were spun off in a company named Solutia.)

Union Carbide

In 1984 a gas leak in the isocyanate plant of Union Carbide India in Bhopal, was responsible for the death of over ten thousand people living close to the plant. The potentially highly adverse consequences of this accident caused Union Carbide's stock price to fall precipitously. GAF Corporation then announced its intention to take over a weakened Union Carbide and began to acquire its stock. Defending itself against this serious takeover threat, Union Carbide made an offer to buy back up to 55 percent of its shares for a total price of $85 per share, about $10 over the GAF offer, and to raise its dividend from $3.50 to $4.50 per share. To finance this purchase, it announced a series of divestitures to raise cash for paying back the massive debt incurred. Over the next two years UCC sold its films packaging business, battery business, specialty polymers and composites business, home and automotive products business (including the Prestone brand), and agricultural chemicals business. Total proceeds for these businesses were on the order of $3.5 billion.

As a result Union Carbide stayed independent and became a considerably smaller, more focused firm, with a major commitment to petrochemicals (ethylene, ethylene oxide, ethylene glycol, ethylene oxide derivatives, vinyl acetate, and acrylates), polyolefins, and specialties (for example, Amerchol [a personal care chemicals subsidiary], resins, solution vinyls, and waxes). It also started an important joint venture investment program in western Canada to make ethylene glycol destined for Asia. The company further restructured itself into a holding company, with Union Carbide Industrial Gases (later named Praxair) and UCAR Carbon Company as separate subsidiaries slated for divestment.

The changes at Monsanto and Union Carbide must be considered as among the most important restructuring moves made by the industry during the 1980s. Several of the traditional, old-line chemical companies (Monsanto, Allied, DuPont, and ICI) became disenchanted with petrochemicals and adopted new strategies involving greater concentration on specialties. The chemical landscape had changed considerably, with a much greater presence of foreign companies, the creation of a number of LBO firms, and the decision by several petrochemical firms to look for areas of the world (Canada and the Middle East) where they could, usually in joint venture with local firms, become low-cost global producers. As the industry entered the 1990s, another downturn would present new challenges that would result in more dramatic change.

Endnotes

1. "Rising Oil Prices Spur Growth of New Technology," *European Chemical News, European Review '79*, supplement to 23 July 1979 issue of *Chemscope*, 18–30.

2. Darrell Aubrey, a consultant at Chem Systems, developed a historical correlation that showed ethylene demand to have a consistent long-term demand growth trend of 3.2 percent annually.

3. Martin B. Sherwin, "Chemicals from Methanol," *Hydrocarbon Processing* (Mar. 1981), 79–84.

4. "Rising Oil Prices" (cit. note 1), 30.

5. William A. Brophy, "New Routes to Chemicals from Synthesis Gas," Chemical Marketing Research Association conference, San Antonio, 15–18 March 1982.

6. "Costlier Crude Oil: No Snag for Synthetics," *Chemical Week* 124 (28 Mar. 1979), 19–20.

7. R. L. Grandy, "The Promise of the 1980s," Chemical Marketing Research Association conference, New York, 5–7 May 1980.

8. "It Was a Grim First Quarter," *Chemical Week* 130 (28 April 1982), 10.

9. "Coping with the Cost-Price Squeeze," *Chemical Week* 130 (20 Jan. 1982), 40.

10. "For CPI Leaders: Pre-dawn Darkness," *Chemical Week* 132 (2 Mar. 1983), 66.

11. "Capital Spending: First Drop Since 1968," *Chemical Week* 132 (12 Jan. 1983), 41.

12. Chem Systems had a process research laboratory from the late 1960s through the mid-1980s dedicated to developing new catalytic routes to large-scale petrochemicals. Amoco Chemical's maleic anhydride process used technology originally developed in this laboratory, and BASF licensed but never used commercially a propylene oxide process patented by Chem Systems.

13. Dick Carlson, personal communication.

14. Joseph R. Bower, *When Markets Quake* (Boston: Harvard Business School Press, 1986), 98–99.

15. Chem Systems, personal communication.

16. Keith Chapman, *The International Petrochemical Industry: Evolution and Location* (Oxford: Basil Blackwell, 1991), 236.

17. H. Kamata and M. Hongoh, "Reconstructing the Japanese Petrochemical Industry—The Japanese Petrochemical Plan," European Chemical Marketing Research Association conference, Venice, 10 Oct. 1983.

18. J. A. Philpot, "Climbing the Barriers," European Chemical Marketing Research Association conference, Amsterdam, Oct. 1984.

19. Bower, *When Markets Quake* (cit. note 14), 180–184.

20. "Restructuring of the European Chemical Industry," Chem Systems Multiclient Study, 1983, personal communication.

21. Chapman, *International Petrochemical Industry* (cit. note 16), 241.

22. Ibid., 243.

23. Ashish Arora, Ralph Landau, Nathan Rosenberg, eds., *Chemicals and Long-Term Economic Growth* (New York: Wiley Interscience, 1998), 400.

24. C. M. Doscher, "Restructuring: What Still Needs to Be Done," European Chemical Marketing Research Association conference, Antwerp, 1986.

25. Italo Trapasso, "Rationalization in Western Europe," European Chemical Marketing Research Association conference, Venice, 10 Oct. 1983.

26. Fred Aftalion, *A History of the International Chemical Industry* (Philadelphia: University of Pennsylvania Press/Chemical Heritage Foundation, 1991), 326.

27. Chapman, *International Petrochemical Industry* (cit. note 16), 248.

28. J. D. DeWitt, "The Needs and Progress in U.S. Chemical Rationalization," European Chemical Marketing Research Association conference, Venice, 10 Oct. 1983.

29. Under global trade agreements, importing countries can initiate antidumping

actions against countries that export products at what must be shown as a price that is lower than the exporting country's production costs.

30. "A Craving for Chemicals Changes to Aversion," *Chemical Week* 130 (21 Apr. 1982), 32–33.

31. The story of this acquisition is told in Gordon Cain, *Everybody Wins! A Life in Free Enterprise*, 2nd ed. (Philadelphia: Chemical Heritage Foundation, 2001).

32. A typical "take-or-pay" contract is an agreement between a buyer and seller, whereby the buyer agrees to purchase a certain quantity (usually a maximum and a minimum amount) at an agreed-on formula price and incurs a financial penalty if a decision is made to take less product.

33. Cain, *Everybody Wins!* (cit. note 31), 220–221.

34. Information from brochure published in 2000 by Huntsman Corporation, Salt Lake City, Utah.

35. John Akitt, "The U.S. Chemical Industry—New Directions," Society of Chemical Industry meeting, Florence, 10 Oct. 1983.

36. Harold Sorgenti, "Restructuring towards a New Age of Vitality," Society of Chemical Industry meeting, Madrid, 6 Oct. 1986.

37. *U.S. Chemical Industry Statistical Handbook* (Arlington, Va.: Chemical Manufacturers Association, 1994), 29.

38. Kamata and Hongoh, "Reconstructing the Japanese Petrochemical Industry" (cit. note 17), 59.

39. As plant size increases, the investment generally rises by only the six-tenths power of the increase. Thus, doubling the size of a new plant may increase its construction cost by only 60 percent. As a result the fixed production cost per pound from the larger plant is considerably less than that from the smaller plant, making the large plant more competitive.

40. Ryota Hamamoto, personal communication.

41. Sumio Takeichi, "The Nature of the Chemical Industry in Japan," Chemical Management and Resources Association meeting, New York, 9 May 1990.

42. Michel Hartveld, personal communication.

43. Ibid., based on statistics from Abiquim (Association of Brazilian Chemical Producers).

44. Chemicals are either sold under contracts or in the spot market, where normal laws of supply and demand prevail. Thus, during periods of high demand, the spot market price can be substantially higher than the market price, which reflects typical contracts then in force. The reverse is also true.

45. Chem Systems U.S. Annual Report, "Price Forecasts," Oct. 1999, and personal communication.

46. *U.S. Chemical Industry Statistical Handbook* (cit. note 37), 27.

47. Bryan Burroughs and John Helyar, *Barbarians at the Gate* (New York: Harper & Row, 1990).

48. Lewis Koflowitz, "Set for a Sparking Second Half," *Chemical Business* 10 (June 1988), 43–49.

49. Andrew Wood, "No Easy Balance in Petrochemicals," *Chemical Week* 146 (28 Mar. 1990), 23.

50. Andrew Wood, "Polyethylene Prices Bottom Out, but Oversupply Looms," *Chemical Week* 148 (8 May 1991), 12.

51. Chemical Data Inc., "Monthly Petrochemical and Plastics Analysis," Feb. 1999, p. 81, and personal communication.

52. Wood, "Polyethylene" (cit. note 50), 13.
53. Andrew Wood and Ian Young, "Oversupply Dogs the Petrochemicals Arena," *Chemical Week* 148 (10 April 1991), 24.
54. "The Reworking of Monsanto," *Chemical Week* 132 (12 Jan. 1983), 42–47.
55. Lyn Tattum, "Monsanto Is Putting the Accent on Europe-Africa Operations," *Chemical Week* 147 (1 Aug. 1990), 20.

Petrochemical Technology Developments

Jeffrey S. Plotkin

During the last two decades of the twentieth century little new breakthrough process technology was commercialized. Petrochemical process technology and catalyst developments had progressed rapidly in the 1950s and 1960s but had slowed down somewhat in the 1970s, a turbulent decade with depressed industry conditions both before and after the first oil shock in 1973–74. The industry's situation worsened shortly after the second oil shock in 1978–79, and companies began retrenching in the early 1980s, with much more emphasis on cost cutting than on research and development.

But there were other reasons for the decline in successful process research. Companies recognized that plants for producing most commodity chemicals were usually based on quite efficient processes. These plants could, with proper maintenance, operate for decades without fear of being made obsolete by plants using new higher-selectivity catalysts, as had been the case in the 1960s and 1970s. It also became clear that any new process would have to decrease production costs by 25 percent to 35 percent to create "shutdown economics" over existing plants: a new plant using newly developed technology based its economics on earning an acceptable return on the new investment (over and above cash costs), while largely depreciated existing plants would keep running if they just recouped cash costs. So although a number of new, "elegant" processes, often based on less expensive feedstocks, were patented in the United States, Japan, and Europe from 1980 to 2000, few of them reached full commercialization, and none had shutdown economics.

Technology has often been characterized as a "strategic weapon," a means of dominating the production of a given chemical using patented technology. Arco Chemical used this technique with propylene oxide, and Amoco

Chemicals used it with purified terephthalic acid. Monsanto could also have emulated this strategy if it had not decided to license its methanol carbonylation technology for producing acetic acid, and there are other examples as well. In the last two decades of the twentieth century, technology's role as a strategic weapon declined markedly as the availability of world-class process technologies became much more prevalent through licensing arrangements, thus creating competitors: Union Carbide's Unipol polyethylene process is a good example. The period was characterized by technology dissemination, with few technology breakthroughs that could be exploited by the inventing firm. With fewer companies staying the course as commodity chemicals producers, the amount of process research carried out on potentially "game-changing," or breakthrough, technologies declined markedly, with much more emphasis placed on improving existing technologies.

The chemical industry has always been characterized as a leader in technology development and commercialization. Organic and inorganic chemists invented new molecules, physical chemists optimized reactions, and chemical engineers designed complex manufacturing plants, scaling up laboratory results. Over the hundred or so years of the so-called modern chemical industry, inventors—in many cases working for chemical firms—have filed more patents than in almost any other industry. These patents cover new reactions, specialized equipment, and other technological advances. They have allowed companies to build plants that, for a time at least, enabled the firms to establish competitive advantage over peer companies that could not produce the same product or enjoy the same economic benefits conferred by the patented technology. Proprietary technology thus became an "entry barrier," allowing the firm holding patent protection or confidential know-how to gain financial benefits over its competitors.

That such new, unique technology has been playing a less important role, despite continued fairly substantial R&D spending, is a significant recent trend. The reasons for this are examined in some detail in this chapter. A contributing factor is that a significant part of companies' R&D budgets has been spent on process and product improvements rather than on process research, with much less on "blue sky" fundamental research (often defined as research for research's sake or research with long-term time horizons) of the type carried out in earlier times, which sometimes led to breakthrough products and processes. Companies now generally insist that R&D efforts be tied to company strategy, particularly to business-level strategy. Such strategies tend to involve supporting product lines, making incremental process improvements, and tweaking existing products and processes, in some cases only to gain small advantages over competitors, thus enhancing market position.

Process research of the earlier kind is still being carried out, but recent history has shown that the odds for probable success of a really novel technology that might change the way the industry makes a certain large-scale chemical have grown very long. In the 1950s and 1960s new plants were often sized in the range of 5,000 or 10,000 tons per year. The much lower cost of building such small "first-of-a-kind" plants greatly lessened the financial risk for commercializing a new process. Scientific Design, arguably the most successful independent developer of petrochemical technologies, stubbed its toe a couple of times when a small new plant based on a process the company developed in its laboratory eventually had to be scrapped because the plant would not replicate laboratory results or was built with unsuitable materials of construction. But the next plant was usually a great success.[1] New plants are now often built with capacities of 500,000 tons per year, which raises the bar for the commercialization of new technologies to a level few if any companies are willing to scale.

This chapter covers the state of petrochemical technology at the turn of the millennium, with some coverage of other chemical technologies.

Are We at the Limit of Petrochemical Process Technology?

Is virtually everything in catalyst development and process design for petrochemical process technology already invented? Are there opportunities for any truly new developments in the major petrochemical building blocks, intermediates, polymers, and plastics? Looking back over the last twenty years, an unbiased observer might answer these two questions with "just about" and "relatively few." While some may disagree with these answers, there is little doubt that new process technologies with shutdown economics that might have reached full commercialization between 1980 and 2000 are virtually nonexistent. Why was nothing new created, and what have all the R&D dollars spent over the last two decades achieved? Where has the emphasis been placed, and what are the prospects for new developments in petrochemical process technology over the next twenty years?

Nexant, Inc./Chem Systems, has been tracking the performance of the major petrochemicals and plastics for over twenty-five years. Table 1 lists the major petrochemicals and plastics and their corresponding process technology, feedstocks, and yields. The yield data are subdivided into "leader" and "laggard" categories. The leader figures reflect performance for the top 25 percent of plants currently operating in the United States, while the laggard data are representative of the bottom 25 percent of plants and can serve as a proxy for technology and performance of, say, twenty years ago.

The yields overall are strikingly good for most processes for both leader and laggard plants. This reflects a mature industry.

Oxidation process technology is one area in which the most potential for process improvement through more selective catalysts might be expected; however, with the exception of ethylene oxide–ethylene glycol and acrylonitrile, oxidation catalyst yields were already very good by 1980, and only small to moderate incremental improvements were achieved over the last twenty years.

Table 1. Feedstocks, Processes, and Yields for Key Petrochemicals

Product	Feedstock	Process Type	Yield Laggard	Yield Leader
Ethylene	Ethane	Steam cracking	83.8%	85.6%
Ethylene	Naphtha	Steam cracking	30.4%	30.6%
VCM	Ethylene	Balanced oxychlorination	94.9%	96.5%
Ethylene oxide	Ethylene	Direct oxidation	72.2%	81.5%
Ethylene glycol	Ethylene	Direct oxidation/ hydrolysis	66.4%	69.8%
Acrylonitrile	Propylene*	Ammoxidation	77.5%	81.1%
Styrene	Ethylene	Alkylation/ dehydrogenation	87.9%	94.7%
Cumene	Propylene†	Alkylation	94.4%	96.3%
Phenol	Cumene	Cumene peroxidation/ rearrangement	91.8%	96.0%
PTA	p-Xylene	Oxidation	94.6%	95.3%
LDPE	Ethylene	Laggard—autoclave/ leader—tubular	96.1%	98.5%
LLDPE	Ethylene‡	Butene comonomer	98.9%	99.5%
HDPE	Ethylene	Injection mold grade	97.5%	98.5%
Polypropylene	Propylene§	Laggard—slurry/ leader—bulk loop	97.0%	98.8%
PVC	VCM	Suspension process	98.4%	99.5%
Polystyrene	Styrene	Bulk polymerization— GPPS	98.7%	99.4%
PET	PTA	Continuous/bottle grade	98.9%	99.8%

Source: Nexant, Inc./Chem Systems.
GPPS = general-purpose polystyrene; HDPE = high-density polyethylene; LDPE = low-density polyethylene; LLDPE = linear low-density polyethylene; PET = polyethylene terephthalate; PTA = purified terephthalic acid; PVC = polyvinyl chloride; VCM = vinyl chloride monomer.
* Chemical grade, 93%.
† Refinery grade.
‡ 7% butene comonomer.
§ Polymer grade.

Of course, even small improvements in the production economics of commodity petrochemicals can add up to significant cost savings.

In the production of ethylene oxide (EO), yields based on ethylene starting material were around 70 percent (average over the time the reactor was in operation before replacing the spent catalyst) in 1974. EO yields have increased incrementally over time to the current state-of-the-art EO catalyst enjoying an average selectivity of about 81.5 percent. Start-of-run selectivities of the latest EO catalysts are above 90 percent. Future gains in selectivity will be ever more difficult to find, and the question of whether the effort and expense will outweigh the benefits is legitimate. Nevertheless, the two leading EO catalyst licensors, Shell and Scientific Design, are continuing to press ahead with their EO catalyst improvement programs, probably driven mostly by competitive pressures to gain new EO process licensees. And further intensifying competitive pressures in this area is another EO process technology licensor, Dow, which obtained this expertise through its recent acquisition of Union Carbide.

Acrylonitrile is another example of a product with a catalyst selectivity still significantly less than 95 percent. And, as shown in Table 1, while some progress in improving selectivity has been made over the last twenty years, yields are still only slightly more than 80 percent. However, even in this case, the situation is not as poor as the low selectivity might indicate because the main by-product, about 9 percent to 10 percent by weight, is hydrogen cyanide (HCN). HCN has value as a feedstock in methyl methacrylate, hexamethylene-diamine, methionine, and glyphosate production and in other specialty chemical products and intermediates. Acrylonitrile producers, depending on their own internal needs or contractual obligations may actually choose to run their acrylonitrile units "sloppily" and intentionally increase HCN production. The next important change in acrylonitrile production will probably be the use of propane instead of propylene as feedstock. This approach will be discussed later in the chapter.

In brief, process yield improvement, with just a few exceptions, was not an area of great progress or even need over the last twenty years with regard to most of the largest-volume petrochemicals and plastics. Despite this, many catalyst advancements have been commercialized in the last twenty years. Excellent reviews by John Armor have detailed the catalyst advancements developed in the United States from 1980 to 2000.[2] Armor cites more than 250 examples of catalyst developments over the last twenty years in the United States alone. Many of these examples are in the areas of specialty and fine chemicals synthesis, emission control, and petroleum processing. However, Armor does cite examples of new catalyst developments in regard to large-volume petrochemicals production; and in the new process-technology examples that follow in this chapter, catalyst innovation was the key enabler for virtually all of them.

Size Matters

So if feedstock efficiency gains were not an overriding issue during the last twenty years, what was the major driver in the quest for continual cost reduction? Table 2 helps to answer this question and reveals that the other driver was success in finding out how to build larger "single-train" plants to reduce fixed costs per pound of production.

As discussed above, from 1980 to 2000 there was relatively little difference, with some exceptions, between leader and laggard technology in terms of catalyst selectivity and feedstock efficiency. Thus, for operators to gain competitive advantage, other differentiators had to be used. Large plants can reduce capital costs per unit of product by virtue of "economy of scale." If plants could be enlarged by making equipment larger (instead of just ganging existing equipment), then capital costs per unit of product could be reduced by an exponential factor. In addition to lowering capital costs per unit of product, large scale also reduces fixed operating costs per unit of product.

A convenient rule of thumb often used by engineers to estimate quickly the capital cost of a proposed larger plant is to use the known costs of a smaller existing plant using similar process technology and to scale up the capital by an exponential factor ranging typically from 0.55 to 0.7. It is an interesting and informative exercise to "back out" scale-up factors retrospectively by comparing sizes and investments for existing leader and laggard plants for the same products. The magnitude of the implied scale-up factor is a strong indicator of whether or not fundamental process improvements were implemented or the larger plant simply benefited from economy of scale.

An ethane-fed olefin steam cracker is a good example. The implied scale-up factor between a leader (571 thousand metric tons [KMTs] per year) and a laggard (367 KMTs per year) plant is 0.67. This is within the range of the 0.55 to 0.7 scale-up factor commonly used by cost engineers and infers that ethane steam cracking has not undergone any significant change in technology and is mainly benefiting from economy of scale. No change in the basic approach to olefin production, such as catalytic ethane or naphtha cracking, is foreseen in the short to medium term, although producing olefins using methanol is a possibility and will be discussed later.

Catalyst improvements have contributed to lowering the implied scale factor to less than 0.55 in such products as ethylene glycol and cumene-phenol. In the case of ethylene glycol the implied scale-up factor is 0.504. Catalyst selectivity improvements in EO production, as discussed above, have led to increased feedstock efficiencies, allowing more product throughput per unit volume of reactor. In cumene, use of new zeolite-based alkylation catalysts has, in comparison to older solid phosphoric or aluminum chloride catalysts, increased selectivities only modestly but, much more important, boosted product throughputs by as much as 50 to 100 percent. This has allowed existing plants to switch to new zeolite catalysts and boost capacities with relatively low additional capital investment or allowed new plants to be less capital intensive. In addition to improved throughputs and yields, solid zeolite catalysts avoid issues of disposal often encountered with solid phosphoric acid and aluminum chloride catalysts. This trend toward the use of solid catalysts is likely to continue for reasons of environmental friendliness but only if increased operating efficiencies and hence lower costs are also part of the equation. Another example of this is in bisphenol A production, where new ion-exchange catalysts offer very high throughputs in comparison to conventional mineral acid catalysts.

In the cases of polyvinyl chloride, polystyrene, and polyethylene terephthalate, simple economy of scale was the major factor for decreased capital costs

Table 2. Technology and Unit Investments for Key Petrochemicals

Product	Feedstock	Process	ISBL/Unit Laggard ($/MT)	ISBL/Unit Leader ($/MT)	Implied Scale Factor
Ethylene	Ethane	Steam cracking	597	517	0.674
VCM	Ethylene	Balanced oxychlorination	389	272	0.618
Ethylene glycol	Ethylene	Direct oxidation/hydrolysis	512	387	0.504
Acrylonitrile	Propylene*	Ammoxidation	791	651	0.566
Styrene	Ethylene	Alkylation/dehydrogenation	412	288	0.618
Cumene	Propylene†	Alkylation	203	87	0.507
Phenol	Cumene	Cumene peroxidation/rearrangement	1,648	681	0.497
PTA	p-Xylene	Oxidation	762	578	0.600
LDPE	Ethylene	Laggard—autoclave/leader—tubular	678	418	-2.152
LLDPE	Ethylene	Butene comonomer	630	189	0.294
HDPE	Ethylene	Injection mold grade	492	211	0.324
Polypropylene	Propylene‡	Laggard—slurry/leader—bulk loop	673	313	-0.803
PVC	VCM	Suspension process	586	347	0.552
Polystyrene	Styrene	Bulk polymerization—GPPS	389	295	0.789
PET	PTA	Continuous/bottle grade	1,156	809	0.558

Source: Nexant, Inc./Chem Systems.
GPPS = general-purpose polystyrene; HDPE = high-density polyethylene; ISBL = inside battery limits; MT = metric ton; LDPE = low-density polyethylene; LLDPE = linear low-density polyethylene; PET = polyethylene terephthalate; PTA = purified terephthalic acid; PVC = polyvinyl chloride; VCM = vinyl chloride monomer.
* Chemical grade.
† Refinery grade.
‡ Polymer grade.

per ton of product. On the one hand, no major change in catalyst or reactor technology was commercialized for these materials in the last twenty years. On the other hand, it is clear that some important developments in polyolefins took place between 1980 and 2000, based on the low (even negative) implied scale-up factors. A negative implied scale factor indicates that owing to an important change in technology building a new, larger plant requires less capital than building a smaller plant using older technology. For example, a tubular low-density polyethylene (LDPE) plant with a capacity of 141,000 metric tons per year requires an ISBL (inside battery limits) investment of about $59 million, while a smaller 121,000-metric-ton-per-year autoclave plant requires a greater investment of $82 million.

Perhaps no area of petrochemical technology received as much attention and R&D spending as polyolefins from 1980 to 2000. The following section highlights the major issues and results of polyolefin research activity over the last twenty years.

Polyolefins—Technology Driving Change

No other segment of petrochemical processing technology has seen as much change and innovation over the last twenty years as the polyolefins area. However, these developments have come with a high price tag in terms of R&D expenditure. Nexant, Inc./Chem Systems estimates that over $3.5 billion were spent in technical and commercial R&D for second-generation polyolefin technology in the 1990s alone (this figure does not include capital investment for any new plants). It is valid to ask whether the benefits of these new technologies—for example, lower costs, higher throughputs, and improved polymer properties—have accrued to the technology developers. In the short term the answer seems to be "no," as producers struggle to pass along the development costs by way of higher polymer prices. Evidence of this inability to recoup development costs is the intense industry consolidation that polyolefin producers have engaged in over the last several years.

Union Carbide's UNIPOL Process

Through 1977 the polyethylene business had matured in a fairly smooth curve, with two general process types making up the bulk of the industry output: LDPE produced using autoclave and tubular reactors and HDPE produced using slurry reactors. In 1977 Union Carbide announced that it had adapted its fluidized-bed gas-phase process for HDPE to make a polyethylene copolymer with a density of 0.925 grams per cubic centimeter. This new material, linear low-density polyethylene (LLDPE), supposedly offered many property advantages over conventional LDPE. In addition to these advantages Union Carbide claimed that the process offered cost savings over the cost of high-pressure LDPE processes. This development breakthrough therefore seemed to promise higher resin selling prices, owing to superior performance, and lower pro-

Table 3. Polyethylene Technology Licensing Competition (1998)

Company	High-Density	Linear Low-Density	Low-Density
Basell*	✔	✔	✔
Borealis	✔	✔	—
British Petroleum (BP)	✔	✔	—
Chevron Phillips	✔	✔	—
Dow†	✔	✔	—
DSM	✔	✔	✔
Enichem	—	—	✔
Equistar/Maruzen	✔	—	✔
ExxonMobil	—	—	✔
Mitsui	✔	✔	—
Nippon PC/Japan Polyolefin	✔	—	—
Nova	✔	✔	—
Simon Carves	—	—	✔
Univation‡	✔	✔	—

Source: Nexant, Inc./Chem Systems.
* Joint venture of BASF and Shell.
† Does not license but operates major competitive technology.
‡ Joint venture of Dow and ExxonMobil (formerly Union Carbide/Exxon).

duction costs. Because of these claims Union Carbide decided to license the process, called UNIPOL, broadly. Although history tells us that gas-phase LLDPE resins sold at a discount to LDPE, Union Carbide sold many licenses and generated significant revenue from up-front license fees and running royalties.

Owing to Union Carbide's success, many other technology developers also started licensing polyolefin process technology. Table 3 presents an overview of polyethylene licensing technology competition. While the UNIPOL process has become the dominant technology, intense competition has caused lower licensing fees for the licensors and greater access to world-class process technology for producers, which in turn leads to more competition and lower prices for better resin—clearly not a recipe for success. In order to differentiate themselves in the marketplace, licensor-producers initiated in the early 1990s R&D programs to examine so-called second-generation polyolefin technology.

Second-Generation Polyolefin Technology

In the mid-1970s a new type of polyethylene polymerization catalyst was discovered. This new catalyst type was based on a combination of a metallocene (a transition metal sandwiched between cyclopentadienyl rings) and a cocatalyst, methylaluminoxane. The development of metallocene catalysts, also more broadly termed single-site or second-generation catalysts, did not start in earnest in commercial laboratories until the late 1980s and early 1990s. The number of patents and patent applications for new catalyst compositions, polymerization processes, and new products using single-site catalysts exceeds 2,700 and were assigned to more than fifty companies around the world. From 1994 to 1998 patent activity exploded, reflecting the intense research effort in this area in the early 1990s.

In the late 1990s nonmetallocene single-site catalysts began to appear in the patent and trade literature. So-called bimodal resins were also being developed at that time, as it was recognized that metallocene-catalyzed polymerization was too "perfect" and led to resins with a very narrow molecular-weight distribution. While narrow molecular-weight distribution is desirable for certain end-use applications, it causes difficulty for processors' equipment. Researchers hoped that bimodal resins could solve this problem and lead to a combination of tailored properties and "easy processing" characteristics. Significant commercialization of second-generation polyolefins did not occur until 1997 to 2000, although owing to the higher prices for these new resins, market demand for the second-generation polymers was weaker than hoped for. In short, second-generation polyolefins development has followed a typical new product development life cycle. From 1990 to 2000 there was a very high level of R&D activity, followed by commercialization and capital investment. The post-2000 period is expected to be the market development and growth period.

In addition to the effort in second-generation catalyst development, other engineering innovations were also commercialized during the same period. These developments have allowed producers to lower costs and reduce capital intensity. These innovations include implementing "condensed mode" operation in gas-phase plants to increase reactor productivity, gas-phase octene-1–based resins, in-situ production of comonomer, and development of mega-scale reactors. Perhaps the most noteworthy part of the development of polyolefin technology is the amazingly large number of companies participating in the development effort and the large number of approaches that have been investigated and invested in. This is clearly illustrated in Table 4, which lays out the various companies and polyolefin polymerization techniques that have been developed. The high level and breadth of industry participation is indicative of the importance of polyolefins to the petrochemical industry.

A number of technology developments for polypropylene production also occurred between 1980 and 2000. The polypropylene industry is only about forty years old because the original Imperial Chemical Industries (ICI) high-pressure polymerization technology for polyethylene does not work for propylene polymerization. In the 1950s Giulio Natta, a chemist working for Montecatini Chemical Company, recognized that Karl Ziegler's polyethylene catalyst could also cost-effectively polymerize propylene to polypropylene. Ziegler, who worked at Hoechst, had earlier produced a linear form of poly-

Table 4. Companies Involved in Second-Generation Polyolefin Development and Commercialization (1998)

Company	Linear Low-Density	High-Density	Polypropylene
North America			
Dow	Commercial	Commercial/ semi-commercial	Commercial/ semi-commercial
Equistar	—	Developmental	—
ExxonMobil	—	—	Commercial/ semi-commercial
Nova	Commercial/ semi-commercial	—	—
Chevon Phillips	Commercial/ semi-commercial	Developmental	—
Western Europe			
Atofina	—	—	Commercial/ semi-commercial
Basell	Commercial/ semi-commercial	Commercial/ semi-commercial	Commercial/ semi-commercial
Borealis	Commercial/ semi-commercial	Commercial/ semi-commercial	—
British Petroleum	Commercial/ semi-commercial	Developmental	—
DSM	Commercial/ semi-commercial	—	—
Japan			
Chisso	—	—	Developmental
Japan Polychem	Commercial/ semi-commercial	—	Commercial/ semi-commercial
Japan Polyolefin	Commercial/ semi-commercial	Developmental	—
Mitsui	Commercial/ semi-commercial	Developmental	Developmental
Ube	Commercial/ semi-commercial	—	—

Source: Nexant, Inc./Chem Systems.

ethylene. Many advances, primarily in catalyst activity, have been made in the last forty years. For example, early polypropylene catalysts in the 1960s had a polymerization activity of 1,000 tons of polymer per ton of catalyst. In contrast, the latest fourth-generation catalysts now have a polymerization activity up to 60,000 tons per ton of catalyst. Metallocene and nonmetallocene single-site catalysts have also been under development for polypropylene. Single-site catalysts allow syndiotactic polypropylene (a specific type of branching of polypropylene produced with certain newer types of catalysts) to be made, but commercial acceptance of this material has been slow. Much progress has also been made in reactor technology and reactor scale, as discussed earlier.

C₁ Chemistry: Chemicals from Coal

The second oil shock of the late 1970s and early 1980s caused petroleum prices to soar as high as $30 per barrel by 1980. Forecasts of $50 per barrel of oil were common during this period. The oil shock caused a flurry of research activity by many U.S. petrochemical companies in the early 1980s to develop new process technology to make chemicals from feedstocks other than oil. Coal, abundant in the United States, was a natural choice as a feedstock for obtaining the necessary carbon atoms for synthesizing "petrochemicals" traditionally made from petroleum-based feedstocks.

The roots of coal-based liquid fuels and chemicals date back to pre–World War II. Germany, having found its sources of petroleum cut off, developed techniques to convert coal to synthesis gas (syngas), which consists of mixtures of carbon monoxide and hydrogen. In turn, syngas was converted into liquid fuels using a process called the Fischer-Tropsch reaction. Syngas was also useful for making certain chemicals. This type of reaction chemistry using syngas is broadly termed C_1 chemistry.

While such chemicals as methanol, ammonia, urea, formaldehyde, and acetic acid were all commonly made using syngas-based chemistry, new C_1-based routes to other chemicals were being explored by a number of oil and petrochemical companies, particularly in the United States, that were hoping to escape the impact of rising crude oil prices.

Direct routes to ethylene glycol from syngas (or direct derivatives such as methanol and formaldehyde) received a lot of attention from research laboratories in the early 1980s. A good review article by John Kollar presents a summary of the various routes explored.[3] Very severe reaction pressures and temperatures were usually required, especially when attempting to convert syngas directly to ethylene glycol. Unfortunately, robust catalysts able to achieve good selectivities were never developed.

During 1981 and 1982 Halcon International, the parent company of Scientific Design, a technology development company with a long history of successful petrochemical developments, announced an elegant multistep, totally C_1-based route to vinyl acetate monomer (VAM). The Halcon route was a multistep sequence beginning with the reaction of methanol and acetic acid. In contrast to the conventional route to vinyl acetate where only two of the four carbon atoms are derived from syngas, all four of the carbon atoms in the Halcon process were derived from syngas. However, as oil prices returned to more typical pricing, research on this route was abandoned.

One significant C_1 chemistry process that was successfully commercialized during the early 1980s was Eastman Chemical's coal–to–acetic anhydride plant in Kingsport, Tennessee. Eastman, then a subsidiary of Eastman Kodak, was already a major producer of a number of chemicals. The Eastman plant is back integrated into coal, and the key aspect of the process is the non-noble metal-catalyzed carbonylation of methyl acetate to acetic anhydride. This process technology was jointly developed by Eastman and Halcon.

Other candidates for syngas research during the early 1980s were direct routes to acetic acid, ethanol, propanol, methyl acetate, and ethyl acetate. Texaco, Union Carbide, and Exxon Chemical were particularly active in this type of research during this time. Unfortunately, owing to the challenge of developing selective catalysts and the severe reaction environments needed, none of these approaches were ever commercialized. As oil prices returned to more normal levels, this line of research was put on the back burner, not to be picked up again in earnest until about the mid-1990s. This renewed effort in C_1 chemistry was and is driven not by the availability of low-cost coal, but by the desire to monetize low-cost natural gas in remote, or "stranded," areas (see later under "C_1 Chemistry Is Visited Yet Again").

No discussion of C_1 chemistry would be complete without mention of Sasol's innovative technology developments. Sasol, in South Africa, the world's technology leader in converting coal or gas feeds to synthetic fuels, has also developed technology to extract valuable olefins and certain oxygenates out of their reactor product streams. In the last decade Sasol has been particularly successful in extracting comonomer-range alpha-olefins, first hexene-1 and pentene-1 and later octene-1, by developing proprietary extraction processes.

The Synthol product streams from Sasol's Secunda complex in South Africa contain a range of even and odd carbon-atom alpha-olefins. This, of course, is very different from conventional ethylene oligomerization technology for producing alpha-olefins in which only even carbon-number olefins are produced. Availability of the alpha-olefins from the Synthol streams generally decreases with increasing carbon number. Sasol now has a dominant position in comonomer-range alpha-olefins. The company has leveraged its access to low-cost alpha-olefins and has moved downstream into detergent-range alcohols and linear alkyl benzenes by acquiring Condea. Sasol has also developed extraction technology for isolating various oxygenated chemicals and solvents, such as ethanol, acetic acid, propionic acid, and various ketones and alcohols, and has set up a new division, Sasol Solvents, to produce and market these materials.

New Processes over the Last Twenty Years: Challenges and Opportunities

During the last twenty years very few major new process technologies with widespread applicability were commercialized in the field of large-volume petrochemicals, with the exception of polyolefins technology. Nevertheless, there were a number of innovative accomplishments, many of which might be called "niche" technologies because they will probably not shut down existing processes but instead will satisfy a need unique to the user of the new technology.

Much of the technology referred to above, as commercialized from 1980 to 2000, can be classified into three categories, or driving forces: the need to overcome or exploit regional supply-and-demand imbalances; the need to reduce, avoid, or recycle an environmentally noxious feedstock or by-product; and technology as a means to enter or create new businesses. Examples of technologies in each of the above categories follow. While this discussion is not intended to be a complete listing of all new process technology developed over the last twenty years, it is an attempt to give the reader a flavor of the type of processes commercialized and the driving forces behind their development.

Regional Supply-and-Demand Imbalances

This group of processes provides examples of how new technology was developed in response to real or forecasted local supply-and-demand imbalances.

Decoupling Propylene from Ethylene Production:
"On-Purpose" Routes to Propylene

Propylene is typically considered a by-product of ethylene production, and there is some concern that the supply of propylene, especially in certain regions, could fall short of demand, owing to the strong global demand for polypropylene. Because of this concern some effort has been devoted to so-called on-purpose routes to propylene. These routes cover a number of approaches.

Propane Dehydrogenation: The chemistry of propane dehydrogenation involves catalytic abstraction of hydrogen from propane to give propylene. The major process conditions are determined on the basis of thermodynamic limitations, reaction kinetics, and the economics of the conversion-selectivity relationship.

The first propane dehydrogenation plant was the National Petrochemical Company (NPC) plant in Thailand, built in 1990 using the Oleflex process from Universal Oil Products (UOP). Since then a total of six or so propane dehydrogenation plants have been built around the world. Propane dehydrogenation process technology is offered by several licensors. The processes vary among the licensors in terms of reactor design, pressure, temperature, heating method, catalysts, and catalyst regeneration techniques.

Propylene via Olefin Metathesis: Phillips Petroleum Company discovered olefin metathesis in 1964. This technology, the interchange of olefins, while quite astounding from a theoretical point of view, has found only a few niche applications. A small plant was built in Canada by Shawinigan using the Phillips' triolefin process to transform propylene into ethylene and butenes. This plant operated only a short time, from 1966 to 1972. Late in 1977 Shell incorporated metathesis chemistry as part of its Shell higher-olefins process (SHOP). In the mid-1980s Lyondell began using metathesis chemistry to make propylene. Lyondell achieved this by effecting the metathesis of ethylene and 2-butene. While using ethylene to make a lower-priced product (propylene) is questionable, two-thirds of the produced propylene molecule is derived from even lower-valued 2-butene.

Recently, ABB Lummus bought the licensing rights to the Phillips metathesis technology. BASF and AtoFina are currently jointly building a new world-scale steam cracker in Port Arthur, Texas, and are planning to incorporate ABB Lummus's olefin metathesis process for converting ethylene and butenes to propylene. When run with the metathesis unit, the new BASF-AtoFina cracker will have the capability of producing more propylene than ethylene, an unusual situation for a petrochemical complex.

Propylene via Deep Catalytic Cracking: Deep catalytic cracking (DCC) is a process for producing light olefins from heavy feedstocks. DCC was developed by the Research Institute of Petroleum Processing (RIPP) and Sinopec International, both located in the People's Republic of China. DCC uses fluid catalytic cracking principles combined with a proprietary catalyst to produce propylene and other light olefins from vacuum gas oil. RIPP has worked on the development of a zeolitic catalyst for DCC for over ten years. Four plants are currently operating in China, with three more under construction. The first application of the technology outside of China was a plant built by Thailand Petrochemical Industries (TPI) in Rayong, Thailand, in 1997. Stone and Webster, an engineering firm, has the licensing rights to DCC outside of China. This development highlights the high quality of petrochemical research that is being carried out in China, and it is likely that other process developments will be coming out of that country in the future.

Aromatics via Nonconventional Routes

Several processes that convert either light olefins, light naphtha, or liquefied petroleum gas (LPG) to aromatics have been commercialized. These types of processes have the potential to upgrade relatively low-valued feedstocks to aromatics. They also provide a new source of aromatics to those regions of the world that are typically short on conventional sources of benzene-toluene-xylene (BTX) aromatics, owing to limited availability of either pyrolysis gasoline, as produced in naphtha cracking, or catalytic reformate from naphtha reforming. These new aromatics technologies may also have growing importance in the United States as regulatory requirements for excluding light components from conventional reformate for gasoline increase.

The British Petroleum (BP)-UOP Cyclar process produces aromatics from propane and butanes (i.e., LPG). The process is best described as dehydrocyclodimerization and works best at temperatures above 425°C. The catalyst is a proprietary zeolite formulation containing a non-noble metal promoter. (Most reforming processes making aromatics use a platinum or other "noble metal" catalyst.) The aromatics product is rich in BTX. In 1999, Saudi Basic Industries (SABIC) built the first commercial plant using the Cyclar process.

The Chevron Aromax process is similar to conventional catalytic reforming, with the exception of extra sulfur removal facilities and the high paraffin level of the feedstock. The catalyst is an L-type zeolite with a high proportion of large crystals. One of the disadvantages of this catalyst is that it affords a relatively high level of toluene, the lowest-valued product of the three major aromatics. Of course, the toluene could be further processed via disproportionation to give benzene and mixed xylenes. The product mix depends on the composition of the feedstock. Chevron's Aromax process is operating at its Pascagoula, Mississippi, refinery and in Saudi Arabia.

Environmentally Driven Process Technology

New Approaches for Methyl Methacrylate

The next group of technologies discussed involves processes that avoid or mitigate the use of environmentally noxious materials and do so economically.

The conventional route to methyl methacrylate (MMA), first used in 1937, starts with the reaction of acetone and hydrocyanic acid (HCN) to give acetone cyanohydrin (ACH), which is then converted to MMA. Although this classic reaction has been used as a commercial process for nearly sixty-five years, it involves two unavoidable safety and environmental problems: the need to use very toxic HCN and the need to dispose of the ammonium bisulfate by-product safely (about 1.5 pounds per pound of MMA produced).

Two approaches have been employed over the last twenty years to solve or mitigate both of the problem areas of the conventional ACH route to MMA. One approach can be characterized as an "end-of-pipe" solution; that is, it does not avoid use of these noxious materials but rather finds ways to convert them back to their original form for reuse. Mitsubishi Gas Chemical (MGC) has used this type of approach for making MMA. In the MGC process the HCN is reconstituted at the end of the process, thus avoiding the need to make HCN and disposing of ammonium bisulfate. The second approach to making MMA in a more environmentally friendly manner is to avoid the use of HCN completely. It was logical to surmise that since methacrylic acid contains four carbon atoms, a four-carbon starting material might be ideally suited for MMA production. In fact, processes using either isobutylene or its derivative, tertiary-butyl alchohol, have been commercialized by several Japanese companies.

It was the lack of adequate HCN supply from acrylonitrile plants that drove Japanese companies to develop new C_4-based technologies that did not require the fourth carbon atom to be derived from HCN. Thus, such companies as Sumitomo Chemical, Mitsubishi Rayon, Tosoh, and Nippon Shokubai developed effective oxidation catalysts for oxidizing either isobutylene or t-butyl alcohol to methacrolein or methacrylic acid, or both. The catalysts for this transformation are similar to the catalysts used for oxidizing propylene to acrylic acid.

In 1984 Asahi Chemical converted a conventional ACH process MMA plant to a process that ammoxidized isobutylene to methacrylonitrile similar to the way propylene is ammoxidized to acrylonitrile. Asahi Chemical has recently developed alternative isobutylene-based MMA technology called the "direct metha" process. This process does not use ammoxidation but rather is a novel mixed-phased oxidation of methacrolein, methanol, and air to give MMA. Asahi's new plant using the direct metha process was started up in Japan in 1999.

BASF has taken a different tack to make MMA. The BASF process is ethylene-based and is believed to involve hydroformylation of ethylene to give propanal. The propanal is condensed with formaldehyde to methacrolein that in turn is oxidized and esterified to give MMA.

Nonphosgene Routes to Polycarbonate

The classical route to polycarbonate resin is through the reaction of bisphenol A with phosgene. Phosgene is a highly poisonous gaseous material that was used as a chemical weapon during World War I. To avoid the use of this gas, General Electric Plastics developed a nonphosgene route to polycarbonate and built a plant in 1990 in Japan that used this new technology. The new approach uses diphenyl carbonate as the source of carbonate functionality and a transesterification process. The GE plastics route, sometimes called the "melt process," has an advantage over the conventional phosgene-based solution process in that the product polycarbonate is undiluted and may be directly pelletized. Furthermore, the melt product has no chloride impurities and has better properties for the fast-growing optical disk market. Disadvantages of the nonphosgene route include the need to use equipment capable of tolerating high temperatures and low pressures and the limitation on molecular weight imposed by the high melt viscosities encountered.

In addition to GE Plastics, Bayer, Asahi–Chi Mei, and Mitsubishi Chemical have proprietary technologies for nonphosgene polycarbonate. The nonphosgene routes to polycarbonates now appear to be preferred, marking a major change in polycarbonate technology.

Adipic Acid and Nitrous Oxide Emissions: Turning a Debit into a Credit

The conventional route to adipic acid involves oxidation of ketone-alcohol oil (a mixture of cyclohexanone and cyclohexanol) with 50 percent to 60 percent nitric acid. A by-product of this reaction is nitrous oxide (N_2O), which is particularly problematic as it is a radiatively chemically active trace gas that is believed to contribute to global warming. The global warming potential of N_2O is estimated at about three hundred times that of carbon dioxide. Because of this negative environmental impact most adipic acid producers have developed catalytic or thermal processes to destroy the nitrous oxide. For instance, Alsachimie, a Rhodia subsidiary, has developed a process that burns the N_2O at a high temperature in the presence of steam to produce nitric acid. In this approach the benefits are twofold: the nitrous oxide emission from adipic acid production is safely eliminated, and the nitrous oxide that would have been produced from additional nitric acid production is obviated.

Solutia has handled the problem of nitrous oxide emission in a creative and potentially cost-effective manner. The company, working with the Boreskov Institute of Catalysis in Russia, found that nitrous oxide in the presence of certain metal-containing zeolite catalysts is able to transfer the oxygen atom selectively between one of benzene's carbon-hydrogen bonds to give phenol in almost quantitative yield. Since Solutia is a large producer of adipic acid, and hence nitrous oxide, this finding gives a double benefit: elimination of nitrous oxide going into the environment and a potentially low-cost route to phenol.

Solutia had announced plans to build a commercial plant using this novel technology but has now decided to shelve its plans owing to the poor business environment. No matter how clever and environmentally friendly a new process might be, if it does not compete on a cost basis, it will not succeed.

Technology as a Means to Enter or Create New Businesses

Proprietary technology can serve as a strategic weapon. Technology can serve as a lever to allow new entrants into a previously tightly held business area. A good example of this is the case of the 1,4-butanediol (BDO) business.

BDO and THF: New Technologies Finally Smash the Barriers to Entry

BDO process technologies are excellent examples of how the power of proprietary technology first establishes a barrier to entry into a business and then later provides the means to hurdle the barrier. The production technology for BDO was first developed in Germany in the 1930s by Walter Reppe of I.G. Farben; it is based on the copper-catalyzed addition of formaldehyde to acetylene followed by hydrogenation of intermediate butynediol and butenediol. Up until the late 1970s, BASF, International Specialty Products (ISP) (a subsidiary of GAF, itself an original U.S. arm of I.G. Farben), and DuPont were the exclusive producers of BDO because of the formidable challenges of handling explosive acetylene at the high pressures needed to carry out this reaction chemistry. BASF served most of the European needs for BDO production, while ISP served the Americas, and DuPont used most of its material for its own downstream derivatives.

In 1979 the technology barrier to entering into the BDO business started to crumble owing to Mitsubishi Chemical's development of a butadiene-based route to BDO. This development did not upset the market too much because Mitsubishi mostly served the small but growing Asian market for BDO. The business changed abruptly in 1990, however, with the entry of Arco Chemical (now Lyondell). Arco Chemical, primarily a propylene oxide–driven company, developed a new route to BDO based on propylene oxide. Owing to Arco's low-cost position in propylene oxide manufacture, its BDO process was also low in cost and enabled Arco to compete on price and capture the BDO business necessary to fill out their plant.

The BDO business continued to evolve, and in 1992, Davy, an engineering firm, licensed its newly developed butane- and maleic anhydride–based BDO technology to Shinwha in Korea and Tonen in Japan. For the first time BDO technology was available for license. In the last six years or so there has been a flurry of new technologies and licensors by such firms as DuPont, BP-Lurgi, Sisas, and Linde-Yukong.

It appears that now, after nearly sixty years, the technology barrier to entry to the BDO business has finally been surmounted. However, as is typical when technologies are freely available, commercial barriers must be erected, or the business will be unprofitable. New producers often mean overcapacity, which leads to lower prices; then the market must undergo some shakeout to survive. Thus, in 1999 and 2000 we saw ISP, one of the original market participants, close all its U.S. BDO capacity. In addition, Sisas, an Italian firm in the business just two years, declared bankruptcy, and the BDO assets were bought by BASF in 2001. Sisas may have underestimated the resolve of BASF to compete in this market on pricing, and Sisas, in spite of using a low-cost butane- and maleic anhydride–based route to BDO, was competing against BASF's highly integrated and depreciated plant in Ludwigshafen, Germany. And while the market for BDO is growing apace, the available merchant market—that is, the segment of demand that is not captive or tied up by long-term contracts—is rather small. All of these factors contributed to Sisas's exit from the BDO business.

A New Monomer Makes Its Commercial Debut: 1,3-Propanediol

The power of technology as an enabler not only to enter existing industry segments but also to establish a whole new business is illustrated by the commercialization of process technology to make 1,3-propanediol (PDO).

PDO is a rare example of a small molecule that had eluded commercialization because of lack of low-cost production technology. For over fifty years polymer scientists knew that polyesters based on PDO had some commercially interesting properties. But it was not until 1998 that Shell Chemical figured out an economical route to PDO and built a fifty-million-pound-per-year PDO plant in Geismar, Louisiana, based on new and proprietary chemistry—the hydroformylation of ethylene oxide.

PDO's primary use will be as a new monomer to react with purified terephthalic acid to make the corresponding polyester, polytrimethylene terephthalate (PTT). PTT has some interesting properties not necessarily expected from a polyester that has an intermediate chemical structure between that of polyethylene terephthalate (PET) and polybutylene terephthalate (PBT). This property is sometimes referred to as the "odd carbon" effect. PTT seems to combine the resiliency of nylon with the inherent stain resistance of polyethylene. These properties make PTT an ideal polymer for use as a fiber in carpets and stretch apparel. Shell is constructing its first PTT plant in Quebec in cooperation with Societé General Federation (SGF).

Of course, when speaking in terms of carpets and stretch fibers, the name DuPont comes to mind. Not surprisingly, DuPont has now also begun development efforts in PDO and PTT using biotechnology as the key enabler to defend the threat to its core businesses by PTT. This is discussed later in the chapter.

A Look into the Next Twenty Years

The major advances in petrochemical process technology over the last century have been driven by the relentless pursuit of lowering costs primarily by changing process feedstocks—from coal tar to acetylene and now mainly petroleum-based olefins and aromatics. Can feedstock costs be reduced further? If so, what will be the feedstocks of the next century? And what chemistries and skill sets will be required to develop this next generation of process technologies? Insights into the answers for these questions can be gleaned by looking back over the last ten years, and even before, to find the seeds of the new process technologies for the next twenty years.

Three future feedstocks are currently being investigated, each at a different pace of development: C_1 chemistry (chemicals from stranded natural gas); alkane activation (chemicals from fuel, for example, low-cost ethane, propane, and butane); and renewable resources (chemicals from agricultural products and by-products).

The drive to use each of these different sources of carbon atoms is fueled by a confluence of factors, including cost-competitiveness, politics, societal pressures, sustainability, and availability. Furthermore, in order to bring each of these three areas to commercial fruition, ever-more-powerful R&D tools will be needed. Some of these new approaches are already on the scene and yielding results. These new tools can be called "enabling" technologies: biotechnology consists of very selective, productive, and robust biocatalysts using directed evolution techniques; high-throughput experimentation uses materials discovery via combinatorial techniques; and nanotechnology uses nanoscale (one dimension in the one-billionth of a meter size) particles to improve the performance of existing materials.

C_1 Chemistry Is Visited Yet Again: Methane Refineries of the Future

As discussed in the beginning of this chapter, C_1 chemistry had its roots in prewar Germany. Later, because of the second oil shock in the late 1970s, C_1 chemistry enjoyed renewed interest, driven by the need to convert coal to liquid fuels and chemicals to eliminate our dependence on "high-priced" crude oil. However, with the exception of Sasol's unique situation, R&D efforts in C_1 chemistry were largely abandoned as oil prices returned to historically normal levels in the mid-1980s.

In the mid-1990s, C_1 chemistry once again became the object of research of R&D laboratories around the world, an interest driven by the availability of low-cost stranded natural gas. Unlike previous ventures into C_1-based devel-

Table 5. Global Natural Gas Reserves (2001)

Location	Trillions of Cubic Feet	Percentage, Global
North America	267	5
South America	253	5
Western Europe	172	3
Eastern Europe	1,983	36
Africa	395	7
Middle East	1,975	36
Asia/Pacific	433	8
World Total	**5,478**	**100**

Source: BP Statistical Review of World Energy (June 2002).

opment efforts, which were driven by dislocations, World War I, and later, embargo-stimulated oil shortages, the current emphasis on C_1 chemistry is being driven by good economics stemming from technological innovation and market expansion into developing regions. These factors are the basis of a more sustainable development effort.

Almost 60 percent of the world's proven natural gas reserves are in stranded locations. This is illustrated in Table 5. While liquefied natural gas (LNG) technology provides a means for transporting this resource to market, capital investment in the LNG infrastructure is very high and can be prohibitive. Moreover, there is significant economic opportunity to add even greater value to gas if new technologies can be developed. This notion is supported in Figure 1.

In order to monetize stranded natural gas and exploit these downstream opportunities, the economics of any new process technology must be very cost-effective to offset the transportation cost of delivering the product from a remote location to the consuming market and to compete with the locally produced product.

Improved technologies for methane reforming, gas-to-liquids (GTL) technologies, megascale methanol plants, and emerging methanol-to-olefin (MTO) processes have prompted the idea of a "methane refinery." As shown in Figure 2, the idea is to build huge natural gas reformers to feed syngas units for making liquid fuels and megascale methanol plants. The methanol could be used for fuel or power production and to feed a chemical complex. This parallels the way petrochemical complexes have grown alongside petroleum refineries to take advantage of fuel production's huge economy of scale.

Megascale Methanol Technology

The prospect of low-cost natural gas; the potential for huge new methanol demand for power generation, use as a hydrogen carrier for fuel cells, and

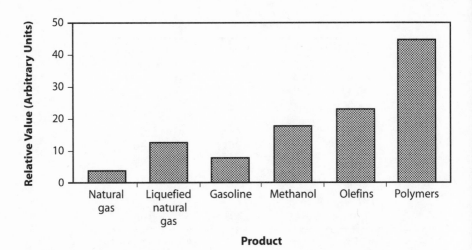

Figure 1. Relative value added to natural gas. *Source:* Universal Oil Products (UOP) LLC, Des Plaines, Illinois, by permission.

conversion into dimethyl ether for use as a cleaner diesel fuel alternative; and the possibility of MTO conversion technology moving closer to commercialization has stimulated the development of single-train, so-called megascale (that is, a huge single reactor) methanol process technology. Current world-scale methanol plant capacity is about 2,000 to 2,500 tons per day. Several technology developers, including Lurgi, Toyo Engineering, Foster Wheeler, Kvaerner, and Synetix, are now developing the capability to offer megascale single-line plants as large as 5,000 to 10,000 tons per day. In December 1999, Lurgi was awarded a contract by Atlas Methanol of Trinidad and Tobago to build the world's first 5,000-ton-a-day single-train methanol plant.

The huge economy of scale of such plants combined with low-cost remote-gas pricing will result in very low-cost methanol production. The availability of such low-cost methanol will undoubtedly stimulate development efforts aimed at adding further value to methanol. One way to accomplish this is to use MTO technology.

Methanol to Olefins (MTO)

In the mid-1980s Mobil brought on stream in Montonui, New Zealand, a new process that converted methanol to gasoline. Key to this process was the use of a shape-selective zeolite catalyst, called ZSM-5. Mobil found that by modifying reaction conditions a substantial amount of light olefins could also be produced. This finding prompted Mobil to develop a process for the intentional conversion of methanol to olefins.

In pilot plant testing, Mobil's catalyst gave, as a percentage of total hydrocarbon produced, approximately 43 percent ethylene. One of the problems with the MTO approach to olefin production is that even if the reaction proceeds with a 100 percent yield to olefins, 56 percent of the total product is water. Although this process was never commercialized by Mobil, it demonstrated a potential C_1 route to olefins and perhaps an economic method to derive commercial value from stranded gas.

In the early 1990s UOP and Norsk Hydro together picked up this line of MTO research. The UOP-Hydro MTO process is based on a different catalyst than that used by Mobil. The UOP-Hydro catalyst is a silica-aluminophosphate molecular sieve (SAPO-34). The size of the molecular pores of silica-aluminophosphates and the strength and distribution of acid sites are better suited to giving light olefins than ZSM-5. However, this material is not as thermally and physically robust as ZSM-5, and one of the key challenges successfully overcome by UOP was to formulate SAPO-34 and later-generation catalysts with the strength and stability to withstand fluidization and continuous thermal regeneration. This feature is important because coke lays down on the catalyst surfaces during the MTO reaction.

Lurgi has recently offered a version of an MTO process with a slightly different twist. The Lurgi process makes mostly propylene instead of ethylene and consequently is called methanol to propylene (MTP). The Lurgi MTP

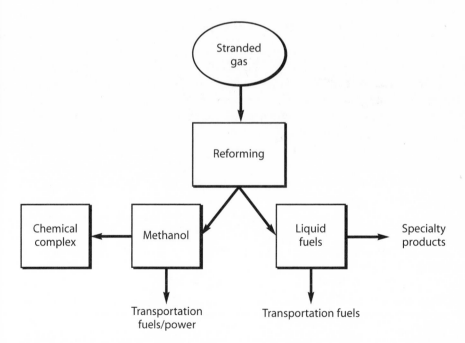

Figure 2. Natural gas refinery of the future. *Source:* Nexant, Inc./Chem Systems, by permission.

process is based on a pentasil-type zeolite catalyst and gives propylene in 46.6 percent molar selectivity—only 4.6 percent selectivity to ethylene. Because of the small amount of ethylene produced, it is not purified but rather burnt for fuel. The products that are heavier than propylene can be sold as a gasoline fraction.

Other companies besides UOP, Norsk Hydro, and Lurgi that have been active in MTO research, based on examination of patent activity in the 1990s, are Exxon, BP-Amoco, Phillips, and Dailim. The economics of MTO processes are highly dependent on the cost of methanol production, which in turn is dependent on the price of natural gas and the scale of the methanol facility.

If low-cost stranded natural gas stimulates real commercialization of mega-scale methanol plants and MTO, then the next logical step is to consider making large quantities of methanol derivatives. Of course, one does not have to look far to imagine the idea of extending the methane value chain. Much of the conceptualization and early laboratory and pilot plant work was carried out in the early 1980s as the result of the second oil shock. Thus, it will be "back to the future," and the research done in the early 1980s may be dusted off and used as a platform for further development. Figure 3 maps out a potential methane value chain based on low-cost natural gas and the development of innovative C_1 chemistry.

Alkane Activation—Feedstocks of the Future

The petrochemical industry has always striven to lower production costs by using the least-refined feedstocks possible. This goal allowed the industry to replace acetylene with olefins as the primary feedstock. The next natural phase of using cheaper feedstocks will be the use of such alkanes as methane, ethane, and propane to replace olefins. This effort has been the subject of R&D efforts for the last forty years.

The technical challenges for the direct conversion of alkanes to chemicals are formidable. Unlike acetylene and olefins, in which chemists exploit the high electron density of the triple and double bond, there is no ready "handle" in alkanes for catalyst designers to grab hold of. Moreover, the engineering challenges will be even greater than the issues that were successfully addressed in the switch from acetylene to olefin-based feedstocks. Conversions per pass (that is, the percentage of raw materials converted on each pass through the reactor) will probably be lower than that in olefins-based processing owing to the need to maintain acceptable selectivities. Lower conversions per pass will require extensive recycling loops with the necessary separation techniques and associated equipment. Significant levels of overoxidation to carbon oxides could occur, creating an environmental burden. Exotic materials of construction may be needed because activation of the alkanes may require high reaction temperatures and corrosive catalyst promoters. All these factors will probably lead to increased capital cost. Successful development of alkane activation process

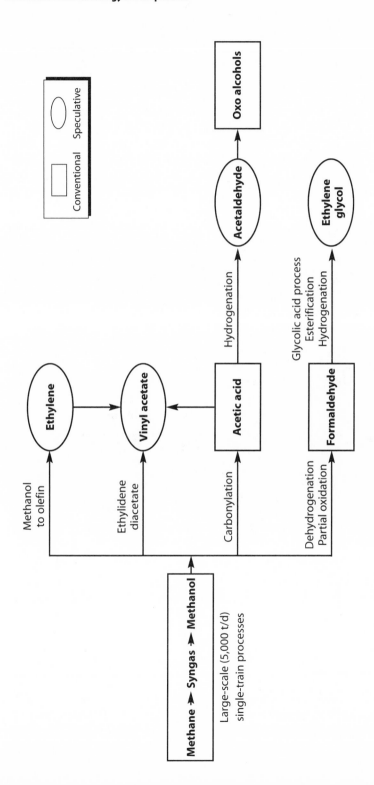

Figure 3. Extending the methane value chain. *Source:* Nexant, Inc./Chem Systems, by permission.

technologies may therefore hinge on whether or not the feedstock cost advantage will offset the probable higher capital investment required.

Butane to Maleic Anhydride

The first example of selective alkane activation was the transition metal-catalyzed oxidation of n-butane to maleic anhydride. The butane–to–maleic anhydride process was first commercialized in the 1970s by Monsanto and Amoco (the latter using a Chem Systems process). Thirty years later this is still the only example of a broadly used commercial alkane activation process. Of course, various process improvements have been implemented since 1975, including improved catalysts, the use of fluid-bed reactors, aqueous or solvent-based recovery solvents, and butane recycle schemes.

Ethane to Acetic Acid

SABIC announced a new alkane activation process for commercialization in 2003: its process for converting ethane to acetic acid. SABIC's subsidiary, Ibn Rushd, is planning to build a 30,000-metric-ton-per-year acetic acid "semiworks" plant. SABIC also plans to build a full-scale 200,000-metric-ton-per-year plant once the technology is proved.

Several chemical companies are highly interested in direct ethane oxidation to acetic acid. During the last five years an increasing number of U.S., European, Japanese, and other patents have been granted on this approach to acetic acid production. The most active companies have been Hoechst, SABIC, Mitsubishi Chemical, and BP. In the estimation of Nexant, Inc./Chem Systems, ethane to acetic acid can compete with the methanol carbonylation process if low-cost ethane is available, as is the case in Saudi Arabia. Earlier, in the mid-1980s, Union Carbide announced development of a new process, Ethoxene, that converted ethane to acetic acid and created fairly large quantities of ethylene as a coproduct. This process was never commercialized.

Ethane to Vinyl Chloride Monomer

Another alkane activation technology that is reportedly close to commercialization is the ethane–to–vinyl chloride process created by the European Vinyls Company (EVC). The company has tested the process in a 1,000-metric-ton-per-year pilot plant in Wilhelmshaven, Germany. The EVC process is believed to be based on reaction of ethane with chlorine and oxygen over a copper-cerium-lanthanide-potassium catalyst. Several other companies previously tried to develop an ethane-based vinyl chloride process. However, these attempts failed because the temperatures required to activate ethane resulted in low VCM selectivities and severe corrosion problems. Key to the new EVC process is discovering a catalyst that is active at relatively low temperatures, avoids the severe corrosion problems, and significantly reduces reaction by-products previously encountered. INEOS has recently acquired EVC, and the status of the ethane-to-VCM process is not known.

Propane to Acrylonitrile

Propane to acrylonitrile is another process that has received considerable attention over the years, and it now seems close to commercialization. BP, Mitsubishi Chemical, and Asahi Chemical have been most active in developing this technology. BP has in fact claimed that its next acrylonitrile plant may well be propane based. Nexant/Chem Systems' analysis of propane-to-acrylonitrile technology, based on patent information, indicates that the propane route can be economical depending on the spread in cost between propane and propylene. Propane and propylene prices do not move synchronously. The fuel market drives propane pricing, while polypropylene demand is the main determinant of propylene pricing. Thus any assessment of the feasibility of alkane activation technologies must include a forecast of alkane and olefin price movements. Of course, in a location that does not have a propylene source, the analysis would be between the new propane-based technology and propane dehydrogenation coupled with conventional propylene ammoxidation technology.

Biotechnology and Renewable Resources

Interest in using biotechnology in the chemical industry has increased rapidly over the last five years. Biocatalysts provide, at least in concept, a number of attractive features to designers of chemical process technologies: use of renewable feedstocks, selective introduction of chemical functionality under relatively mild conditions, introduction of chiral centers not otherwise easily accomplished using conventional catalytic techniques, the potential for environmentally cleaner processes, and favorable public relations by using "green" chemistry.

Table 6 presents a summary of application areas for biotechnology in the chemical industry. The combined effect of market development and technological advancements has led to the creation of a number of biotechnology-based processes in several different industry segments. Each of these technologies is moving at its own pace and has different prospects for commercialization.

Although there are now a number of examples of the use of biocatalysis for making high-priced, molecularly complex pharmaceutical intermediates or fine chemicals, there are relatively few cases in which biotechnology has made a significant impact in large-volume chemicals or plastics. The challenges for chemical and mechnical engineers involved in the scale-up of large-volume chemicals using biotechnology differ from the challenges typically experienced when working with conventional catalyst systems. Biotechnological processes generally must deal with low-concentration aqueous solutions of microbes,

Table 6. Commercialization Status of Biotechnologies for Chemical Production

Industry Sector	Application	Maturity of Technology	Potential for Commercial Growth
Energy and refining	Starch/sugar to ethanol	Mature	Medium; linked to regulation
	Biomass to ethanol	Developmental	Medium; linked to regulation
	Biodiesel fuels	Growth	Medium
	Diesel desulfurization	Developmental	Medium-high
Environmental	Cleaner technology	Growth	High
	Bioremediation	Growth	Medium
Specialty organics	Novel intermediates	Growth	Medium
	Chiral intermediates	Growth	High
	Oleochemicals	Growth	High
Pharmaceuticals	Therapeutic proteins	Late growth	High
	Chiral drugs	Late growth	High
Polymers	Biodegradable polymers	Late growth	Medium
	Xanthan polymers		
Crop-produced chemicals	Oleochemicals	Mature	Medium
	Carbohydrates	Developmental	High
	Polymers	Developmental	Medium
		Developmental	Medium
Commodity chemicals made by fermentation	Citric acid	Mature	Low
	Lactic acid	Developmental	Growth linked to new applications
	1,3-propanediol	Developmental	Growth linked to new applications
	Lysine	Mature	Medium

Source: Nexant, Inc./Chem Systems.

feedstocks, and end products. In addition, reaction rates are often slow compared with the speeds used in conventional petrochemical processes. The low conversion rates require process developers to cope with separating products from large volumes of water without exposing the bioorganisms to harsh conditions—and to do so economically.

Despite the challenges, many large and historically petrochemical-based companies are currently extending their reach into biotechnology. DuPont, Dutch State Mines (DSM), Dow, the former Hoechst (recently split into Aventis and Celanese), Monsanto, Rhône-Poulenc, and others have declared that biotechnology and the life sciences will be the major new technology platform for them in the twenty-first century.

Perhaps the first example of biotechnology having an effect on a large-volume chemical (other than ethanol via fermentation) was Nitto Chemical's commercialization in 1985 of a new biocatalyst for the conversion of acrylonitrile to acrylamide. Nitto, a Japanese chemicals producer, developed an "immobilized" enzyme (bacterial cells attached to a polyacrylamide support) that allowed the hydrolysis of dilute aqueous acrylonitrile at near room temperature to give acrylamide in an almost 100 percent yield. This biocatalytic approach is now the preferred method of acrylamide manufacture.

A recent exciting development is DuPont's biocatalytic process for converting glucose to PDO, the new polyester monomer. As discussed earlier, Shell had brought on stream in 1998 a new petrochemical technology for making PDO via the hydroformylation of ethylene oxide. DuPont, working with the biotechnology company Genencor, has developed genetically altered microbes that can convert corn-derived glucose selectively to PDO at relatively high reaction rates. Based on information from DuPont's patents, Nexant/Chem Systems has assessed and compared the economics of a speculative glucose-based process to PDO to the Shell ethylene oxide–based process as well as an older petrochemical route starting with acrolein. The biotechnological route is potentially at least as competitive as the best of the petrochemical routes.

Since the analysis by Nexant/Chem Systems, DuPont has reported that productivity of the microbes has improved five-hundredfold. This development should further improve the economics of the biotech process.

Another new and promising development is the Dow Chemical and Cargill joint venture, Cargill Dow Polymers (CDP). CDP has commercialized a biotech route to lactic acid, which in turn will be polymerized to polylactic acid (PLA) through intermediate formation of lactide dimer. The feedstock to the lactic acid is corn-derived glucose. PLA not only has the advantage of being derived from a renewable resource, but the polymer is also biodegradable. PLA's properties should allow it to compete with certain large-volume petrochemical-based plastics. CDP is developing application data for PLA for use in injection-molded bottles, films, coatings, and foams. PLA may also be used as a fiber, where it is claimed to have qualities that are a good compromise between synthetic fibers and such natural fibers as silk, cotton, and wool. CDP believes that a world-scale PLA plant will produce polymer in the fifty-cent-per-pound range, allowing competition with such large-volume petrochemical-based polymers as polystyrene and polyethylene terephthalate.

Full-scale commercialization of either of these glucose-based processes, PDO or polylactic acid, will be an important achievement because it would be the first time a renewable resource–based process technology showed competitive or even better economics than a petrochemical route for a large-volume monomer or polymer. Celanese and Diversa together reportedly have an interesting project under way: the conversion of glucose to acetic acid. Celanese is a world leader in acetic acid production via conventional methanol carbonylation technology and is fully aware of the cost elements of acetic acid production.

Acetic acid production technology illustrates well the evolution of science and engineering developments that have occurred in the chemical industry over the last hundred years. The first technology for acetic acid manufacture was via oxidation of ethyl alcohol from various sources. This biotech approach is still in use in small-scale plants in developing regions. The first synthetic route to acetic acid was by oxidation of acetaldehyde. The acetaldehyde was initially made from acetylene and then later produced using ethylene-based chemistry developed by Wacker, a German firm. This technology was in operation before World War I. Next, noncatalytic oxidation of paraffins, such as n-butane or light crude oil distillate, was developed in the United States and Europe as a means to make acetic acid from low-cost feedstocks. This approach exemplifies nonselective alkane activation technology since this reaction results in a variety of coproducts, such as propionic acid, formic acid, acetone, and methyl ethyl ketone. Methanol carbonylation was not commercialized until 1960, when BASF used a cobalt catalyst operating at high pressure. Then in 1970 the well-known and highly cost-effective Monsanto methanol carbonylation process using a rhodium catalyst was commercialized. Methanol carbonylation technology has been significantly improved over the last ten years by both Celanese with its acid optimization technology and by BP with its iridium catalyst–based process. The latest technologies are based on direct ethylene oxidation (Showa Denko in Japan) and ethane oxidation (SABIC), and now, as mentioned above, work is under way to convert glucose to acetic acid using genetically engineered biocatalysts. Thus, acetic acid technology has come full circle, beginning with naturally derived ethyl alcohol oxidation, followed by nonselective oxidation of hydrocarbons, and then moving through various stages of C_1-based (methanol carbonylation) chemistry, to direct ethylene oxidation, and finally to selective alkane activation and high-tech biocatalytic approaches using glucose as a renewable resource.

The development of new approaches for acetic acid production will not necessarily shut down existing routes. Different feedstocks and process technologies can be brought to bear to produce a particular chemical optimally depending on the situation of any one manufacturer. This is illustrated in Figure 4.

Other biotechnologies for large-volume chemicals or polymers that are being developed include polyhydroxyalkanoates (Metabolix), methane conversion directly to methanol (Chevron-Maxygen), succinic acid–1,4-butanediol (work sponsored by several U.S. government laboratories), and propylene glycol from corn starch (jointly under development by the National Corn Growers Association, Archer-Daniels-Midland, and the U.S. Department of Energy). Biomass to ethanol is another technology being developed by a number of

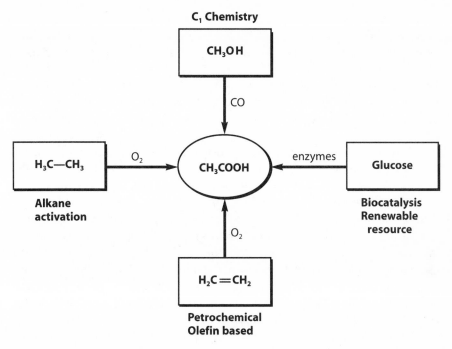

Figure 4. Various routes to acetic acid.

entrepreneurial companies; however, the ethanol product is directed to fuel application.

Enabling Technologies

Second-Generation Biocatalysts

As described above, the key challenge in the development of biotechnology-based processes for the production of large-volume chemicals and plastics is the ability to design biocatalysts that not only selectively carry out a desired biotransformation but also are hardy and robust enough to carry out the chemistry at commercially viable rates and survive the more severe conditions (for example, higher temperatures) of a commercial reactor. To solve this problem, new companies using powerful new technologies have been established. Companies at the forefront of this new area include Maxygen, Diversa, and Genencor.

The approach used for enzyme design in the 1980s was called "rational design," which used the classical scientific method and entailed an understanding of the structure-properties relationship of a protein. It then systematically modified the protein structure through targeted mutations in the hope of improving a particular useful property. However, this approach proved too difficult and too time consuming and led to disappointing results.

A newer approach to the same problem uses decidedly nonclassical methods by joining combinatorial chemistry techniques with the concept of "directed evolution." Combinatorial chemistry methods, sometimes called "high-throughput experimentation," rely not on an understanding of structure-property relationships but rather on trial-and-error experiments conducted at speeds much faster than what can be achieved using traditional laboratory wet methods. Instead of experiments being run several at a time, high-throughput experimentation methods allow hundreds or even thousands of experiments to be run at one time. These high speeds are a result of techniques created for the rapid preparation of microsamples and new methods and techniques for the rapid screening of these samples. Advanced informatics are also employed to handle and interpret the tremendous amount of data that are quickly generated. Combinatorial techniques were first used by pharmaceutical companies as a means to prepare and screen chemicals rapidly for pharmacological activity. Later, it was recognized that high-throughput experimentation could be extended to other areas, such as developing improved biocatalysts. Maxygen, a spin-off of Affymax, one of the companies that pioneered use of combinatorial techniques for drug discovery, was established to extend the use of combinatorial methods for biocatalyst development.

Directed evolution combines techniques for manipulating genetic material with high-throughput experimentation techniques to accelerate the natural-selection process of enzymes. By mixing and matching hundreds of snippets of genes ("gene shuffling"), thousands of variants or mutant enzymes are rapidly produced. Using high-throughput screening methods, these mutants are rapidly assessed for such desired properties as chemical transformation specificity, activity, and stability. Once promising mutant enzymes are identified, repeated rounds of directed evolution are performed on the "best of breed"—enzymes and organisms with the desired property traits are then bred in a relatively short time.

Diversa, a new biotechnology company, uses a similar method but believes it has a critical advantage because it starts with organisms that already have some of the desired features. For example, if high-temperature stability is a key feature needed to carry out a particular transformation, then it stands to reason that microbes isolated from hot springs or geysers would be genetically predisposed to withstand heat. Diversa begins with these "extremophiles" and then uses various proprietary genetic manipulation techniques to effect further rounds of directed evolution. This approach combines natural selection and high-tech directed evolution.

High-Throughput Experimentation to Speed Catalyst Discovery

Combinatorial techniques and high-throughput experimentation are also ideally suited for the discovery of such new materials as catalysts, electronic chemicals, and new plastics and coatings. In the last five years several new start-up

Table 7. R&D Costs Using Traditional and High-Throughput Experimentation (HTE)

Variable	Traditional Approach	HTE
Team	1 chemist + 1 technician	1 chemist + 1 technician
Cost/year	$500,000	$500,000
Experiments/year	500 to 1,000	20,000 to 50,000
Cost/experiment	$500 to $1,000	$10 to $25
Timeline to discovery	Many years	0.5 to 2 years

Source: www.symyx.com, 8 Aug. 2001.

companies have specialized in using combinatorial techniques for new material discovery. The leading companies in this area are Symyx (United States), Avantium (Holland), HTE (Germany), and very recently Torial (United States, a spinoff of UOP). Internal efforts in using high-throughput experimentation to speed catalyst development have been initiated at many companies, including Albemarle, BP, Dow, BASF, ExxonMobil, and Degussa.

A concise step-by-step description of the combinatorial approach for materials discovery has been published by Symyx:[4]

- *Imagine and define:* A new material with desired characteristics and qualities is defined.
- *Select likely elements:* From the whole periodic table, a chemist selects the combination of elements most likely to yield the desired material.
- *Create a library:* Using robotics and other automated devices, a library composed of thousands of different chemical combinations is rapidly created.
- *Process in parallel:* The library is an example of parallel processing, allowing up to twenty-five thousand variations of material to be tested at one time.
- *Miniaturize:* By greatly reducing sample size, miniaturization facilitates processing, saving time and money.
- *Process:* Processing can include any number of variables, including heat, high or low pressure, and time.
- *High-throughput analysis:* The library is screened by detectors that quickly scan various optical, magnetic, electrical, or other chemical or physical properties of a material, and the results are entered into a massive database.
- *Discovery and information:* Scientists apply this analysis to identify the most successful new materials and the process used to produce them.

Combinatorial methods, besides increasing the odds of discovering new materials and products, also reduce the cost of R&D. Research can be vastly increased with the same number of people. Symyx has published an illustration of cost reductions that can be enjoyed by implementing high-speed experimentation, which is shown in Table 7.[5]

Nanotechnology

The field of nanotechnology (the science of ultra-small materials) is broad and has implications in medicine, microelectronics, micromechanics, and the chemical industry. For the purpose of this chapter we will limit the discussion of nanotechnology to the subarea of nanocomposites, as this area has seen some recent commercial applications in the plastics business. Nanocomposites are composed of a polymer matrix intimately combined with nanoscale particles either via in-situ polymerization or postpolymerization compounding. Nanocomposites were first investigated in the late 1980s by Toyota's Central Research Laboratory. Researchers at Toyota found that commonly used reinforcing materials (fillers), such as glass fiber and clays, were not homogeneously dispersed in the plastic matrix when viewed microscopically. They hypothesized that use of nanoparticles (one dimension in the one-billionth of a meter) might lead to composites with improved or new properties. Their experiments focused on developing nanoparticles of Montmorillonite clays and using these as fillers for nylon. Indeed, this intuition paid off, and the Toyota workers found the polymer nanocomposites to have improved tensile strength, gas barrier properties, heat stability, abrasion resistance, and flame retardance. However, the most interesting feature was that the reinforcement efficiency was quite high, requiring less filler and affording minimal alteration of impact strength and surface appearance.

Nanocomposites are finding commercial acceptance in such end-use application areas as films and coatings. The improved barrier properties makes these films interesting for use as bags for processed food or boil-in bags. Another area, and potentially a very large-volume application if the economics prove viable, is plastic beer bottles, in which carbon dioxide and oxygen barrier properties are critical. The good structural stability properties of nanocomposites make such application areas as automotive products (for example, fan shrouds, timing belt covers, and engine covers), power tool housings, and portable electronics attractive candidates for commercialization.

Endnotes

1. Peter H. Spitz, *Petrochemicals: The Rise of an Industry* (New York: John Wiley & Sons, 1988), 328–329.
2. John N. Armor, "New Catalytic Technology Commercialized in the USA during the 1980s," *Applied Catalysis* 78 (1991), 141–173; and Armor, "New Catalytic Technology Commercialized in the USA during the 1990s," *Applied Catalysis: A, General* 222:1 (2002), 407–426.
3. John Kollar, "Ethylene Glycol from Syngas," *Chemtech* (Aug. 1984), 504.
4. www.symyx.com, 8 Aug. 2001.
5. Ibid.

Chapter 3

Specialty Chemicals

Andrew Boccone

The specialty chemicals industry, which was often thought of in the 1970s and 1980s as the miracle solution to restore growth and profits to a maturing industry, began to change dramatically after World War II. The industry blossomed as new markets and applications were developing rapidly. During that period, growth rates for many end markets, such as catalysts, adhesives, and polymer additives, were in the double digits, and gross margins were healthy. New specialty chemical companies emerged to meet the growing demands of the marketplace. All in all, it was a good time for the industry.

Specialty chemicals were traditionally the province of small or medium-sized private companies, since much less capital is needed to produce specialties than to produce commodity chemicals, which require a much larger asset base. However, by the 1970s internationalization of trade, funds for expanded marketing activities, and the adaptation of new technologies created the need for more capital, forcing a number of smaller companies to sell out or to go public. Around that time many commodity chemical firms decided to go into specialties.[1]

A good example of a private company that eventually went public is Rohm and Haas, which had its origins in Germany in the early 1900s as a partnership for producing leather-tanning chemicals.[2] Some of the specialty chemical producers still active today started as spin-offs of commodity companies, including Hercules, which was created in 1913 when DuPont was forced to divest part of its explosives business. Hercules acquired a naval stores company with plants located in Gulfport, Mississippi, and Brunswick, Georgia, soon after and became one of the largest producers of specialty chemicals for the paper industry.[3]

In the 1970s and 1980s W. R. Grace was considered one of the best specialty chemical companies. It was formed in 1954 when Peter Grace decided to diversify the family shipping business by entering into production of specialty chemicals. In the same year he bought Dewey and Almy Chemical Company, an adhesives producer, and Davison Chemical, which had started as a producer of silica gels and later became the leading firm for the production of cracking catalysts.[4] W. R. Grace also went into the production of construction chemicals, sealants, irradiated plastic films (Cryovac), electronic chemicals, and other specialties, becoming the model for a broad-scaled, diversified specialty chemicals producer.

What Are Specialties?

Despite the fact that the term *specialty chemicals* has been in the chemical industry's lexicon for more than fifty years, it still is used loosely, which can cause great confusion about just what specialties are. The terms *specialty chemicals*, *fine chemicals*, and *performance chemicals* are often used interchangeably. Recently, a number of companies that produce mostly commodities, including Eastman Chemical and Union Carbide, labeled a relatively large segment of their portfolio as "specialties" on the basis that these products were much less cyclical and tended to maintain some or most of their margins in downturns. But some true specialties are contained in these companies' portfolios, as is also the case with Dow, BASF, and other such companies.

In the late 1970s Kline & Company, a leading consulting firm for the specialty chemicals industry, developed a simple two-by-two matrix to segment the global industrial chemical industry, which was valued at $1.6 trillion in 2000 (Figure 1).

Specialty chemicals, once the source of high sustained profits that every chemical company wishes for today, are located in the upper right-hand quadrant of the figure. Specialties were historically described as differentiated performance products offered for what they do, not for what they are: they are produced in relatively low volumes and sold at relatively high prices for their effect. In other words, they are made in small quantities and therefore tend to represent a small cost element for the end user, but they are essential in producing the desired effect or performance of the end products. Many specialty chemicals today do not necessarily fit this description because they have become more and more commoditized (that is, produced by a number of firms and therefore not particularly profitable). But the description still rings true for many other specialties. For example, in papermaking, small amounts of drainage-aid chemicals added to fine paper pulp almost magically allow water to be drained from the papermaking machine screens while maintaining the fine pulp particles forming the paper up on the screen to become part of the paper felt. Performance-oriented specialty chemicals typically require a high

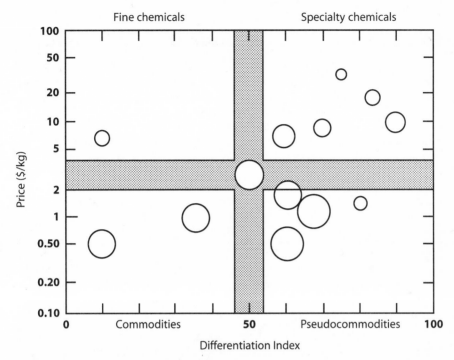

Figure 1. The Kline differentiation matrix. Bubble size is relative to the size of the market for various categories of industrial chemicals. *Source:* Kline & Company, Inc., by permission.

level of technical support, including substantial R&D expenses. Marketing is also essential as brand names tend to be important. The impact of branding will be discussed later in the chapter.

Although specialty chemical businesses have demand cycles linked to global economic activity, they are less volatile than commodity chemical companies because pricing is usually more stable. However, like most sectors of the chemical industry today, specialties are finding that their customers are resisting price increases more than they have in the past, in part because of changes in the structure of the major customer businesses. In today's global and highly competitive business environment, the industrial chemical industry cannot be defined by just a two-by-two matrix. As profit margins shrink and raw material costs escalate, segmenting the industry along the value chain creates a more meaningful distinction between categories (Table 1).

Basic chemicals include such products as caustic and chlorine, soda ash, and ethylene and propylene. The primary emphasis of companies producing these products is in achieving economies of scale to drive down production costs and in developing superior process technologies. Differentiated commodities are typically large-volume products that possess a degree of functionality and

Table 1. Chemical Industry Value Chain

Value Chain	Basic Chemicals	Differentiated Commodities	Technical Specialties/ Formulated Products
Attribute			
Chemistry	Production of molecules	Development of molecules	Modification of molecules
Focus	Economy of scale	Operations	Customer/market
Structure	Centralized	Some decentralized	Decentralized
Technology	Process technology	Process and product technology	Product and application know-how
Management	Internal process development	Internal/external	External—sales and marketing
Capital intensity	High	Moderate	Low

Source: Kline & Company, Inc.

performance attributes that differentiate them from true commodities. Epoxy resins, large-volume surfactants, and some thermoplastic resins are examples of differentiated commodities.

Specialty chemicals are divided into two categories: technical specialties and formulated or market specialties. Agricultural chemicals, biocides, and plastics additives are all examples of technical specialties. Success in technical specialties is typically driven by a company's skill in process and synthesis know-how. Fine chemicals could also be defined similarly, which is one reason why fine and specialty chemicals are often confused. However, the key difference is that fine chemicals are typically intermediates that are further reacted, while technical specialties are generally finished products that are either used "as-is" or formulated with other products into the final end product.

The critical skill of manufacturers of formulated specialties is the ability to blend additives uniquely into a single formulation to achieve the optimal performance characteristics. A high degree of field and technical service is associated with formulated specialty businesses. Examples include Ondea-Nalco in water treatment and Ecolab in industrial and institutional cleaning compounds.

True specialties are products that combine the attributes of both technical and formulated specialties. Companies that manufacture true specialties combine superior technical competence with strong formulation and service skills. Examples include specialty adhesives, electronic chemicals, and certain construction and oil field chemicals.

Through the late 1980s specialty chemical companies were usually small, with many concentrating in one or two markets or technologies, such as electronics (MacDermid), bromine chemistry (Great Lakes Chemical), and fuel and lubricant additives (Lubrizol). Others covered several areas, including Grace (construction chemicals, catalysts, and electronic chemicals), Hercules (paper chemicals, water treatment, and water-soluble polymers), and Witco (surfactants, plastics additives, and urethane intermediates). There are still many small specialties companies, with some trend toward growth and consolidation, but no consensus on whether size alone makes that much difference. At least fourteen specialty firms serve the paper chemicals market alone, some a part of very large firms like BASF and Bayer and some quite small independent firms like Buckman Laboratories (Table 2).

Differentiation: The Key to Success

The recent wave of restructuring and portfolio rationalization among specialty chemical companies is an attempt to restore the once strong and consistent levels of profitability. However, mergers and acquisitions alone cannot solve the underlying causes of the declining financial performance of specialty

Table 2. U.S. Firms Serving the Market for Paper Chemicals

Supplier	Strength Additives		Process Aids	Sizing Agents		Water	Waste
	Wet	Dry		Internal	Surface	Treatment*	
Hercules	✔	✔	✔	✔	✔	✔	✔
Akzo Nobel	✔	✔	✔	✔	✔	—	✔
Ondea Nalco	—	✔	✔	✔	✔	✔	✔
Vulcan	✔	✔	✔	—	—	✔	✔
Bayer	✔	✔	—	✔	✔	—	—
Georgia-Pacific	✔	✔	—	✔	✔	—	—
Drew Industrial†	—	✔	—	✔	—	✔	—
Buckman Laboratories	—	—	✔	—	—	✔	✔
Ciba Specialty Chemicals	—	✔	✔	—	—	—	✔
Geo	—	✔	✔	—	✔	—	—
Vinings	—	✔	✔	—	✔	—	—
Plasmine Technologies	—	—	—	✔	✔	—	—
BASF	—	—	—	—	✔	—	—
Borden Chemicals and Plastics	✔	—	—	—	—	—	—

Source: "Mill Closures, Price Cuts Put a Tear in Profits," *Chemical Week* 163 (14 Feb. 2001), 29–30.
* Boiler and cooling.
† Unit of Ashland Chemical.

chemical companies, which has been brought on by maturing of the end-use markets, increased global competition, and an inability to regain pricing leverage. Further, as the customer base of specialty chemical producers consolidates, the need to differentiate through innovative products and services has become even more critical.

Differentiation has long been the defining characteristic of specialty chemicals. The extent to which specialty suppliers can create differentiation through superior technical know-how, a unique product delivery system, or some other value-added service can often determine commercial success or failure. The degree of differentiation has been one of the distinguishing characteristics separating specialty chemicals from other classes of industrial chemicals. In its simplest terms, undifferentiated products are fully described by their chemical formulas or by a statement of their chemical content. Examples are sulfuric acid and tall oil fatty acids. In all cases undifferentiated products are sold to composition specifications for what they contain, not for what they do.

Differentiated products are described by their performance characteristics. There are usually real or imputed differences among the suppliers. Differentiation can be achieved in several ways, but perceptual research conducted at Kline & Company on consumers of specialty chemicals suggests that the traditional keys to creating differentiation include creating market awareness and developing a strong brand franchise and product and service innovation. Al Schuman, the CEO of Ecolab, had it right when he said that the idea was to "sell the account with our differentiated product and keep the account with our service."[5] It is what Ecolab calls their "circle the customer" approach. The goal is to create a superior value proposition for the customer—sometimes easier said than done. This can be achieved by new product development, but it takes lots of time and money to go from pilot plant to problem solving. For example, PQ Corporation got into the zeolites business in 1981, but it did not begin to get a return on its investment until the early 1990s. Alternatively, adding value can be achieved by providing customers with tailored services and products to meet specific needs. Fernand Kaufmann, vice president of strategic development and new business at Dow, said that "adding value to products is a matter of finding ways for the customer to do his thing cheaper, better and faster so that there still is something left for you."[6]

Innovation in the chemical industry began to steamroll right after World War II. At that time, new product development was emphasized, which lasted through the 1960s. Such major chemical product innovations as nylon plastics, silicone, high-density polyethylene, polypropylene, and Aramid fibers were all discovered between 1940 and 1970. Although these are not specialty chemicals, they are popular examples of "breakthrough" chemical products born of innovation in the laboratories of leading companies.

Gerhard Mensch, an economist, found that major commercial innovations as distinct from original inventions occur in cycles of roughly fifty years, each beginning in a recessionary period. Kline & Company found that the frequency of chemical product innovations, albeit somewhat subjective, peaked during 1945 to 1955 and has declined sharply since then. If Mensch's observation is correct, it raises the question, "Is the chemical industry poised for another wave of innovation?" There is some reason to think so, owing to major advances in molecular screening, biocatalysis, combinatorial chemistry, nanotechnology, and computer sciences of various kinds.

In any case the pace of true chemical innovation slowed and has resulted in chemical companies pursuing other strategies to achieve growth and sustained profitability. The decline in innovation can be traced back partly to a corporate shift in R&D strategy, which took the responsibility away from the business sectors and placed it at the strategic business unit level. This resulted in short-term investment decisions. In the 1970s the focus shifted from chemical innovation to applications know-how. The goal was to understand more about the application of the product than the customer did, which, among other things, enabled the specialty company to maintain the perceived proprietary nature of its products.

In the 1980s, as the number of true chemical innovations continued to slow, companies again shifted focus and emphasized the marketing of existing technologies. This decision created the proliferation of new and improved products that were modifications of a core technology with new functional groups added to impart unique performance characteristics. In the late 1980s and throughout the 1990s companies began to look down the value chain for opportunities. Specialty companies took on a systems approach by combining their chemical products with equipment and fashioning a unique delivery system to create an alternative kind of differentiation. Today the industry has gone one step farther, by taking a customer-centric solutions approach to the marketplace: the companies work directly with customers to develop new products specifically needed by that customer. This is a logical extension of a supplier-customer partnership. Closer cooperation of this type has also been driven by the consolidation of the customer base, resulting in larger, more powerful end users who are exerting pricing and service pressures on their chemical suppliers. End users frequently expect their preferred suppliers to provide total chemical management service, which may include order entry, inventory management programs, and waste management.

Many believe that new product innovation in the specialty chemical industry has faltered to a large extent because of pressure from Wall Street for quick hits. This could be explained in part by the industry's desire to achieve higher earnings through the reduction of research spending and other cost-saving steps. In fact, EBIT (earnings before interest and taxes, noted on a typical income statement) margins on average increased from 10.2 percent in 1988 to 13.1 percent in 1999. "Leaders had a large commitment to new products and technology. . . . We whipsawed ourselves," says Tom Reilly, chairman of Reilly

Industries. "We have done too much financial maneuvering and do not have enough technological know-how."[7] Today the life cycle for newly developed specialties is probably five years or less, reducing the time for realizing the profits of innovation and resulting in earlier commoditization. This condensed timeframe is partially because of the effects stemming from globalization and single-vendor sourcing. Thus, customers have initiated supply-chain management programs that can lessen the applicability of traditional methods of differentiation, such as service and custom formulation. Further, many so-called new products appear to be variations of traditional products.

With companies shifting their focus to growth strategies, much of the growth has come about through mergers and acquisitions. In fact, according to a survey on sustainable and profitable growth, cosponsored by the consulting firm Cap Gemini and the Economist Intelligence Unit, the chemical industry expects to achieve more of its growth from acquisition (42 percent) than any other industry. An additional 21 percent will come from strategic alliances, and the balance, 37 percent, from internal or organic growth, including new product development. R&D budgets have been slashed as companies look to trim expenses to help offset the lack of top-line growth. R&D, which was once the growth engine, appears now to have run out of steam.

In the future, superior innovation will be achieved through a multidisciplinary approach. Joseph Miller, senior vice president and chief science and technology officer at DuPont, puts it this way when describing DuPont's initiatives in discovering new processes and materials: "This will be accomplished though the collaboration of chemical, biotechnology and information technology engineers, thus creating a new engineering discipline."[8] Chemistry will be an enabling technology, a part of a multidisciplinary approach of science coupled with process know-how. Successful specialty chemical companies will be able to react quickly to market needs, maintain a strong customer interface, practice a multidisciplinary approach, and have a strong market-oriented strategy.

Company branding has long been a mainstay of consumer product companies. Branding is now becoming a more important part of the marketing and awareness programs in the business-to-business sector, and the chemical industry is no exception. Given the number of mergers and acquisitions, including corporate name changes, it is no wonder that there is confusion in the marketplace. For some chemical companies branding is not a new phenomenon, but it is gaining more attention now as a basis for differentiation as specialty markets are maturing. For example, Dow, 3M, and General Electric have had branding campaigns for twenty years or more. We associate silicone sealants with GE and adhesives and stain and water repellants with 3M. These associations come from years of print and media reinforcement. Although no spending statistics are generally published for the chemical industry, we believe that typical spending on branding is 0.5 percent to 1.0 percent of annual sales. Many companies spend much less. In 1999 DuPont's branding expenditures were $15.7 million compared with $5.9 million for Dow.[9]

Branding can be specific to either a product or a corporation. According to Al Ries, a marketing expert and author of books on branding strategy, "One can be just as effective as the other, depending on the company, the marketplace and the competition. The problem is that most companies are in the middle and compromise is the wasteful way of doing this."[10]

Air Products and Chemicals, under the leadership of its new CEO, John Paul Jones, launched its first branding campaign with the tag line "Tell me more." The purpose is to create market awareness of all the resources and capabilities at Air Products. Similarly, when DuPont divested Conoco, it realized that it needed to better define itself, and so abandoned the slogan "Better things for better living," replacing it with "The miracles of science." Other companies have different slogans for different parts of the world, each tailored to the particular culture and issues of the region. In the United States we became familiar with BASF's slogan, "We don't make a lot of the products you buy. We make a lot of the products you buy, better." However, in Europe, BASF uses "Innovative thinking—responsible action" to convey its concerns for the environment. In Asia the tag line is "Bring out the best." Branding will play an increasingly important role as specialty companies seek to distinguish themselves from their peer-group companies and try to create awareness among the customer base of all the products and services they have available. But branding will never replace innovation: sophisticated customers will continue to buy according to price, product performance, and quality rather than depending on brands.

With customers becoming more knowledgeable and economy minded, specialty companies are now emphasizing the cost savings that customers can achieve when using a specific specialty chemical. The customer is asked to pay a bit more for the product because savings will be realized elsewhere; if the customer agrees, the specialty chemical seller will regain a higher margin, a margin that shrank owing to the generally greater buying power of today's customers.

Chemical management programs are another way to add value and create differentiation. Such programs typically consist of some or all of the following services: application expertise, process expertise, inventory management, process management, monitoring and testing, procurement outsourcing, on-site support, and waste management. The main driver for implementing chemical management programs is cost reduction. Cost savings are typically realized through such factors as consolidation of vendors, inventory reduction, decreased use of indirect products, and increased productivity. Clearly, environmental issues are another strong driver to embrace chemical management programs as companies strive to be "green."

The automotive industry has led the support for chemical management programs since the early 1980s, when the driving force was the desire to reduce the number of vendors and to bring knowledge and resources in for plant and support service activities, which were reduced because of downsizing. Since then chemical management programs have spread throughout the metalworking industry into various other specialty chemical markets.

Quaker Chemical, Cognis, and Milacron are examples of companies that have implemented chemical management programs targeted at the metalworking industry. Similarly, BetzDearborn and Ondea Nalco provide chemical management programs for water treatment through the use of proprietary computer systems and equipment. These systems allow both companies to provide monitoring and testing services as well as automated reordering and inventory management control.

With true innovation slowing and commoditization increasing, specialty firms are finding it difficult to gain competitive advantage. An Accenture poll of forty chemical industry executives in 1999 showed that these leaders believed that customer relationships and service still represent the most effective means to gain advantage (50 percent). Innovation ranked second (20 percent) and organizational flexibility third (15 percent). The capacity to achieve global reach was ranked fourth (10 percent), while operational excellence was last (5 percent).[11]

Financial Performance

Specialty chemical companies have since the early 1950s generated higher and more predictable levels of profitability than their commodity or diversified chemical company counterparts. Historically, specialty producers in the United States have been able to increase prices by 2 percent to 4 percent a year without significant resistance from the customers. End users have been reluctant to switch from one additive to another simply on the basis of price when this represents only a small fraction of the end users' total costs. Products are therefore not easily substituted, and there has been no global pricing for true specialty chemicals. Nevertheless, cyclicality for specialty chemicals does exist, in volumes and in prices. Volumes are affected by swings in end-user demand, and prices are mainly affected by cheaper imports or the introduction of new, improved products. However, because specialty chemical companies have traditionally benefited from lower expenditures for feedstocks, they have been better able to adjust to these cyclical effects, and their earnings performance surpasses that of commodity companies in periods of slow economic growth.

Until recently, these paradigms all held true. However, since the mid-1990s we have seen global deflation in specialty chemical prices owing to more intense competition and shortening product life cycles, resulting in faster commoditization. These outcomes result from the slowing of the underlying end-user market growth, customer consolidation, increasing customer leverage and globalization, and increased competition from producers in emerging countries. The advent of a multitude of information systems, search engines, and available databases, among other things, has made customers smarter and has also contributed to a loss of pricing leverage.

Table 3. Growth Rates for Selected Specialty Chemicals

Category	Average Annual Increase (%/Year)	
	1981–1986	1999–2002*
Electronic chemicals	12	7
Diagnostic aids	12	5
Food additives	8	6
Oil field chemicals	8	4
Paper additives	7	3
Specialty elastomers	8	5

Source: Kline & Company.
* These projections, made in 1999, are considered too optimistic currently.

Specialty chemicals, valued at $95 billion in the United States and about $300 billion worldwide in 2000, benefited from a growth engine spurred by increasing demand in existing and emerging end-market applications that supported high single-digit growth for the sector as a whole and moderate-to-high double-digit growth for certain sector components in the 1980s. Such specialty categories as electronic chemicals, diagnostic aids, and synthetic lubricants had growth rates of 12 percent a year between 1981 and 1986. However, these chemicals and such others as additives for lubricants and fuels, colorants, paper chemicals, and specialty surfactants have since matured, with individual growth rates for most specialties now ranging from 2 percent to 4 percent a year. Today the weighted average growth for all sectors of the specialty chemical industry is about 3 percent a year compared with 7 percent a year in the early to mid-1980s (Table 3).

Because of the slowdown in new product and service innovation, both the U.S. and European specialty chemical industries have seen a decline in demand growth as a result of maturing end-use markets, such as refining, automotive, textiles, and appliances. What was once a predictable business model for growth is now being driven by the economic cycles of the customer base, resulting in a more commodity-like picture.

The relative financial performance of specialty chemicals companies, which earlier had led to considerable investor interest, deteriorated in the late 1990s, when compared with the companies making up the Standard & Poor 500 index. While specialty chemical firms were improving their margins somewhat as a result of reengineering and other cost-reduction moves (Table 4) and were on average consistently earning returns over and above their cost of capital (Figure 2), the S&P 500 index was moving up rapidly. This index became heavily weighted toward technology companies, such as Cisco and Microsoft, whose stock had been appreciating rapidly. Investors became less interested in owning specialty chemical firms, not only because it was clear that their growth

Table 4. Specialty Chemicals Industry, NOPAT Margins

Year	Percentage
1990	7.5
1991	6.8
1992	7.5
1993	7.2
1994	8.0
1995	8.3
1996	8.5
1997	9.1
1998	9.1
1999	9.3
2000	9.3
2001 (estimated)	8.3
2002 (estimated)	9.1

Source: ABN Amro Chemical Equity Research.
* NOPAT = net operative profit after taxes.

had slowed but also because of their small market capitalization, which created both investment and liquidity problems for large funds. Investment problems were created because a number of the larger funds could not invest a typical amount of money (for them) in such firms, since that amount of investment would give them more than the maximum percentage they could invest in any given firm, according to their bylaws. Liquidity problems were created because the small market caps of many of these firms would make it difficult to liquidate their positions, if so desired, without an excessively negative effect on the firm's share price as the shares were sold.

The New Specialty Chemicals Industry

Perhaps the single most important issue on the minds of specialty chemical executives today is the loss of selling price leverage. There are four main reasons for this.

- Information systems have made real-time knowledge available globally. Pricing has become more transparent for end users. There are Internet Web sites for plastics additives, specialty lubricants, and other fine and specialty chemicals. The Internet over time will probably lead to the commoditization of certain specialty sectors and eliminate the previously claimed points of differentiation for a number of specialties. (However, specialty producers benefit from these systems in the purchasing of their raw materials, which may negate some of the adverse consequences for their businesses.)

- Customer consolidations, as in the automotive and petroleum industries, have put greater power in the hands of the buyers, who negotiate contracts involving lower pricing; suppliers are often selected for their ability to provide extra services. The number of suppliers to any given customer is decreasing, with sole-source supply contracts becoming more common. As the customer base consolidates and moves more toward sole-source supply contracts, relationship selling at the local or plant level will become unnecessary and is already being replaced in some cases by master contracts at the corporate level. If this trend continues, sales and technical service forces will be pared, since the customer is no longer willing to pay for this luxury. This scenario has already become evident in selling water-management chemicals to the pulp and paper industry and in selling specialty metalworking fluids to the automotive industry.

- To counter this trend, some specialty companies have implemented chemical management programs to further differentiate themselves. The larger specialty chemical companies have begun to act like the larger global organizations they are, reducing the number of products offered, focusing on branded products, and managing the business globally with a

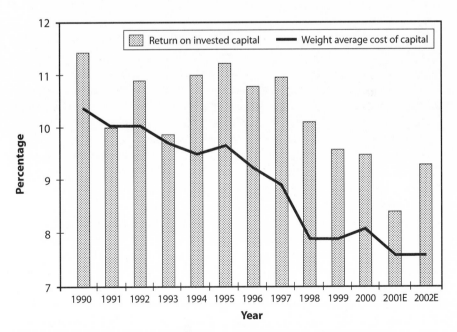

Figure 2. The specialty chemicals industry, ROIC-WACC. ROIC = return on invested capital; WACC = weight average cost of capital. *Source:* Deutsche Bank Securities, Chemicals/Specialty Chemicals, by permission.

market focus instead of a regional one. Smaller producers are left out because global supply-chain implementation favors the larger, more global suppliers.

• Globalization and increased competition from suppliers in emerging countries are resulting in additional pricing pressure for suppliers of specialty chemicals. Specialties are more easily transportable, because logistics typically add a small fraction to the cost of the formulated product. Specialty suppliers from India and China are placing increased emphasis on the production of technical specialties as opposed to formulated specialties and have made some market penetration in the less critical end-use applications, such as dyes and complex fine chemicals molecules, where their quality of product has improved to competitive levels.

As a result of these initiatives the industry has found it more difficult to price to value and virtually impossible to pass along raw material price increases sufficient to offset costs. The result is generally lower margins across the board. Top-line growth has also stagnated. Facing this rather bleak picture, the industry looked for miracle solutions and sought to regain value, using its traditional formula of product tailoring and service. This has led to product proliferation, with some companies now claiming to have over a thousand products in their portfolios, and to increased costs, which results in lower earnings growth.

Then in the early 1980s the business was further commoditized by commodity players entering into the specialty arena, drawn by the promise of expanding their portfolios into businesses with higher valuation multiples and higher growth. Such companies as Exxon and Union Carbide entered the business principally through acquisition. These firms approached the market with a different business model and cost structure. This strategy has failed, and commodity firms have now generally abandoned the idea of seeking growth through specialties. Thus Exxon Chemical sold off its specialty acquisitions in paper chemicals and surfactants, and Occidental Chemical similarly abandoned a once-heralded specialty chemicals effort. BP Chemicals, after acquiring Burmah Castrol, sold off Castrol's entire specialty chemicals portfolio and its own specialty plastics businesses.

Sustained profitable growth is the core issue facing the specialty chemical industry. The business continues to be plagued by slow to moderate growth in its key end-use markets; customer consolidation; increased competitive rivalry, especially at the commodity end of the business; rising raw materials costs; and investors' preference for large capitalization companies with greater liquidity. In these economically uncertain times companies with significant portions of their portfolios in the automotive, electronics, and paper and packaging end markets could be in for the roughest ride. Such companies as Lubrizol, Engelhard, and Cabot, each with a significant portion of their sales tied to the auto industry, would be the most threatened by slowing auto sales in North America and by changes in consumer maintenance practices.

The slowdown in the paper and packaging industry is posing challenges for companies like Minerals Technologies and Hercules and has already resulted in Cytec's sale of its paper chemicals business to Bayer. Some companies have responded to the downturn in the traditional way by expanding their product portfolio and by bundling service-based solutions with products.

Companies serving the noncyclical, consumer-oriented end markets of personal care, food and beverage, and health care should be able to maintain better operating results even in a slowing economic environment. Such companies as International Flavors and Fragrances, Sigma-Aldrich, International Specialty Products, Cognis, and Cambrex, each with more than 50 percent of their revenue generated from these sectors, should not suffer as much from a slowing economy. Sectors that depend on petrochemicals or petrochemical derivatives will probably continue to be affected by rising raw material prices. These sectors include specialty lubricants, lube additives, adhesives, coating resins, surfactants, specialty polymers, and elastomers. Rohm and Haas, Ashland Chemical, and a number of others cited the surge in natural gas prices in the first quarter of 2001 as the prime reason for recent poor financial performance.

A number of specialty chemicals firms have shifted some or a substantial part of their portfolio to "life sciences," which in their case means production of fine chemical intermediates primarily for the pharmaceutical sector. Examples of such companies are Lonza, Great Lakes Chemical, and Cambrex. Some larger firms, such as DSM, Clariant, and Rhodia, have made similar moves through acquisition of, respectively, Catalytica Fine Chemicals, BTP, and Chirex. With drug companies outsourcing some or much of their required upstream intermediates, the firms listed above and others saw an opportunity to use their synthesis capabilities to make higher value-added chemicals. Whether this strategy will pay off is not clear because the outsourcing candidate field is becoming crowded.

In the late 1990s it was thought that size would become an important criterion for success in specialty chemicals. But based on research conducted at Kline & Company in analyzing publicly traded U.S. specialty chemical companies, size alone has little correlation with profitability. It is more the ability to dominate the market or to be a leader that directly relates to profitability. Companies that lead the market or hold a strong second position are thus rewarded by the marketplace, as exemplified by General Electric, Ecolab, Sigma-Aldrich, and Avery Dennison.

A key to developing a strong market position is achieving a high level of differentiation in a product or business unit and segmenting the market in a manner that facilitates such differentiation. Furthermore, Kline's analysis indicates that companies that focus on a single product or technology, or both,

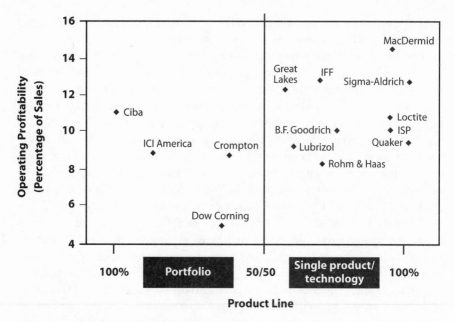

Figure 3. Product line versus apparent profitability, 2000. *Source:* Kline & Company, Inc., by permission.

typically achieve a higher level of profitability than diversified firms (Figure 3). These firms are able to centralize various activities, including sales and marketing, R&D, and administration. They are also better able to realize synergies between business segments and can more effectively provide technical service to their customers. The most profitable specialty companies are the organizations that have successfully achieved and managed three critical issues: focus, culture, and market leadership.

Growth prospects, profitability of customers served, and portfolio focus are additional contributors to the overall quality of a business. Kline's research has found that companies that possess a high degree of differentiation, technology, or market focus and strong growth prospects have higher market valuations than peer-group companies, with rare exceptions. Recently, market capitalization has been identified as a contributing factor to shareholder value because investors favor companies with large market caps. However, not everyone believes that bigger is better. While large cap stocks have greater liquidity and resources to compete in a global economy, they are not truly a measure of the underlying quality of the business. Liquidity has become a positive factor almost by default, since many specialty companies have lost the edge that made them special to begin with. Vince Calarco, chairman, president, and CEO of Crompton Corporation, says that "in the final analysis if performance is there, the street tends to look past size."[12] Certainly this is true for such companies as Ecolab and Cambrex. Both companies have a high market valuation principally driven by the quality of their respective businesses, not by their size, since annual revenues for Cambrex are around $500 million and just over $2 billion for Ecolab, much lower than for firms like Clariant, Ciba Specialties, and Degussa. Both Cambrex and Ecolab possess a high degree of market focus and have high growth business models.

Companies such as Clariant and Ciba possess sizable market capitalizations and strong product and technology portfolios. However, their financial performance has not been good, with Ciba suffering from the very high acquisition premium it paid for Allied Colloids. Other participants, such as HB Fuller, serve mature end markets and need to consider portfolio realignment to drive growth and earnings to increase their market valuation. However, as mentioned previously, few areas now represent solid growth opportunities in specialties. Electronic chemicals, chemicals and materials for batteries and alternative power generation, such specialty chemicals and delivery systems for pharmaceuticals and cosmetics as microfine titanium dioxide, ultraviolet absorbers and microcapsules, and products of biotechnology currently represent attractive portfolio positions from a growth perspective.

Cabot Corporation, for example, recently completed its spin-off of Cabot Microelectronics. Cabot is a leading producer of two commodity-oriented products: carbon black and fumed silica. Since the late 1980s the company has been investigating growth opportunities outside of its core markets. It eventually discovered a new opportunity and became an early merchant provider of consumables for the chemical mechanical planarization (CMP) process, which is used to flatten chip surfaces in preparation for photolithography. The main business of Cabot is to provide slurries for the CMP process. The CMP market is expected to grow 20 percent to 25 percent a year between 2002 and 2007.

Rohm and Haas has expanded its position in electronic chemicals, and it now has sales of over $1 billion annually in this sector. And Cambrex has totally repositioned its portfolio from when it was Caschem, a specialty chemicals company, and now has leading positions in life sciences and molecular biology.

All these businesses serve high value-added end markets, although in the case of electronic chemicals, business cyclicality is an issue, which needs to be considered when evaluating the potential opportunity.

Restructuring

As discussed earlier, specialty chemical companies traditionally provided investors with sustained earnings growth and in earlier years traded at a premium over both other segments of the industrial chemical industry and the broad market indices in general. However, beginning in the early to mid-1990s, this pattern of superior performance slowed, as both top-line and bottom-line growth was starting to falter. The factors contributing to this downturn have been discussed earlier.

Prior to the mid-1990s the specialty chemical companies were divided into two tiers, each with its own business model for value creation. One tier, the diversified or portfolio companies, typically included small to mid-cap companies that possess a diverse product and market mix. Such companies as Crompton, Hercules, and BFGoodrich Specialty Materials (now called Noveon) are examples of diversified specialty companies. Morton International was a diversified company and was acquired by Rohm and Haas after Morton's management recognized the problems inherent in the firm's diversified model. The second tier comprises the focused or niche players, which typically target one or two areas with a strong technology platform and market position. Their portfolios leverage their common market, customer, and technology positions. Examples of these companies include Dow Corning, International Flavors and Fragrances, and Lubrizol. Focused companies consistently outperformed their diversified company counterparts, partly because of the higher cost structure of diversified companies resulting from having to service a disparate group of end users and being unable to leverage technology platforms across end-use segments. RPM remains a firm dedicated to the diversified model, but it has a unique structure of family-run subsidiaries in somewhat related businesses, as well as absolutely minimum corporate overheads.

Today it is clear that the overdiversified business model is not working. Although to some extent the new industry models may appear similar, there are significant structural changes. The declining margins and lack of top-line growth are forcing industry executives to rethink their growth strategies. The publicly traded companies can no longer command the premiums they enjoyed up to the early 1990s. This factor, along with the emergence of the large European-based mega-specialty companies, are the two biggest contributors to the structural change the industry is undergoing today. As a result the specialty industry is now evolving into a global structure with three models: mega-specialty companies, small to mid-sized diversified companies, and focused niche players. Of these three the second model will be the most difficult to manage and maintain successfully, a typical case of being "caught in the middle."

To date, the mega-specialty company has been a Eurocentric phenomenon. These companies have emerged principally as an outcome of shareholder pressure on the large European pharmaceutical and chemical companies, but also because of a move to diminish the fragmentation in the European specialty chemical industry. Creation of larger players resulted, focusing on pharmaceuticals, agricultural chemicals, and specialties. Imperial Chemical Industries (ICI), Ciba Specialties, Degussa, and Clariant are the four most visible examples of a mega-company. The events that transformed two of these companies are briefly described below.

With ICI, the management and shareholders wanted to eliminate the cash flow volatility owing to the cyclical nature of its commodity markets and to refocus the portfolio on profitable growth markets. The first step was to spin off ICI's life sciences businesses and some specialties into Zeneca. The restructuring continued in 1997 with ICI's acquisition of Unilever's National Starch and Chemical, Quest, and other specialty businesses. It was completed with the sale of its polyurethanes and titanium dioxide businesses to Huntsman. Today ICI, which was for most of the twentieth century one of the world's largest commodity chemicals companies, supplies a range of specialty products to such end markets as food and beverage, paper and packaging, coatings, home and personal care, catalysts, and electronics. ICI is a significantly smaller company today, with sales of £6 billion down from the fairly recent range of £9 billion to £10 billion.

Ciba Specialty Chemicals was spun off from Ciba-Geigy, which merged the rest of its business into Novartis to become a pure life sciences player. The driving force behind the transaction was the parent company's strategy to focus on health care and agrochemicals and to achieve an independent focus for its chemical businesses. Ciba subsequently acquired Allied Colloids and sold the epoxy chemicals business unit (now called Vantico) to Morgan Grenfell in June 2000, along with its 49 percent stake in Hexcel. Today Ciba has five business units: plastics additives (25 percent), coating effects (26 percent), water and paper treatment (20 percent), textile effects (23 percent), and home and personal care (6 percent). Sales in 2001 were around 7 billion Swiss francs. Among the most globalized specialty companies, the firm obtains 38 percent of its revenues in Europe, 36 percent in the Americas, and 26 percent in Asia.[13]

Mega-specialty firms differ from other models of specialty chemicals principally in terms of the breadth of their product line and services and their enhanced global footprint. However, they run the risk of having too many small businesses with insufficient differentiation.

Sales of specialty chemicals for these European mega-companies are at least $5 billion annually, with Degussa much higher (Table 5). This size should provide for greater economies of scale, an attribute that historically has been of less value to specialty chemical businesses. Although these mega-companies are relative newcomers, they have not generated the fundamental growth that was expected and therefore have experienced no particular benefit on market capitalization. The benefits and value creation have mostly come from cost reduction and supply-chain management initiatives and from expanding their market positions outside of Europe. On the market side Ciba significantly expanded its presence in the United States with the acquisition of Allied Colloids. This acquisition gave Ciba a new platform in flocculants and also expanded its existing position as a supplier of coating additives. The cost enhancements for the mega-firms have come from reductions in head count, facility rationalization, improved procurement practices, and other internal measures. Annual estimates of savings from these activities are in the tens of millions of dollars, or even substantially more. However, unless the innovation pump is primed, it may be difficult for these firms to reach both managements' and shareholders' expectations.

The pressures generated from the formation of these mega-companies,

Table 5. Lineup of Companies Producing Specialty Chemicals (in Descending Order)

Specialty Chemical Revenues		
$4 Billion to $12 Billion	$2 Billion to $4 Billion	$1 Billion to $2 Billion
Degussa	Sumitomo Chemical	Lubrizol
Atofina	OM Group	IFF
Bayer	Eastman Chemical	Grace
Clariant	Honeywell	Cabot
ICI	Engelhard	Ferro
BASF	Cognis	Cytec
Rohm and Haas	Crompton	Fuller
Ciba Specialty	PolyOne	Great Lakes
Rhodia	Hercules	Nippon Shokubai
Akzo Nobel	Shinetsu	Noveon
Dow Chemical	DuPont	Sigma Aldrich
Dainippon Ink	Dow Corning	Lonza
DSM	Ecolab	Avecia
	Johnson Matthey	Gentech
		Hexcel
		Schulman
		Arch
		Albemarle
		Merck KGaA
		Sensient Technologies
		ISP
		MacDermid
		Ethyl

Source: *Chemical Week* 164 (26 June 2002), 23.

coupled with investors' belief that size and liquidity are important success factors for specialty chemical companies, have forced smaller companies to merge or to acquire or seek out strategic partners. But size and liquidity have never been among the important success factors for specialty companies. Nevertheless, these attributes have become a defensive mechanism for investors, who want to offset their risk owing to a lack of perceived innovation and top-line growth. Although investors may find this logical, it ignores the fundamental underlying problems of poor earnings growth.

Nonetheless, small to mid-sized companies are consolidating. Crompton and Knowles acquired Witco to form Crompton Corporation, Rohm and Haas acquired Morton, Geon and M. A. Hanna have merged into PolyOne, and Hercules earlier acquired Betz Dearborn. Crompton acquired Witco to en-

hance its position as a supplier of plastics additives. Rohm and Haas acquired
Morton to strengthen its presence in both electronic chemicals and adhesives
and coatings. Hercules acquired Betz Dearborn to add a strong service-related
business to its existing paper chemicals additives. (Hercules bought this business at its peak and overpaid for it, just as the paper companies were instituting
programs to reduce the number of specialty suppliers and put their specialty
requirements out for bid. To reduce its high debt burden, Hercules recently
sold most of Betz Dearborn to General Electric.) The hope is that these new,
larger firms will benefit from a range of advantages, including market and channel overlap, sufficient strength to compete with global players, and large enough
cash flows to reinvest in the business. However, it does not address the fundamental problem of how to create differentiation. Allan Cohen of First Analysis
Corporation, an investment banking and equity research firm, says: "Diversified companies produce less new technology, provide lower value-added products, are more cyclical and grow at slower rates."[14] This can certainly be said
for suppliers of coatings, dyes, and pigments and surfactants, all businesses that
have a more commodity-like profile. So while these mergers have created larger
new enterprises, they also raise a big question: are these new companies better
positioned now to drive internal growth, or will growth come from acquisitions and market share expansion at the hands of the smaller companies? If the
latter, then it is a short-term fix for a more serious systemic problem. Further,
these second-tier companies will probably continue to be squeezed, not only
by the larger companies but also by the smaller niche players that tend to dominate one or two specific market segments.

The third model, or tier, of companies consists of the smaller, more regional
firms that typically focus on one or two product or technology platforms. Examples include Quaker Chemical in metalworking fluids, Lubrizol in additives
for transportation vehicles and other fluids, and the OM Group in specialty
metal salts. As the specialty industry continues to take on more commodity-
like characteristics, it may be a challenge for some of these companies to survive as smaller, focused firms. The key determinant will be whether they can
continue to innovate and differentiate themselves sufficiently to create a unique
value proposition. If they cannot, they may have to seek a merger partner or be
acquired. If they merge or are acquired, they risk losing some or much of their
unique culture and brand franchise. On the other hand, being part of a larger
organization can provide them with the necessary muscle to compete in a global marketplace and to offset the pressures resulting from customer consolidation.

The Next Phase

There does not appear to be any near-term relief for specialty chemical companies. End markets are continuing to mature, product life cycles are shortening, customers are demanding more for less, and innovation is probably at an

all-time low. It appears that the recent strategies used by players in specialty chemicals have been relatively unsuccessful from an investor's standpoint. Those who decided to acquire other specialty companies in a move to grow have been particularly unsuccessful. The premiums paid for acquired companies were too large, as mentioned earlier. While reductions in overhead costs were achieved, the market and technical synergies have yet to be realized. The current notion that a higher market cap should confer a higher price-to-earnings (P/E) ratio needs to be reexamined as smaller companies that have not combined with others—for example, Avery Dennison, Engelhard, and Cytec—are outperforming the larger firms and enjoy higher P/E ratios.

In *Profit Patterns*, Adrian J. Slywotzky, David J. Morrison, and their colleagues detail a laundry list of businesses in the "no-profit zone," including parts of consumer electronics, homeowners' insurance, and environmental remediation. On the waiting list for a place on the no-profit zone are "many classes of chemicals." The authors cite the following as characteristics of competing in the no-profit zone: globalization of competition, which has eliminated price management and pricing discipline, and customers becoming more aggressive and more professional in their procurement activities.

Does any of this sound familiar? The authors then go on to report that in times like this managements find "new crutches" to lean on to avoid the tough job of business model redesign:

- In recent economic cycles, the ratio of good years to bad years has improved. [*Interpretation:* We will be OK and benefit from the majority of overall economic good times.]
- Given that growing competition threatens our profit pattern, we can resort to acquisitions, mergers or strategic partnering to lessen the intensity of competition. [*Interpretation:* These are mechanisms to slow down the advent of the no-profit pattern without rethinking the business.][15]

To rely on either of these crutches as the way out is at best a risky decision in the long term.

The specialty industry needs to go back to the basics that made it the darling of the investment community for forty years and apply those tools and techniques in the context of the new global marketplace. Business and innovation models are needed that will rekindle the internally driven growth initiatives. External growth through mergers and acquisitions has a limited life span: there are only so many companies that can be merged or acquired.

If we accept the commoditization of the specialty chemical business as the ultimate outcome, then the industry structure is predictable. To date, industry restructuring has focused on the benefits gained from plant rationalizations, administrative cost reductions, and market share growth. Little if any mention is made of the contribution of organic (informal) top- and bottom-line growth through innovation and outside-of-the-box strategy development. Under the continuation of the current three model industry structures, the mega-companies will continue their quest to create value by exploiting their larger scale and scope and their global reach. In this rapidly changing global environment size matters, and it is important to take advantage of the good things size brings. However, size alone without the fundamentals of differentiation will ultimately lead to a commodity chemical business model. The risk for these mega-companies is having too many portfolio businesses with insufficient focus and differentiation and reduced flexibility. Nonetheless, further consolidations are likely because the need to develop additional skills to support innovation and rapid commercialization requires deep pockets.

The second model (small to mid-sized diversified companies) includes firms in a "catch-22" situation: they are too big to be small and too small to be big. Pressure on these companies will come from above—the mega-companies—as well as from below—the smaller, more focused companies, which often compete effectively with larger companies in their niche markets. As mentioned earlier, this condition was described by strategy guru Michael Porter (see Chapter 4) as being caught in the middle,[16] a poor strategic place to be. Over time we may see fewer of the mid-sized companies, since they will probably have to merge with or acquire other specialty companies to remain competitive. Some of these companies will, because of their value-chain fit, be desirable acquisition candidates for some of the large players.

The fate of the smaller, niche players is not so clear. On the one hand, if the investment community continues to reward size, it will be difficult for many of these companies to achieve the earnings growth necessary to offset the market discount for lack of size. On the other hand, these companies possess many of the cultural and business attributes that define true specialty chemicals: speed, flexibility, service, innovative business models, and superior technology. Those companies that are able to differentiate through either service or technology that possess or achieve sufficient critical mass, and that are able to maintain focus can compete effectively with larger companies in their particular market niche. Firms that cannot achieve critical mass or are not able to reinvest sufficiently in their business to maintain their competitive advantage will become acquisition targets or be owned by private investors, who are not subject to the vagaries of the stock market and institutional investors.

Regardless of the structure, the specialty chemical industry needs to think differently. The industry has lost some of its drive to be unique, but in fairness it is more costly today to be different. Successful specialty companies in the future will not only need to seek higher growth markets, but they will also need to target industries that possess a customer base that values service and innovation. Customers are increasingly demanding and willing to pay for value-added products. Specialty companies that have traditionally been good technologists and less adept innovators and marketers are now aligning their technology and market platforms.

Innovation can provide the greatest payoff for an industry that is maturing, but companies desiring to succeed in a rapidly changing environment, where new discoveries can have an enormous payoff, will need to have diverse skill sets in material sciences, chemistry, physics, and life sciences. Equally as important as the quality of the product portfolio and the customer base, however, is to think differently about adding value and redefining the power shift between suppliers and customers. How the shift in power has affected the pharmaceutical industry is an excellent example. Pharmaceutical companies held most of the power in the 1970s and 1980s. On the other side was the fragmented customer base, which allowed pharmaceutical companies to earn better-than-average profits because they held most of the power. Then along came managed care, which altered the power structure, and so the industry lost some of its clout.

To some extent the chemical industry now finds itself in a similar situation. And like the pharmaceutical industry it needs to increase the rate of innovation. The pharmaceutical manufacturers consolidated not only to counterbalance the power shift but also to enhance the prospects for new product discovery.

There are, of course, some bright spots. *Chemical Week* magazine highlighted some of these success stories, which included the following:

- Cabot Corporation's development of mechanical planarization slurries, which was so successful that the company decided to split off its electronics business to the public in 2001. Sales of this material rose from $33 million in 1997 to $227 million in the 2001 fiscal year.
- Air Products and Chemicals' nitrogen trifluoride business, also used in the semiconductor industry, which generated sales of over $100 million in 2001.
- Crompton's and Degussa's commercialization of sulfur silanes for "green" tires, which generated sales of around $200 million in 2001.
- Sensient Technologies' success in developing new pigments for ink-jet printers, building sales from $3 million to $36 million over a five-year period (1996–2001).
- PPG Industries' joint venture with Essilor, which uses light-sensitive resins to provide eyeglass protection. Sales tripled to $300 million over eight years.

The article also mentions Lubrizol's PuriNOx technology to clean up diesel fuel emissions of nitrogen oxides and Dow Chemical's SiLK dielectric resin for latest-generation semiconductors, where Dow forecasts sales of $200 million per year.[17]

The specialty chemical industry also needs to think differently about all its business processes as well as the value proposition (that is, the reason to buy specific offerings) it brings to the customer. For example, when GE Plastics had difficulty in cracking open an account, it implemented a "plan B." It charged its Solution Engineering Group to come up with systems-wide cost-saving ideas in plastics, lighting, and electrical systems. Then armed with a new and expanded value proposition, Jack Welsh went on a CEO-to-CEO sales call.

Purchasing is no longer the sole decision maker. As an example, the OM Group offers its customers superior cobalt powder technology combined with a high-quality cobalt recycling program.

Businesses can also differentiate value and achieve more rapid innovation by combining their separate capabilities with other companies. Recombination of differentiating capabilities outside the company boundaries greatly enhances the value proposition and the growth potential. Although much has been written about alliances and strategic partnerships, much more can and needs to be done.

A good example in the specialty sector was Lubrizol's venture with GE Transportation Systems, which focused on providing lubrication, diagnostic, and maintenance services to optimize service intervals and improve fuel consumption. Dow's alliance with Alchemia combined the latter firm's carbohydrate synthesis technology with Dow's pharmaceutical manufacturing capabilities to develop a lower-cost synthesis route with a greatly accelerated development time. An interesting collaboration case was described in an article in which a large firm selling personal-care products to chains of retail stores purchased a block of manufacturing time from a specialty chemical supplier to make sure it would receive on-time deliveries of the surfactants or other chemicals it required, having previously experienced substantial delivery problems.[18]

In the future, successful specialty companies will need to combine the benefits of size and scope with a differentiated approach to the marketplace. Differentiation and resurgence in innovation are critical to restoring both top- and bottom-line growth and effecting a shift in the supplier-customer power base. Although some investors currently hold the chemical industry in disfavor, the science and business of chemistry is vital to the world's economy.

Novel, unexplored organic molecular families and combinatorial chemistry can contribute to drug discovery, while biocatalysts can produce some products more cheaply with less impact on the environment. Chemistry combined with physics will aid in the development of next-generation electronic chemicals and such materials as fiber optics. Inorganic chemicals, often not considered specialty chemicals, represent an enormous opportunity for product development and innovation. Major inorganic chemicals are already used in such specialty applications as electronics, ceramics, pharmaceuticals, and plastics. And natural products such as biodegradable polymers are expected to see wide use in orthopedic and dental devices and in drug delivery systems.

Endnotes

1. Fred Aftalion, *A History of the International Petrochemical Industry*, 2nd ed. (Philadelphia: Chemical Heritage Press, 2001), 346.
2. Sheldon Hochheiser, *Rohm and Haas: History of a Chemical Company* (Philadelphia: University of Pennsylvania Press, 1986), 2–6.
3. Davis Dyer and David B. Sicilia, *Labors of a Modern Hercules* (Boston: Harvard Business School Press, 1991), 10–11.
4. Aftalion, *International Petrochemical Industry* (cit. note 1), 260.

5. Bill Schmitt, Claudia Hume, and Kerri Walsh, "Specialty Chemicals Becoming a Commodity Market?" *Chemical Week* 163 (30 May/6 June 2001), 27.
6. Rick Mullin, "Hoping for a Breakthrough," *Chemical Specialties* 1:4 (Nov./Dec. 1999), 24.
7. Schmitt, Hume, and Walsh, "Specialty Chemicals" (cit. note 5), 27.
8. Mullin, "Hoping for a Breakthrough" (cit. note 6), 23.
9. Esther D'Amico, "Image Adjustment—Jumping on the Brand Wagon," *Chemical Week* 163 (14 Feb. 2001), 24.
10. Ibid.
11. Hans van Doesberg, "Embracing the Service Economy," *Chemical Specialties* 1 (Jan. 1999), 40.
12. Vince Calarco, personal communication.
13. Ciba Specialty Chemicals, "Solid Performance in a Challenging Environment," Salomon Smith Barney Twelfth Annual Chemical Conference, New York, 4 Dec. 2001.
14. Schmitt, Hume, and Walsh, "Specialty Chemicals" (cit. note 5).
15. Adrian J. Slywotzky et al., *Profit Patterns: 30 Ways to Anticipate and Profit from Strategic Forces Reshaping Your Business* (New York: Random House, 1999), 60–61.
16. Michael E. Porter, *Competitive Strategy: Techniques for Analyzing Industries and Competitors* (New York: Free Press, 1980).
17. David Hunter et al., "Leveraging Innovation: Specialties Firms Reembrace R&D," *Chemical Week* 163 (19/26 Dec. 2001), 19–22.
18. Phil Schoepke and Jim Welch, "Five Routes to Deep Collaboration in the Chemical Industry," *Chemical Innovation* 31 (Oct. 2001), 41.

Chapter 4

Strategy Development in the Chemical Industry

Michael Eckstut and Peter H. Spitz

The problems confronting the chief executive officers (CEOs) of chemical firms coping with an increasingly complex business environment in the early 1980s led to a sharp increase in the use of outside consultants. Many firms were unsure of the best way to plan for the future and either believed that their own experience did not provide enough answers or wanted to get a second or third opinion. Corporate planning departments did not always engage in independent thinking, and division heads were preoccupied with day-to-day problems.

Business strategy planning was still a relatively new discipline. Some companies, such as General Electric and Shell, and some management consultants, such as McKinsey and Company, Boston Consulting Group (BCG), and Booz Allen and Hamilton, had been pioneering ways for firms to look at their businesses. These techniques became much more popular and would be used by many chemical firms in the 1980s as a discipline for deciding how to expand and for reviewing their product portfolios. Consultants from academia or from management-consulting firms were more familiar with the new tools and methods used for planning, and they vied with each other to create ever newer and better techniques. The chemical industry became an ebullient market for management consultants.

Consultants were used not only to review the firm's businesses and growth prospects but also to advise how market share could be gained or defended, to develop marketing strategies, to reshape corporate organizations, to reduce costs, to create shareholder value, and to install various types of software. They wrote articles and books and in the case of certain practitioners developed an almost mythical following. Consulting—not just for chemical firms, of course—

became a multibillion-dollar growth industry just when many of their clients' businesses were suffering from maturity or decline in demand.

By the end of the 1990s the chemical industry had undergone dramatic change, much of it occasioned or abetted by consultants and investment bankers. It is difficult to assess whether all the consulting advice made a difference in the financial performance of firms that used consultants copiously compared with those that largely did their own planning. In this chapter we review how chemical firms and their outside advisers approached the development of business strategies they believed would be useful for planning their future in an uncertain environment.

Strategy development and execution in the chemical industry is a fascinating subject. As the industry evolved over the latter part of the twentieth century, many of the firms developed a degree of complexity in their portfolio of businesses that made it exceptionally difficult to adopt rational growth strategies that would also result in reasonably profitable operations. Looking at the many businesses within such firms as Union Carbide, Allied Chemical, and DuPont in the 1970s points to the problems such companies faced when they began to experience serious competition in areas of their business in which they were not particularly strong or when there were economic turndowns. Driven primarily by technology and engineers, most chemical firms from 1950 to 1970 were bent on putting their products on the market to satisfy the surging demand of consumers for the many novel synthetics, most of which were replacing such natural materials as glass, metal, wood, and paper. So when business problems arose and companies found themselves with a large number of small or medium-sized plants, some with older technologies, and with relatively little experience in marketing in competitive environments, companies had no ready means to assess their business prospects and to review and adjust their portfolios. They often turned to outside consultants to find answers. But since the results of steps recommended by the consultants were often mixed, it is not surprising that new sets of analyses and tools kept coming into vogue. Looking back, it was clearly not until "value-based management" (VBM) and "economic value added" (EVA) became widespread in the 1990s that the industry finally had the tools with which to manage their businesses properly.

Strategy is about making choices and developing a framework for decision making. This chapter reviews a number of approaches developed by companies and by consultants and, with the benefit of hindsight, comments on the strengths and weaknesses of these techniques. Unquestionably, the strategists that developed newer ways of thinking about business issues had excellent ideas, which resonated for a time with the managements of their chemical clients. But the path from early strategy development to the techniques now acknowledged to be the most effective was long and expensive. Some might argue that firms like Marakon and Stern Stuart may have delivered the most value in getting companies to accept the principles of VBM. These consulting firms are considered the pioneers in teaching the concept of managing to increase shareholder value.

* * *

The chemical industry entered the 1970s after an almost unprecedented period of growth during the previous two decades. Chemicals had developed into one of the first multibillion-dollar industries, with at least a dozen large and increasingly complex companies headquartered in the United States, Europe, and Japan.[1] The businesses within many of these companies were similar in that they had common raw materials, but they were frequently operating in different parts of the full product value chains. New businesses or divisions within these companies came about as product offshoots, product integrations, or internally developed technology. For the leading companies at that time— for example, DuPont, Dow, Monsanto, and Union Carbide in the United States and Hoechst, Bayer, BASF, and Rhône-Poulenc in Europe—the result was increasingly complicated business portfolios. Real strategy development, as it was to become popularly known in the 1980s,[2] was not used as a planning tool, with most companies focused on financial budgeting and capital planning. Strategy at that time was primarily project based, making large capital decisions for new plants or new plant locations, with plant siting preferably driven by where feedstock was plentiful and cheap. Business development focused on commercializing new product technologies and on building new plants, with minimal acquisition activity. Plant scale was driven by engineering and technology requirements for unit costs and by capital availability along with market research for market identification and customer demand. Competitive differentiation, to the extent it was thought about explicitly, was derived from new product development, proprietary product technology, and low manufacturing cost. DuPont, ICI, Hoechst, and Bayer were prime examples of companies following an R&D approach, developing proprietary products based on in-house technology skills. Cost leadership (low unit production costs resulting from plant-scale effective integration of manufacturing steps or low feedstock cost) was viewed as the best strategy for commodity businesses, with Dow and BASF as the main examples of successful companies following this approach, which included early globalization of their businesses. This strategy was also primarily used by the chemical divisions of oil companies, such as Exxon, Mobil, and Shell, which relied on full-scale integration between chemicals and refineries and practiced this in North America, Europe, and later in Japan and the rest of Asia.

The price spike caused by the first oil shock in 1973–74, which was the first significant business uncertainty the industry had to face after World War II, caught the chemical companies unprepared. The strategic issues went beyond the traditional planning activities the companies were comfortable with. They had to address concerns about feedstock cost and availability and the resulting impact these would have on the industry's cost structure. They were also faced with a slowdown in the economy and a slowing underlying industry growth rate, as materials substitution also abated.

As the effects of the first oil shock receded, the industry began to look at its structure and dynamics. It had learned two important lessons: the need to secure low-cost feedstock and a recognition that much of the industry, especially the high-volume chemical intermediates, plastics, and fibers, was periodically susceptible to large increases in costs and hence profitability swings. Cyclicality had hit with a vengeance just as firms began to face up to declining growth rates for many of their end-use products. Growth in consumption had slowed from a multiple of the gross national product (GNP) to GNP levels. By the early 1980s many of the possible areas in which plastics replaced paper, wood, glass, and metal in a number of products were beginning to be saturated, and substitution rates slowed. New tools were needed to address these issues and concerns.

The industry thus began to focus on its existing businesses, investing in bigger feedstock facilities, finding more geographically diverse plants, and acquiring competitors, using primarily incremental economics and traditional methods of financial analysis. The industry shored up its commodity businesses with downstream integration into plastics and fibers or back-integration into feedstocks. Feedstock ownership was viewed as critical for ensuring raw materials availability and for maintaining a low-cost position. But the industry was not looking at full costs and total business chain profitability. As capital costs increased, owing to inflation and higher interest rates in the developed economies, and as capital became both scarce and expensive, a total integration strategy was not only costly but also unsustainable for all but the largest firms.[3] The implications of this would not become fully apparent until the mid- to late 1980s.

In the late 1970s scenario planning was introduced by several academics and popularized by Shell, building off the "war gaming" approaches that the U.S. military had begun to use somewhat earlier. Papers covering Shell's work in this area were published in *Long Range Planning* during that period.[4] Shell tried to sidestep some of the shortcomings of traditional long-range planning and strategic planning by moving away from the linear-thinking approach inherent in typical forecasting methodologies and focusing on a more conceptual, qualitative analysis of the factors that affect decision making within a particular business. What Shell and others using this methodology tried to do was identify the key elements—for example, competition, politics, and technology—and develop an overall framework, perhaps a full-blown model combined with financial outcomes that incorporate these factors. Scenarios—generally no more than two or three—were created by considering different combinations of the factors that would influence the outcome. Shell was known to use this kind of approach in highly uncertain situations, such as the effect of oil prices on its business. It examined a variety of technical, economic, and political factors to develop an overall model of potential oil price scenarios and then used the scenarios to develop business plans and actions.

Scenario planning, when used properly, can be powerful, providing insights beyond those available from a pure extrapolation of data and developing a variety of responses to deal with highly uncertain situations. The major shortcoming of scenario planning, however, is that it usually examined an insufficient number of scenarios, with only a limited number of variables for consideration (Table 1). There is also a need to understand the links among the variables in order to create a credible and meaningful assessment framework, and these links and interdependencies present a major challenge. Thus, scenario planning at that time never went beyond very limited use, and when some companies thought they were using scenario planning, they were really doing sensitivity analyses, which are far more restrictive and less thought provoking.

In the late 1980s, Booz-Allen developed a simulation process that added more structure for carrying out these complex scenario-planning exercises using computing capability to handle the data requirements. It involved the management team in a multiday exercise to "live" the scenario development and outcome.

Scenario planning, and a simpler version known as "what if" analyses, became for a time foundations for resource building as an approach to strategy development. In the late 1980s this method focused on building resources within a business (for example, technology, people, and product portfolios) that would be valuable under a variety of external outcomes. The premise was that external outcomes (for example, economic growth, political changes, and technology developments) are unpredictable but that resources can be built to protect competitive position or to exploit change. The fact that these resources are in place in and of itself creates competitive advantage.

Long-range planning had been used by many companies but was found to be inadequate to address the new concerns. Traditional long-range planning is based on a single-line projection of the current situation into the future. Although long-range plans often tried to include a number of variables, such as technology, economic environment, and geopolitical changes, they were aimed at forecasting a single output, not at trying to determine an appropriate direction or a range of outputs. That kind of long-range plan is much like a rifle shot: if it hits the target, it can be deadly, but there is no way of preparing for what to do if the target is missed. Long-range plans have a built-in bias for an extrapolation of the existing situation, and they are often driven by financial models yielding financial outcomes only, because these can be easily determined and measured. Long-range plans do not encompass the range of business complexities executives must deal with as they resolve business uncertainties and changing economic and competitive dynamics. Furthermore, it is difficult to forecast significant changes, or more important, structural discontinuities that can alter business positions being considered. Although many events can

Table 1. Strategic Scenarios Example: Scenario Descriptions

Scenario Dimensions	The Accepted Future	Tough Green	Technological Leap	Slow Systematic Degradation
Technology	Manageable technological innovation	Moderate pace directed toward environmental opportunities created by government intervention	Rapid and "breakthrough" technological advances	Slow technology changes
Environmental regulation	Steady and predictable increases in environmental regulation	Greening of popular perception; developed world willing to sacrifice	Steady and predictable; more options on the table owing to economic growth and technology	Environmental regulatory pace decreases
Economic	Slow to moderate economic growth; free trade within blocks	Low economic growth in the developed world; high in developing world	Fast pace of growth; free trade	Slower economic growth and global basis
Political stability	Basic stability; some political cooperation	Great levels of political stability	Greater levels of political stability	Increased conflict
Energy market conditions	Relationship between oil and natural gas remains the same or changes in a manageable way	Natural gas becomes a premium fuel; natural gas reaches or exceeds BTU parity	Oil and natural gas become less important; no supply price constraints. Coal may become more important	Oil and natural gas prices increase owing to supply interruptions

Source: U.S. utility company.

be successfully forecast from the status quo, in strategy development the changes or discontinuities are important and are what can determine real success.

Development of Modern Strategic Planning

There was clearly a need for a new, strategic approach to address the problems faced by the chemical industry. Many other industries were going through similar changes and had similar needs for new strategic thinking. Corporate strategy therefore began to develop to meet these needs. It proceeded along two paths: a vision path that emphasized the development of a broad or general direction a company needed to take and the specific actions to get there, and a causality path that aimed to determine the underlying factors that cause business success. (This could perhaps be considered the precursor to VBM.) The vision approach had many converts, but this exercise was often done haphazardly, with little data or analysis supporting the outcome. The results were either simplistic vision statements or marketing slogans (valuable internally as a company rallying cry), or they were very general, leading to flawed conclusions and questionable actions. The causality path became the province of the leading consulting firms, exemplified by BCG, with the analytical skills, required data-collection abilities, and chemical industry and general business understanding to be able to carry out major strategy development efforts. Data-based, analytically driven strategy as a separate discipline within management science, and as a practice area in consulting firms, was being born just as the chemical industry started to grapple with the first wave of strategic issues—portfolio complexity, more intense competition, global operations, asset management, and resource allocation. The chemical industry, along with the electronic and automotive industries, became a logical, practical testing ground, as it was one of the first industries to face this multitude of issues. It was also a multibillion-dollar industry, with large companies having the resources and management talent to implement the new ideas and, also important, to pay the large consulting fees charged by the leading academics and consulting firms. A number of these firms began to develop strategic approaches. The traditional firms—at that time McKinsey and Company, Booz Allen and Hamilton, Cresap, McCormick, and Paget—were joined by such industry-specific firms as Arthur D. Little and a number of others, including BCG, Bain and Company, and other smaller or more highly focused firms, such as Chem Systems.

In 1980 Michael Porter, a professor at the Harvard Business School, published a landmark book on strategy development and practice.[5] Porter's book provided a broad framework for structural analysis in various industries. It identified generic competitive strategies (Figure 1), market moves, and competitive moves and then proposed strategies applicable to such specific industry environments as fragmented industries, mature industries, declining industries, and global industries. Perhaps the most enduring of Porter's proposed analyses was the so-called Five Force diagram, which identified the forces driving industry competition (Figure 2). In the early 1980s Porter became the guru of strategic thinking, as planners recognized the validity and reasonableness of many of his ideas. Porter and coworkers also published a number of associated articles in the *Harvard Business Review*.[6]

Many major U.S. and European companies established strategic planning groups at corporate headquarters as a way to internalize the new ideas, develop their own ideas, and train a new generation of strategic thinkers as leaders. These groups, often led by a respected and thoughtful executive, were then frequently staffed with newly recruited MBAs who would work with the businesses to gather information and develop corporate strategy material. They would use mechanisms akin to budgeting processes—perhaps an annual senior management

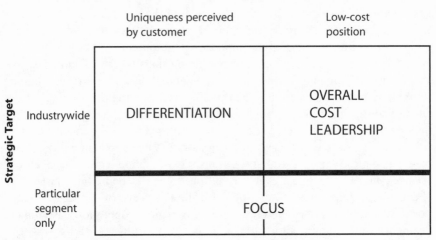

Figure 1. Porter's three generic strategies. Reprinted with permission of the Free Press, a Divison of Simon & Schuster Adult Publishing Group, from *Competitive Advantage: Creating and Sustaining Superior Performance* by Michael Porter. © 1985, 1998 by Michael E. Porter.

meeting or a quarterly review session with the CEO—as a way to develop and maintain a strategic perspective. In the late 1970s General Electric, Allied Chemical (later Allied and then Allied-Signal), W. R. Grace, American Cyanamid, and Shell had been leaders in setting up such groups and using them for these kinds of strategic planning roles. The groups in turn worked with strategy consulting firms and also served as training grounds for strong performers and as a way to attract talented outsiders to the industry. Such corporate planning groups peaked in popularity in the mid-1980s and by the late 1980s had generally been disbanded or integrated into the rest of the organization. By the time their popularity ended, they had often become victims of "staff work gone awry" and of a recognition that strategic thinking, including the strategic concepts often externally developed, needed to be embedded directly into the business. It was also recognized that strategic thinking, rather than strategic planning, took place in the space between the CEOs' ears, as the difference between long-range planning (now called strategic planning) and strategic thinking made clear the role of strategy. While many corporate groups attracted high-quality staff, they could not compete with the consulting firms in terms of salary and opportunity. However, these groups did serve a secondary purpose, since a number of senior executives and CEOs, such as Hap Wagner of Air Products and John Macomber of Celanese, had previously spent time as vice presidents of strategic planning or on planning staffs early in their careers.

Early concepts that appealed to the chemical industry focused on industry assessment and addressed the basic concepts of industry attractiveness and company market or competitive position. Arthur D. Little, Inc., and General

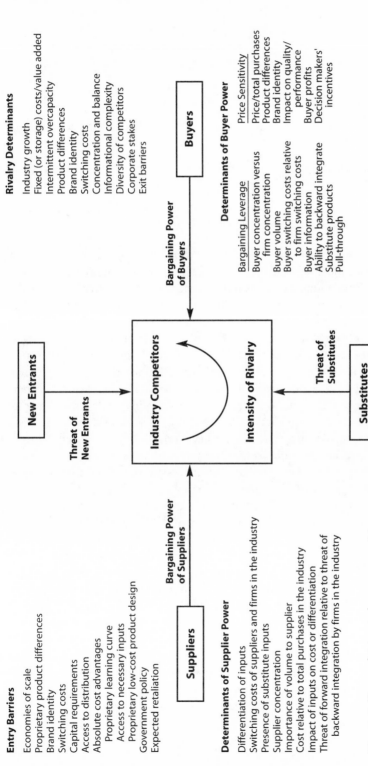

Entry Barriers

Economies of scale
Proprietary product differences
Brand identity
Switching costs
Capital requirements
Access to distribution
Absolute cost advantages
    Proprietary learning curve
    Access to necessary inputs
    Proprietary low-cost product design
Government policy
Expected retaliation

Determinants of Supplier Power

Differentiation of inputs
Switching costs of suppliers and firms in the industry
Presence of substitute inputs
Supplier concentration
Importance of volume to supplier
Cost relative to total purchases in the industry
Impact of inputs on cost or differentiation
Threat of forward integration relative to threat of
    backward integration by firms in the industry

Rivalry Determinants

Industry growth
Fixed (or storage) costs/value added
Intermittent overcapacity
Product differences
Brand identity
Switching costs
Concentration and balance
Informational complexity
Diversity of competitors
Corporate stakes
Exit barriers

Determinants of Buyer Power

Bargaining Leverage   Price Sensitivity
Buyer concentration versus   Price/total purchases
    firm concentration   Product differences
Buyer volume   Brand identity
Buyer switching costs relative Impact on quality/
    to firm switching costs   performance
Buyer information   Buyer profits
Ability to backward integrate Decision makers'
Substitute products   incentives
Pull-through

Determinants of Substitution Threat

Relative price performance of substitutes
Switching costs
Buyer propensity to substitute

Figure 2. Forces driving industry competition. Reprinted with permission of the Free Press, a Division of Simon & Schuster Adult Publishing Group, from *Competitive Strategy: Techniques for Analyzing Industries and Competitors* by Michael E. Porter. © 1980, 1998 by the Free Press.

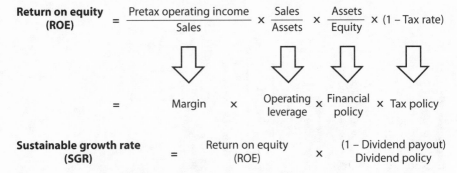

Figure 3. The DuPont equation (simplified form). *Source:* J. Fred Weston and Eugene F. Brigham, *Managerial Finance*, 6th ed. (Hinsdale, Ill.: Dryden Press, 1978), 40.

Electric pioneered the use of an "industry attractiveness" matrix, with some very early assessment of company position. Industry analysis tended to focus on such macroeconomic factors as industry size and growth (measured by demand), market or business segmentation (though generally based on size of buyer or end-use industry only), and some measure of industry profitability (usually focusing on margin or return on sales [ROS] as the key metric). Some form of technology assessment would also be done, from an engineering perspective only, and there would be some attempt at estimating competitive costs.

The financial metrics deserve a word of comment. The industry was generally profitable, interest rates were low, capital was readily available, and margin was used as a measure of profitability. The DuPont equation (Figure 3), which initially used return on assets (ROA) as a key metric and decomposed it into its constituent parts controllable by management, was known and used by many firms, but more as a financial tool to measure asset productivity than as a strategic planning tool to make business decisions. The financing terms were added to create a return on equity (ROE) version of the equation to better model the full financial decisions available to management. Ultimately, it allowed the determination of a sustainable growth rate (SGR), essentially the amount of asset growth supported by the operational, financial, and tax policies of the enterprise. The SGR became a useful measure to allow management to highlight what changes in its business would be required to support greater growth or to determine whether additional capital would be needed. This model was quite useful (though not used as widely as it might have been, perhaps owing to the relative lack of strategic sophistication of chemical industry management at that time) and became a precursor to many future strategic models and metrics, including the "balanced scorecard," which in the 1990s became a very popular measurement method used in conjunction with VBM.

These models were generally targeted at providing information for internal use and considered the business factors only: no attempt was made to think about the financial markets (shareholders or financial analysts) or the drivers of financial value. The concepts of shareholder value or stakeholder value were not yet well established, and stakeholder analysis was generally not considered. ROS was the measure most often used to assess industry profitability and decision attractiveness. Competitor analysis was driven by external data availability, with ROS data for businesses or divisions generally available from public reports. Since for many chemical businesses the dollar value of annual revenues at full capacity is close to the plant's capital investment, ROS is very close to ROA. ROE, a metric with which readers may be more familiar, applies to the entire company and is obviously not a useful concept for running a business—it is a result.

Shell Chemicals had in 1975 published details of a technique for portfolio analysis called the directional policy matrix (DPM). This was a nine-box matrix with two axes, where the horizontal axis indicated market sector profitability and the vertical axis indicated the company's competitive position (Figure 4).[7] This matrix and the work done around the same time by General Electric were forerunners of the BCG matrix, which was used as a tool for portfolio analysis.

The Boston Consulting Group, formed in the early 1970s by Bruce Henderson, a former Arthur D. Little staffer, became a hothouse for the development of strategic planning ideas and concepts. BCG undertook extensive, rigorous studies using published industry data as well as data gathered by its

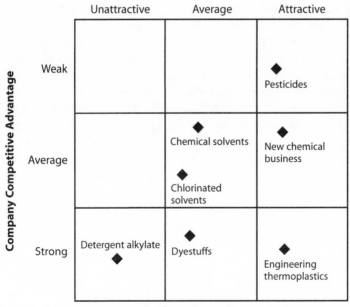

Figure 4. Directional policy matrix for a chemical company. Each company business is positioned with respect to its deemed competitive advantage within its business sector. Reprinted from *Long Range Planning*, 11, S. J. Q. Robinson, R. E. Hitchens, and D. P. Wade, "The Directional Policy Matrix—Tool for Strategic Planning," 8, 1978, with permission from Elsevier.

consultants in an attempt to develop an explanatory model of industry performance. The businesses in the late 1960s and early 1970s were becoming quite complex. Management began to realize that it needed a way to determine whether the firm should participate in all its current businesses and if not, which ones should be kept and which ones discarded. The multitude of businesses in some companies often made it difficult for management to keep track of and determine how to provide resources for all of them. Managements were looking for data and analytically based decision-making (resource allocation) tools to help with these determinations; they needed an easy-to-use, rational way to make decisions and to help mediate the many internal arguments that these kinds of decisions were likely to initiate. Henderson and BCG began by asking the question "What business should we be in?" They were the first to use macroeconomic tools and principles to provide a popular tool useful for management. To answer this question, two other questions had to be answered first: what industry should the company participate in, and what drives success in any particular industry? BCG argued that industry attractiveness could be boiled down to a simple factor: industry real growth. Furthermore, BCG determined in its major conceptual breakthrough that in any industry the single factor that drove success—defined as business performance, measured as a return—was market share and that market share was most valuable in rapidly growing businesses. This concept was later adopted by Jack Welch of General Electric who decided to get out of any business in which GE was not either number one or number two. Although ultimately the concept of market share above all was shown to have flaws, it served to focus a great deal of effort on the part of managers, consultants, and academics. It also became known that Japanese firms drove for high market share before worrying about profitability.

BCG sought to understand why market share was such an important factor, why it may itself be an outcome of success rather than its driver. A company's market share represents its size and often its "experience" in the industry—and by definition its ability to produce the required volumes at the required quality—and it is the combination of these two (embodied in market share) that drives the business to success. BCG was thus able to link market share to its other major breakthrough—the "learning curve," which postulated that costs decline persistently as volume increases. BCG identified a learning-curve factor, the percentage of decline in product costs for each doubling of volume, as its key parameter (known as the "experience curve"). From the perspective of the chemical industry this formulation fit with the well-known chemical engineering concepts of capital equipment costs increasing to the six-tenths power of capacity and with the ongoing process cost reductions that all chemical plants experience over time. These two factors together lower unit costs and follow a model according to the BCG formulation.

The experience-curve concept had considerable validity, but it also had limitations. Companies in a rapidly growing industry could "move down the experience curve" by building new capacity faster than their competitors, using less capital and achieving higher efficiencies owing to their experience with existing plants. They thus could maintain a substantial advantage over competitors. In some industries this also allowed competitors to cut prices ahead of demand, creating large increases in demand. However, in the commodity chemicals arena, there were few places where the experience curve could keep companies ahead of competitors when technology was broadly available. Further, as the industry growth rate slowed, pricing ahead of the curve became a financially disastrous strategy. But where firms controlled technology that would later lose patent protection, the experience-curve concept was valuable as a way to create a head start that competitors could never catch up to.

Arco (now Lyondell) Chemical's unique coproduct propylene oxide process was the basis for the construction of several plants in different parts of the world, each time using the experience gained from previous plants. When patents for the process expired, other firms, such as Texaco Chemical (later a part of Huntsman), could build a plant based roughly on the patent, but such a plant obviously did not include the know-how gained by Arco Chemical and its predecessor (Oxirane) in building and operating its own plants over a number of years. It would therefore not be positioned as competitively as a new Arco Chemical plant, which would be "well down the experience curve" for the process.

The above case was a relatively unusual situation for the chemical industry, where know-how for the construction and operation of most plants tends to be disseminated by contractors or licensors. Exxon Chemical did become a strong proponent of the experience curve,[8] but the company used it not so much for competitive advantage as to show in articles how production costs for a number of processes relentlessly decline as experience is gained, making it difficult for older plants to compete against reduced selling prices, given their narrower operating margins.[9] Exxon Chemical and other firms used the experience-curve concept as a tool to manage productivity improvements and as a means not to get fooled by optimistic price forecasts.[10] The limitations of the concept were published in papers in the 1980s.[11]

BCG combined the experience-curve concept and its market-share leadership findings into its well-known growth-share, two-by-two matrix (Figure 5). The company then took the logical next steps and provided prescriptions for business decisions based on where individual businesses were positioned in the corporate "portfolio matrix"—for example, invest in the "stars" and eliminate the "dogs." So the management task for individual business strategies and for overall corporate portfolio management was simplified, consisting of gathering the necessary data across all the businesses in the corporate portfolio, positioning them on the matrix, and implementing the resultant prescriptions. The matrix and the prescriptions received a great deal of publicity, generated much excitement, and quickly became widely known throughout executive circles across many industries. The concept of industry attractiveness and market share became embedded in both thought and deed for many years thereafter.

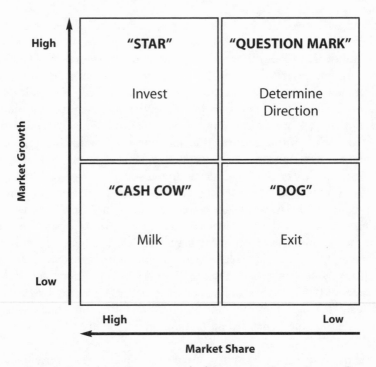

Figure 5. Boston Consulting Group growth-share matrix. *Source:* Boston Consulting Group, by permission.

As expected, the BCG matrix was the starting point for other consultants' and analysts' refinements—analytical, methodological, or formulaic. Business definitions were revised, and each axis underwent scrutiny and redefinition. Growth was variously viewed as industry growth, "real" growth (net of inflation), and relative company-industry growth. The share axis also underwent changes, the most important being "relative" share—the share of the leading competitor compared with that of the next competitor. BCG's concepts even found their way into government policy and regulation, specifically U.S. antitrust enforcement. The Justice Department developed the "Herfindahl index," that is, the sum of the squares of the market shares (as fractions) in a particular industry as a measure of industry concentration, or domination by a single player (a Herfindahl index of 1.0 being a monopoly, anything above 0.30 to 0.40 being a highly concentrated industry where any further consolidations would probably be challenged by the Justice Department).

Many in the chemical industry accepted the underpinnings of the BCG work, especially the experience curve,[12] and built company and corporate strategies in the late 1970s and early 1980s using the concepts for both corporate strategies and business unit strategies. A number of companies restructured their portfolios in response to these kinds of analyses. For example, Monsanto sold its non-chemicals businesses, such as Fischer Controls, an instrument business, while focusing on expanding its life sciences business and using its mature chemical businesses as a "cash cow" to fund its growth programs. DuPont focused on nylon while deemphasizing methanol and petrochemicals.

Nevertheless, use of the growth-share matrix to evaluate businesses generated much controversy. The industry found it difficult to accept fully such a highly simplified, though clearly powerful concept. Chemical industry experts and management felt there was no recognition of the technology differences that existed among the competitors, which in the late 1970s many chemical companies felt they possessed, as well as consideration of other unique aspects of the industry—for example, the measure of integration across various stages of chemical industry value chains. However, the biggest difficulty the chemical industry had was with the growth axis of the matrix, which under the BCG formulation placed much of the industry in the "mature" category. This was a category for retrenchment and business "milking," not a comfortable position for an industry in which management grew up by building plants and expanding the businesses. An often-heard view during this time was that technology could create new markets and thus change the growth outlook for the industry. Therefore many in the industry evaluated the maturity axis from the perspective of technology, not just from the industry growth rate.

A number of articles were therefore written analyzing and critiquing the BCG matrix. Some of the identified pitfalls in the use of this tool are defining the relevant market, negative effects of the product life cycle, unfavorable market structure, inability to factor in market stability, and viewing the portfolio as a closed system.[13] The last point seems particularly important since the matrix views the company's cash-flow cycle as a closed system in which cash generated from "cash cows" goes to cash-requiring businesses ("stars" and "question marks"). This went against the belief that a company should always find adequate funds to finance an investment, provided a project has a net positive value.

The BCG matrix did force the industry to start thinking about strategic business units, or SBUs, the fundamental units that form the basis for strategic analysis—that is, the ability to define a freestanding strategy for a business. Since the outcome of the matrix assessment could result in an SBU divestiture or shutdown, getting the right definition was important, not just for the analytical steps but also for specific actions. For many industries, defining SBUs is a relatively straightforward affair: they are independent, readily separable businesses with a clearly defined set of customers and competitors and focused or dedicated functional support (manufacturing, R&D, and distribution). Many firms started to organize their portfolios around SBUs, with each SBU a profit-and-loss business. For the chemical industry, using this kind of definition is not so straightforward, primarily because of the highly integrated nature of the industry, the substitution of products for one another in specific applications, and the often overlapping customer-supplier-competitor relationships. What is, for example, the right definition for a company that has a plastics business that includes polyethylene (high-density polyethylene and linear low-density polyethylene), polystyrene, polycarbonate, and polyvinyl chloride? Should each broad family be an SBU? Should the feedstock plant (ethylene) be a separate business unit or rolled up into a downstream business or businesses? Should the polymer SBU combine commodity plastics and engineering plastics? Depending on the agreed-on definition, the outcome of the analysis might be different because of growth and share considerations. Any major actions, such as shutdown of a PVC business, would affect the ethylene use and costs for the other businesses. The "right" answer depended to a great extent on the specific company, and chemical firms therefore took different approaches to the problem. According to Fernand Kaufmann, who headed up Dow Chemical's strategy development in the 1990s, VBM "solved this problem for good."[14]

Focusing on Growth as the Industry Approaches Maturity

In the mid-1980s the chemical industry began the first real emphasis on planned diversification and growth, the next stage in strategic thinking. Many different approaches were tried by the leading commodity players in the industry. A comprehensive growth model, covering all the potential modes of growth, is relatively easy to create but much harder to carry out. The industry had been using the simplistic approach of looking at size and profitability in deciding on growth programs and acquisitions as a way to offset the commodity cycles. Some of the large players started to convince themselves—with the BCG portfolio analysis for support—that they could only succeed by moving into "specialties," a broad range of product and market businesses serving multiple end uses as well as new technology developments. Among other things, this approach indicated a move away from commodities, whose manufacture would require a low-cost feedstock position. In addition, globalization—at that time simply defined as investing in other regions—was also pursued with a vengeance, especially by European companies seeking to expand their position in the United States and elsewhere. U.S. companies meanwhile sought to expand in Europe and Asia. Bayer, BASF, Akzo, Rhône-Poulenc, and other leading European producers were actively investing in the United States to increase the U.S. portion of their business to 20 percent to 30 percent of the total group revenues. This thinking continued through the mid-1980s, capped by Hoechst's acquisition of Celanese in 1987.

During this time the attitude of "specialty is good, commodity is bad" really took hold. Many companies investigated, and ultimately acquired, "high-growth" specialty firms. The most popular types were specialty chemicals, advanced materials (performance polymers, advanced composite materials, and ceramics), and separations technologies (for example, membranes), along with the first wave of interest in biotechnology—both for new materials and as a processing technology for producing traditional materials and products. Existing leading specialty chemical companies, such as Nalco, Great Lakes Chemicals, and Loctite, as well as emerging technology companies had the types of businesses the large commodity players were trying to emulate. These firms—for example, Dow, Union Carbide, Monsanto, and ICI—were being criticized by financial analysts and prodded to make over their businesses. No one really understood that specialty and commodity businesses were quite different and that the management tasks and company cultures were significantly different between the two. Unfortunately, few analyses were done at that time to support this point of view, with most analyses focused on the results of a few successful companies rather than on determining the true causes or drivers of success in the business—the real objective of strategy. It was later seen that over a typical eight- or ten-year cycle the returns of "specialty" and "commodity" companies were actually comparable (during commodity cycle upturns, commodity chemical companies could earn as much as 30 percent to 40 percent on capital, making up for poor years at other points in the cycle). It also became apparent that some of the successful specialty chemical producers were serving such markets as refining, mining, and pulp and paper whose growth rate was slowing markedly. This decline in demand had the expected negative effect on the growth rates of products the specialty chemical firms were selling into these markets. The confusion caused by the lack of understanding of "specialties" and "commodities," including the fact that a preponderance of so-called specialties were actually small-volume commodities or pseudo-commodities, was most disruptive.

In the early and mid-1980s there was nevertheless a rash of acquisitions in the industry—many by leading commodity players attempting to shift themselves in part into specialty businesses. For example, BP Chemicals made a series of acquisitions in the materials area, focusing on ceramics, and thermoplastic composites; Rhône-Poulenc purchased a series of food additives and surfactants companies; Olin purchased Hunt and other electronic chemical and materials companies and technologies. Exxon Chemical acquired several small companies in the water-treatment and oil-field recovery businesses (for example, Callaway Chemicals and Tomah), and ICI acquired Beatrice Chemical, which consisted of a number of specialty chemical businesses. The large companies generally did not know how to manage these businesses, and since they paid high prices, they attempted to reduce costs by cutting back on marketing, selling, and application development, the very areas that made the "specialty" companies special in the first place. Some of these acquisition programs were quite unsuccessful, and it did not take long, usually less than five years, for these companies to realize that this strategy was not going to be successful. Overall industry statistics by such firms as McKinsey showed that 70 percent of all acquisitions were failures in that they did not achieve expected financial or strategic objectives. In the chemical industry the percentage was at least at the same level, with only a few successes. By the late 1980s many of the companies that had embarked on these programs earlier in the decade were in the process of selling or otherwise disposing of these companies or businesses.

A brief discussion of the status of strategic thinking (as opposed to strategic planning) during this time will provide a better understanding of the events of the 1980s. Strategic thinking was still thought of primarily as business planning: budgeting, long-term plans, business description, and the concept of strategy as a means of creating sustainable differentiated advantage existed only in a few forward-thinking companies, consulting firms, and academia. Individual companies in the industry used similar tools yielding similar actions and tactics with similar results. This produced the so-called herd effect with respect to steps taken: companies used the same measure of market "attractiveness," similar targets for internal development, and similar information technology tools. Very few chemical companies were bold enough to try something different and to stick with that initiative for any length of time (strategy being, among other things, a long-term discipline). One outstanding example of a company that used a very different approach to implementing a differentiated strategy during the late 1980s was Monsanto, under the leadership of its CEO, Richard Mahoney. He believed in competitive differentiation and became convinced that Monsanto's route to competitive differentiation was through the use of biotechnology. Monsanto fundamentally redefined all parts of its portfolio—which businesses to keep, how to participate in those businesses to serve both existing and new markets, and which businesses to fund the transformation of the company. It was in some respects a good use of the BCG matrix, where cash cows (the traditional chemical business and the profitable agricultural chemicals business) would fund development and growth of Monsanto's putative "stars" and "question marks," the new biotech, nutriceutical (food or food additives with "functional" properties, such as a synthetic sweetener), and pharmaceutical products. Mahoney dedicated all of Monsanto to this vision and invested significant resources over an extended period to execute it. He was succeeded by Robert Shapiro, who strongly embraced Mahoney's vision and strategy and continued to implement it aggressively. In retrospect, Monsanto greatly overspent on acquisitions, which largely contributed to the company's eventual acquisition by Pharmacia Upjohn. The results of Monsanto's bold strategy, viewed from the perspective of almost twenty years later, were only modest. Monsanto developed some successful products, but it was not able to revolutionize the businesses it targeted or to develop and commercialize the breakthrough products it aimed for. In the late 1990s the company split off its traditional businesses into a separate company, Solutia, and focused the remainder on human health, nutriceuticals, and its agricultural chemicals business. For a while Monsanto was able to increase shareholder value to over $40 billion (in 1998), but that did not last. The company was not able to compete in pharmaceuticals against the drug giants and ultimately was split up and sold.

Analytical Techniques: SCP Analysis and Other Tools

McKinsey and Company was a leader in strategy development and worked with a number of chemical companies throughout the 1980s and 1990s. Some

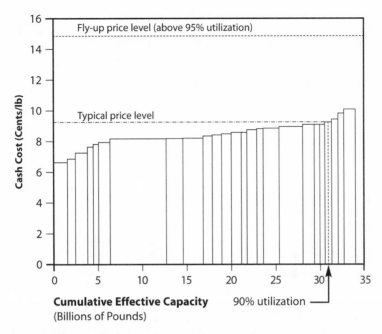

Figure 6. Ethylene cost curve for fiscal year 1987. The effective capacity is 96 percent of the nameplate capacity. *Source:* McKinsey & Co., by permission.

of McKinsey's important chemical clients were Celanese, Union Oil, Shell Chemical, Dow, and DuPont and a number of European firms, such as Bayer, ICI, and Hoechst. McKinsey developed a number of strategic planning tools, some particularly useful for chemical companies. An example was the so-called SCP (structure-conduct-performance) analysis, which could be applied to specific businesses. Chem Systems and McKinsey worked together in the late 1980s, joining Chem Systems' expertise in the chemical industry with McKinsey's analytical techniques and its strategy practice. The SCP analysis had three steps.

Structure

The first step was to assess the attractiveness of the industry structure, for example, the number of competitors, the emergence of new technologies, and the possibilities of product substitution. A key technique used by McKinsey was to construct an "industry cost curve" that plotted the production cash cost of each of the competitors on a graph with total industry capacity as the horizontal axis. Each competitor's capacity was shown as a rectangle on the chart, with the lowest cost competitor on the lower left side and the high-cost marginal player, most often with small capacity, on the far right. Industries with many competitors (for example, ethylene) might have had ten to twenty plants or more shown on the cost curve (Figure 6). McKinsey's research demonstrated that prices in any competitive chemical segment were set by the intersection of

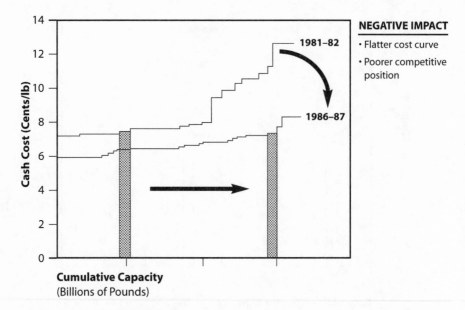

Figure 7. The deterioration of ethylene fundamentals. Each of the cost curves pictures plant ethylene capacity in billions of pounds for the two timeframes. The shaded bars represent the capacity of the same plant in the two timeframes: the capacity of the plant remained unchanged, but the total capacity of the industry increased by 1986–87. *Source:* McKinsey & Co., by permission.

industry demand with the industry cost curve. Consistently, year after year across many commodities, prices were shown to have tracked the cash costs of the marginal producer necessary to meet demand. The margins earned by all other competitors were then determined by their cost level relative to this marginal producer.

An industry with this type of structure is not just highly competitive; it also often exhibits a flattening of its cost curve (Figure 7) as weak players shut down and the remaining competitors strive to lower their operating costs. In Figure 7 the plant singled out for analysis moved from the low end to the high end of the cost curve over a certain period, while its cash cost of production actually decreased somewhat. Meanwhile new, more efficient plants had come on stream, and some existing producers had substantially improved plant efficiencies. When industry cost curves flattened, total industry profits were reduced for all participants, and the less efficient producers on the upper right-hand side of the cost curve no longer provided a "price umbrella" for the more efficient producers.

Conduct

This step involved gathering information on the "behavior" of the business. Was the business "disciplined," meaning was there usually little fighting over market share, generally indicative of a business with few players? Did competitors add capacity in an orderly manner? Were there some traditional price cutters? An example of a notional cost curve for what would be considered an attractive business is shown in Figure 8. In a "disciplined" business the leader usually has the highest capacity and is often happy to see the highest cost player stay in business making a small profit, while the leader is considerably more profitable, being on the low end of the cost curve.

Chemical businesses that had only a few players and were generally more attractive to participants than, for example, polyethylene or ammonia, which had many players, were linear alkyl benzene (LAB), acrylates, propylene oxide, and higher olefins. Technology was an entry barrier in most of these.

Performance

Was the industry especially cyclical? How did the profitability of the players vary over a period of one or two cycles? Was this an industry worthy of reinvestment, or was new investment unlikely to give a good return? The SCP analysis was carried out for a number of a company's businesses to determine which of these should receive new capital and which should be "harvested," or sold. Figure 9 shows profits of a business through a complete cycle. The SCP analysis would show that the detergent business is a better candidate for reinvestment than the PVC business. The latter has periods of high profits, but the cyclical results make PVC less attractive for reinvestment.

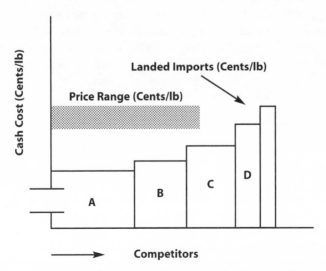

Figure 8. Illustrative notional cost curve for an attractive industry. Four companies (A, B, C, and D) compete, with D providing a price umbrella for the others. Players A and B will generate enough cash for reinvestment in new capacity when required. Player D will probably not reinvest and should not unless cash costs can be lowered significantly for the new plant.

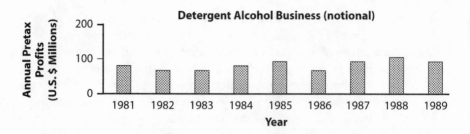

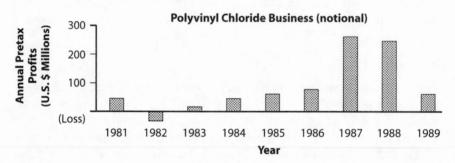

Figure 9. Business profitability over a cycle.

Looking for Competitive Advantage

When companies looked for investment opportunities through forward or backward integration, often through acquisitions, they were told to look at the entire "value chain" for the production and use of the chemical or chemical family. The important thing was to identify those areas in the value chain where profitability was likely to be highest owing to technology advantages, a smaller number of competitors, or barriers to entry. Figure 10 depicts the value chain for some coatings and adhesives businesses. Both PPG and BASF concluded that they should move far down the value chain in automotive coatings by actually building paint shops near car assembly lines and painting the cars themselves. This method was excellent for creating value while building almost impregnable entry barriers.

However, strategy consultants pointed out that a majority of acquisitions ended up creating little or no value, often because the acquiring company had few or no skills (for example, technology and marketing skills) in the businesses they were acquiring. Table 2 shows a case where a company is considering the acquisition of a diverse specialty chemicals company but has few of the skills required to integrate the target company into its operations.

As companies recognized more and more that profitability would not necessarily come from building larger plants or using the latest technologies, they turned inward to look at various aspects of their operations, seeking ways to differentiate themselves from their competitors. Leaders in strategic thinking provided guidelines to help companies make such internal examinations.

The concept of *sustainable advantage* received much attention. Strategists emphasize that for superior performance a company has to outperform the competition, a much more difficult task than one might think. Product innovations can be copied, new processes are hard to protect, learning-curve effects can leak to competitors, and marketing moves can be copied. According to Pankaj Ghemawat, sustainable advantages fall into three categories: size in the targeted market, superior access to resources and customers, and restrictions on competitors' options (for example, technology).[15]

Experience effects are based on size and leadership and can often keep a company well ahead of its competitors. Amoco Chemicals (now part of BP Chemicals) provides a good example with its purified terephthalic acid (PTA) business. Having established patent protection that lasted for a long time, Amoco built large plants alone or in joint ventures in foreign countries, always receiving feedback from technology advances in all these plants. Armed with operating information, including progressive improvements, from many plants, Amoco was able to scale up successive plants to maintain cost advantages against competitors.

Access to customers can be a tremendous advantage in some cases, such as when a specialty chemical firm works inside a paper mill supplying most of or all its processing chemicals and keeping the paper machine line speed at maximum levels. Companies like Hercules and Nalco placed technical service people inside such mills, working closely with the mill superintendent. In such a situation switching costs for a company contemplating a competitor's product can be enormous. Consider a chemical that keeps a paper machine running at

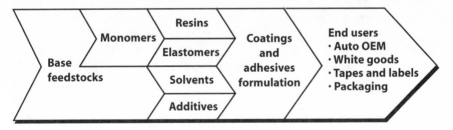

Possible Approaches to Growth

- Pursue related growth opportunities in existing individual business areas, for example, in specific solvents or resins (no links).

- Identify alternatives where the company can pursue an attractive value-chain position (for example, new families of resins, monomers, or new technologies).

- Focus on sectors where value is created through strong links among the company's businesses.

Figure 10. The value chain for coatings and adhesives. OEM = original equipment manufacturer. *Source:* Nexant, Inc./Chem Systems, by permission.

Table 2. Skills Mapping for Assessing Acquisition Fit (Determining Whether Acquirer Skills Match Skills Required for Businesses in Acquisition Target)

Skills	Polymer Additives		Health and Personal Care		Functional Fluids		Energy Chemicals	
	Acquirer skills	Key skills required	Acquirer skills	Key skills required	Acquirer skills	Key skills required	Acquirer skills	Key skills required
Technology								
• Application	✔	✔	✔	✔	—	✔	—	✔
• Product	✔	✔	—	✔	—	—	—	✔
• Process	—	—	—	—	—	✔	—	✔
Distribution system	✔	✔	✔	✔	—	✔	—	✔
Customer position	—	—	—	✔	—	—	—	✔
Marketing position	—	✔	—	✔	—	✔	—	✔

Source: Nexant, Inc./Chem Systems.

several thousand feet per hour by controlling the amount of water that continuously drains from the screens as the paper mat is formed from wet pulp. This drainage-aid chemical could conceivably be sourced for a lower price from another supplier, but it may cause the paper machine to have to operate at lower speeds or cause loss of more cellulose fines, thus slowing down operations or actually costing the company more. Thus, a new potential supplier would have a hard time convincing the mill superintendent to change to another drainage-aid chemical.

Restrictions on competitors' options can create sustainable advantage when a competitor is barred by antitrust laws or is kept from building a plant because of new environmental pollution regulations. Dow and Shell Chemical are the only two U.S. firms making epoxy intermediates, since this process is potentially highly polluting. A new competitor would have a great deal of trouble receiving a permit to build a plant of this type, even next to other plants, given the restrictions now being imposed by the Environmental Protection Agency and communities located next to chemical plants.

In all these situations the companies described established and sustained competitive advantages that would keep their operations more profitable than would otherwise have been the case.

Ghemawat showed that sustainable advantage has a finite time limit in almost all cases. His analysis of seven hundred business units showed that nine-tenths of the profitability differential between businesses that were initially above average versus those below average disappeared over a ten-year period.[16] This finding is relevant to the more recent problems of the petrochemical industry: many of the technologies that once were able to provide competitive advantage to a number of companies have been imitated by or licensed to competitors.

The use of "market segmentation" as a tool for strategic analysis began to take hold in the 1980s and continued in the 1990s. Consumer-oriented businesses had for years been segmenting markets along various dimensions and developing approaches for penetrating specific segments. Most of these segmentation methods were along demographic lines—age, geographic location, and gender being the most common—and were focused primarily on demographic statistics to determine differences in buyer behavior or purchasing methods. Academics, advertising agencies, and consulting firms had long used "psychographic" variables such as buying motivation (for example, impulse buyers) to develop marketing, promotional, advertising, and sales programs targeted at these groups. In the consumer world numerous research firms were formed to construct segmentation schemes and to measure the success of the consumer firms in reaching or penetrating the segments. Industrial companies, including chemical companies, had not generally been using segmentation as a strategy tool, although sales force organizations along geographic lines or for national accounts had been used for a long time and there had been a general interest in serving some customers and segments differently. Leading academics, such as Benson Shapiro at Harvard Business School and Thomas Bonoma at Northwestern University began to stress the value of including industry segmentation as a marketing tool.[17] The consulting firms, always eager to add new "intellectual capital" to their strategy offering, picked up the pace of development in their client work. As businesses became more complex on both the production and the customer sides, chemical companies began to see the value of segmentation as a strategy driver. Segments would have different characteristics, including economics, competitive structure, and success factors. Companies could identify those segments that best met their total offerings and that might give the companies some advantage; then they would build segment-specific strategies. Similarly, segments that did not play to their strengths (for example, a cost-sensitive segment might be avoided if a company did not have a leading cost position) need not be a part of the strategy. Markets served by a particular company could be segmented along dimensions that might determine "cost to serve," such as quantity of purchase (a very simple one) but also product purity, the amount of service required, and the geographic spread of the customer base. Once these segments were identified and assessed for such product requirements as purity or breadth of product line, size of the opportunity, specific customers, price points, and competitive intensity, they could then be addressed as individual businesses.

Some companies used market segmentation to develop strategies to deliver a broad range of products to a particular end-use industry. For example, both Dow and DuPont created "automotive" business groups headquartered in Detroit, headed by an SBU manager, to deliver the full range of products—for example, commodity and engineering plastics, fibers, various solvents, and paints and coatings—produced by those companies to the automotive sector. These automotive strategies were not generally successful because the segment definition was primarily an organizational convenience. The links between the products did not produce sufficient value from a customer perspective because the customers were able to "cherry pick" products they actually needed, and the combination did not provide significant economic benefits to the manufacturers themselves, aside from some shared costs across the businesses and products.

Striving for Lower Costs and Greater Operating Efficiency

By the end of the 1980s, with financial results worsening and pressures for dramatic action by the financial community increasing, new tools were clearly needed to address a new set of strategic issues. Early restructuring had made a number of companies much more efficient, including some of the companies bought by financial investors and operated with minimum overheads, such as Vista Chemical and Georgia Gulf. Chemical industry management was facing challenges in several areas: portfolio composition—beyond the BCG formulation; management processes; industry cyclicality (how to manage and how to offset); and financial results and shareholder value.

New tools were needed to address this new set of strategic challenges, and the next stage of strategy development came at the right time. James Champy and Michael Hammer, two consultants from CSC Index, had been developing an approach to improve a company's efficiency and reduce its costs by focusing on its "business processes," the internal activities of a business defined broadly around "deliverables," which cut across the traditional functional organization. This method was called business process reengineering (BPR).[18] Once these processes are defined, they can be "reengineered" to eliminate the redundancies and greatly improve efficiency; these newly engineered processes are then supported by "enterprise" computer systems that provide the fully integrated data and management information systems needed to implement the results of reengineering. A new organizational structure is then put in place that supports the newly defined processes, leading to lower costs, more rapid organizational responses, and better service to customers. Transformation or more fundamental restructuring of the business, or both, is carried out by integrating various "processes" with an overall vision of the company. The vision is the strategy of the business, and here the vision path identified earlier finds a powerful use—often taken as a given in BPR exercises but clearly a potentially powerful tool for ensuring tight linkage between strategy and operational implementation.

BPR was the technique necessary in the late 1980s and the 1990s to fix a serious problem: companies had too much nonoperating staff, and many of the functions—purchasing, planning, production, sales, logistics, and financial processes—were enclosed in separate "silos" with inadequate communication and cooperation. So BPR was an "emergency measure" needed to bring about greater efficiency. While BPR served to focus managements' attention on fixing internal company "processes," it also created a new way of thinking about the business as a whole—thinking about a company as a set of business processes, both internal and external, and managing along process lines rather than along traditional functional or business unit lines. Early BPR efforts focused on internal processes, often in the administrative areas (for example, people management and order flow) and in the internal supply chain (production, production planning, and logistics). Accordingly, there were only a few early efforts to define growth and innovation processes for repositioning the company in the marketplace. Managing the external side of the supply chain—that is, creating a fully integrated supply chain that included suppliers and customers—is an effort that continues to this day, with only modest success from the standpoints of business process definition, organization, and supporting software. The focus is on integrating the internal systems across all the participants in a supply chain and to do complete product development and production planning. The growth-oriented BPR efforts were generally not successful because the processes in these areas were and still are much harder to define, are less controllable, and require much more cross-enterprise coordination and synchronization.

The overall effect of the BPR wave was to put the classic strategic-planning tools and methodology development on hold. There was a real slowdown in the outwardly focused activities of the industry as seen in the 1980s and early 1990s, while management aimed to improve the company's financial results. The fundamental restructuring enabled by BPR and enterprise resource planning (ERP) systems resulted in significantly improved financial results for a number of companies and enabled them to reestablish credibility with investors and the financial marketplace. A number of strategic investors, most notably Gordon Cain and Jon Huntsman, were able to benefit from the inefficiencies in companies that had not been reengineered, making leveraged, low-cost purchases of inefficient assets at the low point of the cycle, taking costs out, and reaping benefits when the cycle turned up again.

With the failure of many acquisition programs the industry was searching for more effective ways to plan and identify new growth opportunities. C. K. Pralahad and Gary Hamel now introduced a new strategic-planning approach that aimed to link the earlier concepts of competitive differentiation pioneered by Michael Porter with company-specific know-how and skills.[19] This new concept was based on a "core competency" approach, with core competencies defined as unique groupings of organizational knowledge encompassing skills, physical and intellectual assets, technologies, and relationships or other unique aspects that transcend business unit definitions or product lines, which could exist anywhere in the organization, as shown in Figure 11. The key to using this approach was to define a small number—generally no more than three—of the right set of core competencies:

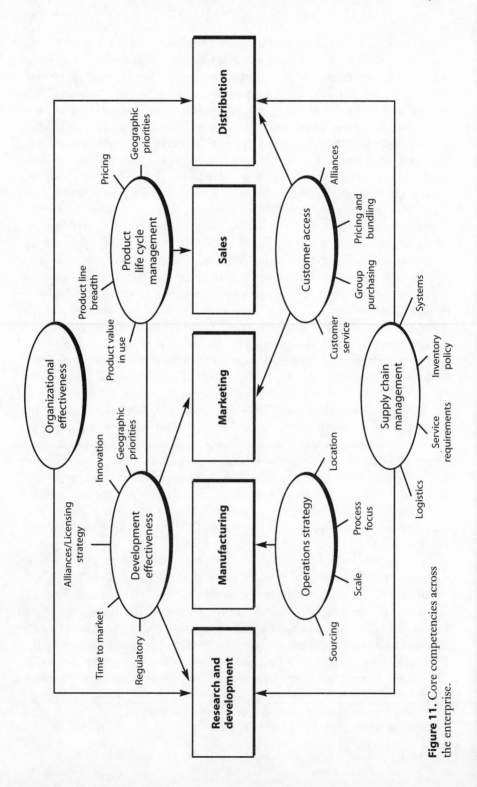

Figure 11. Core competencies across the enterprise.

- Provide competitive advantage—the competency makes a significant contribution to the perceived customer benefits of the end products or services and enables tangible benefits that competitive offerings do not.

- Have broad application—the competency has utility across multiple areas of the firm (products, markets, and businesses), and the cross-business value of the competency provides funding and investment justification even if single-business assessments do not.

- Be difficult to replicate—a unique blending of process, people, and technology that is difficult for competitors to disentangle and imitate.

Core competencies used as examples are Sony's in miniaturization, 3M's in adhesives, and Honda's in building engines.

Joseph Raksis published a paper in the late 1980s that showed how W. R. Grace relied on its core competencies in deciding which new businesses to consider entering.[20] Grace identified its own competencies as designing and manufacturing molecular separation systems, formulating and processing polymeric systems, and formulating and manufacturing alumina- and silica-based materials and catalysts. Grace then targeted a number of new (for Grace) businesses as representing opportunities for their core competencies, including components for high-performance batteries and biomedical devices, smart packaging films and environmentally friendly products for containers and graphic arts, and emission control catalysts for nitrogen oxides (NO_x), sulfur oxides (SO_x), and hydrocarbons.

An overall diversification could be built by linking core competencies to create businesses and working backward by defining what competencies are needed for particular businesses and then seeking to build these competencies from the base of skills already resident within the corporation. With an understanding of its advantaged capabilities, a company can build a vision and a practical path; it can expand business opportunities for future growth by building on and exploiting existing competencies as well as building on others over time.

In some ways the core-competencies concept was related to the process definitions used for BPR, in that they were cross-functional and complex. However, core competencies went further in that they were based on sustainable competitive differentiation. Pralahad and Hamel contended that core competency management should become an integral component not simply of the strategic planning process but also of day-to-day operational management. Most important, competencies would drive the appropriate allocation of resources, people, technology, and capital, with significant investment in the most critical and valuable competencies.[21] In this way management was again given a portfolio management tool, harking back to the BCG matrix: develop a portfolio of competencies and invest in the core competencies.

More important, core competencies became real strategic assets in their own right that could be used as a foundation for growth—new product offerings as well as a foundation for entering a new market and a sharing mechanism to leverage when entering new geographic markets. The focus of an acquisition program could thus be to protect and defend core competencies or to grow by building or adding to competencies. Some companies creatively combined core competencies with market opportunities (specifically by looking at how markets are likely to change over time—either evolving or creating discontinuities) and sought to build growth plans accordingly. It was thought that when markets are about to experience a discontinuity—for example, a technology change—or where a discontinuity can be created—for example, in the Monsanto scenario discussed earlier—such situations present more fertile opportunities if these can use or exploit a company's core competencies. This approach differed from the previous strategic-planning models in that it focused on sustainable market leadership built on skills and "assets" across the organization as the goal. The key metrics are not just historical measures of past performance but are also based on three factors:

- Time—the ability of a company to access demand;
- Value—the ability of a business to shape or create demand; and
- Cost—the ability of business to capture profitability.

By the late 1980s the financial markets were focusing on shareholder returns, as shareholder value became a popular concept across all industries. Shareholders, often just pools of equity capital, were increasingly holding management accountable for achieving acceptable targets. The financial analysts were seeking to go beyond the traditional measures like ROS and ROA and the historical accounting numbers used to calculate returns and value. The inadequacy of ROS as a strategic management tool was discussed earlier. Many analysts believed that historical accounting data did not adequately capture the true amounts and the real costs of the total capital required to support overall business activities. They believed that many capital-intensive businesses were showing profits simply because they had depreciated assets and were, in fact, showing returns that would be inadequate to justify reinvestment. European companies had earlier started to use "replacement cost" accounting for fixed assets to at least begin to reflect more accurately the investment costs companies in an industry were likely to face. Consultants began to use modified interest rates in their corporate assessments (net present value calculations or other calculations, such as calculating business value using a dividend model). A popular method involved taking the difference between actual return on investment and the company's cost of capital (weight average cost of capital, reflecting both debt and equity) and using this difference to determine whether a company or business was truly profitable. Alfred Rappaport at Northwestern was one of the first to use a comprehensive methodology to capture all of the capital, including both fixed and working capital, required to support a business. He also required that the business "pay" the cost of that capital (in the same way that a lender pays interest on debt) first, before any returns are paid to shareholders.[22] He applied this methodology to corporations in total as a first step and then applied it to individual businesses within a corporation's portfolio of businesses. Rappaport coined the term *economic value added* (EVA) for this methodology. He essentially created an ROE model to be used at the individual SBU level. Others used the term *economic value creation* (EVC) to reflect the overall process, and consultants quickly created the term *value-based management* (VBM). There were methodological issues associated with trying to allocate such corporate capital as corporate debt and equity or assets shared by multiple businesses, such as manufacturing plants and distribution centers, but the EVC methodology represented a significant advance over previous tools used by management. By using EVC, management was focusing on measures of prime interest to its shareholders and using a language familiar to observers of the industry—market and financial analysts.

The VBM methodology, used here as a generic equivalent for all of the EVC-EVA methodologies, has major implications for strategic planning for the chemical industry and for the financial analysts following the industry. Historically, the industry had not focused on value creation, thus accounting for its poor performance in the stock market (see also Chapter 9). Companies discovered that, when the industry was looked at from the standpoint of value creation, a number of existing businesses, especially its commodity businesses, were often "destroying value" (and had been for some time). Companies began to concentrate on reducing capital investment and increasing margins (two major drivers of EVC): capital spending was pared back, working capital reduced, and marginal businesses sold off or shut down.

Management in the 1990s had strong incentives to increase shareholder value by having a substantial portion of its compensation tied to the price of their company's stock. The so-called pay–performance ratio increased dramatically between 1980 and 1994, as shown in a recent study. It found that over this timeframe a one percent increase in market capitalization for an average company increased CEO wealth from $18,000 to $124,000.[23]

By the late 1990s most major chemical companies in the United States had adopted some form of EVC, and the concept was taking hold in Europe. The widespread use of EVC as a metric and decision-support tool had the effect of "unblocking" many of the thought processes and actions of major chemical companies, and so for the first time some significant actions were being taken. Major companies, such as Hoechst, ICI, Rhône-Poulenc, and Monsanto, were being dismembered (that is, split or demerged or changing their focus); the German and Swiss chemical industries were consolidated and rationalized; DuPont sold Conoco; Dow sold off its pharmaceutical and consumer businesses; the chemical subsidiaries of some major oil companies, for example, Texaco and Union Oil, were sold; and a number of additional players left the chemical industry altogether.

At the turn of the century the biggest problem facing not only chemical

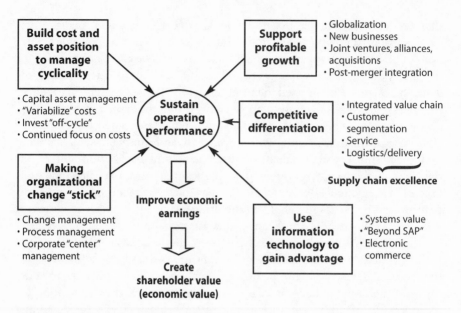

Figure 12. Industry future strategy requirements.

companies but also firms in many other sectors of industry was what might be called a "growth crisis." A number of so-called strategy gurus of the 1980s and 1990s had thoughts and prescriptions on how to tackle this problem, but they differed in their approach.[24] Adrian Slywotzky, a leading consultant and author, pointed out that in the 1990s only 7 percent of U.S. public companies achieved eight or more years of double-digit growth in revenue and operating profits; he believes that a nifty business model is no longer a guarantee of success. In addition, he says that the next strategic wave is what he calls "demand innovation," which involves doing new things for customers to solve their biggest problems to improve their operating performance. Much of this involves offering new services that in some cases act as entry barriers to competitors.[25] Examples in the chemical industry notably include taking over operations at the customer level that involve the use of chemicals. PPG and BASF (in automotive paints), Ecolab (in restaurants and hospitals), and Quaker Chemical (in General Motors' hydraulic transmission plants) have been successful in implementing such comprehensive product-service models.

McKinsey's Richard Foster believes that the main barrier to reinvigorating growth involves what he calls "cultural barriers." Companies need not only to develop new businesses but also must sell old ones and thus avoid what Foster calls "cultural lock-in," where managers get too attached to existing products and processes.[26]

There is no agreed-on formula for setting companies on a new path for growth, and the challenge is formidable. Chemical firms are struggling with this issue, which is closely tied to how analysts and the investing public look at companies' prospects for the near and long term.

* * *

As the millennium approached, the chemical industry in the industrial world, both commodity and specialty industries, had reached a degree of maturity that made it difficult for managers and consultants to come up with exciting, breakthrough strategic concepts. Managers were clearly fixated on their firm's share price, almost to the exclusion of other goals. Value growth became the critical challenge facing chemical companies. The business outlook is as difficult as it ever has been, with continued cyclicality, maturing markets in the developed world, and greater risk and uncertainty in the developing world. Since the early 1990s the industry had relied on cost reduction through downsizing and restructuring and is now facing the challenge of how to create value in the future. The industry has possibly forgotten how to grow with existing organizations, processes, information technology systems, and incentive and measurement systems, particularly when these are not aligned for growth. The key challenge for future strategy development for leading chemical companies is building the capabilities required to support shareholder value growth in the challenging business and economic environment, as shown in Figure 12. The industry's strategy has been to do everything possible to raise earnings per share and to convince analysts that the firms were in a position to achieve future earning growth. This would be done by a combination of acquisitions, cost reductions, and market share growth. CEOs were spending much more time with equity analysts and investment bankers than with strategy consultants. These consultants—firms like McKinsey, Accenture, Booz-Allen, and BCG—were busy developing e-commerce strategies, making ERP systems work, and helping firms with post-merger integrations. Companies, with or without consultants, continued to try to improve their operations, while looking for new ways to grow and to attract investors. Strategy necessarily took a backseat to getting the firms in the best possible shape for renewed growth in the future.

Endnotes

1. While Japanese companies were similar in many ways to those in the other two regions, they were focused mostly on domestic markets, initially using technology licensed from Western firms, with high feedstock costs and relatively small plants. Strategic planning took a different course in Japan and is not covered in this chapter.
2. Some definitions of *strategy:*
   - Tactics is knowing what to do when there is something to do. Strategy is knowing what to do when there is nothing to do. (Polish Chess Grandmaster)
   - Skill in managing or planning; artful means to achieve some end (*Webster's New Collegiate Dictionary* [Springfield, Mass.: G&C Merriam, 1981]).

- Allocating resources across a set of business choices to achieve sustainable competitive advantage and superior financial returns (workable business definition).
3. Robert D. Buzzell, "Is Vertical Integration Profitable?" *Harvard Business Review* 61 (Jan.-Feb. 1983), 92–101.
4. For example, Peter W. Beck, "Corporate Planning for an Uncertain Future," *Long Range Planning* 15:4 (2001), 12–21.
5. Michael E. Porter, *Competitive Strategy: Techniques for Analyzing Industries and Competitors* (New York: Free Press, 1980).
6. For example, Kathryn Rudie Harrigan and Michael E. Porter, "End-Game Strategies in Declining Industries," *Harvard Business Review* 61 (July-Aug. 1983), 111–120.
7. S. J. Q. Robinson, R. E. Hitchens, and D. P. Wade, "The Directional Policy Matrix: Tool for Strategic Planning," *Long Range Planning* 11 (June 1978), 8–15.
8. For example, Mark R. Pratt, "Experience Is a Tough Taskmaster," *Chemical Engineering Progress* 7 (July 1988), 50–55.
9. Darrell C. Aubrey, personal communication.
10. Arnold Lowenstein, Charles River Associates, personal communication.
11. For example, Pankaj Ghemawat, "Building Strategy on the Experience Curve," *Harvard Business Review* 63 (Mar.-Apr. 1985), 143–149.
12. See Peter H. Spitz, *Petrochemicals: The Rise of an Industry* (New York: John Wiley & Sons, 1988), 391–392.
13. Stuart S. P. Slatter, "Common Pitfalls in Using the BCG Product Portfolio Matrix," *London Business School Journal* (Winter 1980), 18–22.
14. Fernand Kaufmann, personal communication.
15. Pankaj Ghemawat, "Sustainable Advantage," *Harvard Business Review* 64 (Sept.-Oct. 1986), 53–57.
16. Ibid. See also Ghemawat, "Competition and Business History in Perspective," *Business History Review* 76 (Spring 2002), 37–74.
17. Benson P. Shapiro and Thomas V. Bonoma, "How to Segment Industrial Markets," *Harvard Business Review* 62 (May-June 1984), 104–110.
18. Michael Hammer and James A. Champy, *Reengineering the Corporation* (New York: HarperBusiness, 1993).
19. C. K. Prahalad and Gary Hamel, "The Core Competence of the Corporation," *Harvard Business Review* 68 (May-June 1990), 79–91.
20. Joseph W. Raksis, "Growth through Focus in Specialty Chemicals," Chemical Management and Resources Association conference, Newport Beach, California, 13–16 Sept. 1992.
21. Pralahad and Hamel, "Core Competence" (cit. note 19).
22. Alfred Rappaport, *Creating Shareholder Value: The New Standard for Business Performance* (New York: Free Press, 1986).
23. Brian J. Hall and Jeffrey B. Liebman, "Are CEOs Really Paid Like Bureaucrats?" *Quarterly Journal of Economics* 113:3 (1998), 653–691.
24. Simon London, "The Growing Pains of Business," *Financial Times*, 5 May 2003, p. 5.
25. Adrian J. Slywotzky, Richard Wise, and Karl Weber, *How to Grow When Markets Don't* (New York: Warner Books, 2003).
26. Richard Foster and Sarah Kaplan, *Creative Destruction: Why Companies That Are Built to Last Underperform the Market—and How to Successfully Transform Them* (New York: Doubleday, 2001).

Chapter 5

Quality and the Reengineering Imperative

Peter H. Spitz

The early 1990s were not good years for the chemical industry—not as bad as the early 1980s, but still fairly miserable for most companies. For the entire industry ("chemicals and allied products"), after-tax income, which had been at the $23 billion to $24 billion level during 1988–90, dropped to $20 billion in 1991 and $13 billion in 1992, while return on equity (ROE), which was between 15 percent and 20 percent in the three preceding years, dropped to between 9 percent and 10 percent in 1991 and 1992.[1] For companies heavily tilted toward commodities, the situation was much worse. Dow Chemical's operating income, which had reached $2.486 billion in the banner year 1989 and was still at a respectable $1.378 billion in 1990, dipped to $935 million in 1991 and became a loss of $496 million in 1992. Union Carbide's income from continuing operations dropped from $978 million in 1988 to $365 million in 1990, became a loss of $147 million in 1991, and reverted to a gain of only $178 million in 1992. At the end of 1991, *Chemical Week* provided a litany of unfavorable statistics for the industry's third quarter relative to that of 1990: diversified companies were off 77 percent, and multi-industry companies with chemical process operations were off 51 percent; industrial chemical profits were down 22 percent; petroleum and natural gas companies' net income was down 46 percent; Lyondell Petrochemical sales were down 20 percent; Hercules sales were down 12 percent; Sterling Chemical sales were down 36 percent; and Cabot Corporation sales were down 15 percent.[2]

In the early and mid-1980s, when the industry had gone through a major slump, some of the petrochemical producers had decided to bail out of these cyclical commodities to concentrate on higher value-added products. Most of the producers had undertaken significant cost cutting, and all were reviewing their long-term strategies. Relatively little had been done, however, to change fundamentally their ways of doing business. The early 1990s, which brought another period of dismal financial performance and high anxiety, provided a second wake-up call for the industry—for commodity and specialty producers alike. This chapter describes how companies in the 1980s and 1990s took a new look at their traditional ways of doing business. In all cases, of course, the first step was to reduce near-term costs. The more important step from a long-term standpoint was to improve internal processes fundamentally and focus additional effort on working more closely with customers.

When companies are trying to reverse a deteriorating bottom line during a period when sales are flat or declining, the only way they can improve profits quickly is by slashing costs. The easy way to do that is to reduce staff, close unprofitable operations, and otherwise eliminate cost burdens. In the early 1990s many companies also initiated comprehensive programs to reduce staff overhead expenses, which usually included incentive programs to make it attractive for middle-level and other staff members to take early retirement. These programs generally reduced the number of years employees had to work to earn full retirement benefits or reduced the retirement age. Of course, such programs had to be offered to all employees in given categories, not just to those the company was content to see depart. So companies lost many key employees, some of whom would go to work for competitors, while presumably observing the noncompetition or confidentiality agreements they had with their previous employer. Years later some companies wondered whether many of these redundancy programs had been sensible, but at the time the emphasis was on cost cutting and not much else. In any case a lot of good employees with valuable experience were lost. Many went into different industries. Many others retired permanently. A few came back and had second careers, consulting to the same companies that had let them go.

Companies did recognize that this kind of cost cutting alone would only solve part of the problem. Individual businesses were also reviewed to determine whether all the components of the company should be kept in the portfolio. For businesses that were retained, later termed *core businesses*, companies concentrated on maintaining or gaining market share. Gaining share is basically a "zero sum" game, especially during low- or no-growth market environments. At such times companies can only try to increase market share at their competitors' expense. Since most chemicals are specification products and one company's offering is essentially the same as another's, except sometimes in degree of purity, the most obvious means of gaining share is by cutting price, slicing margins even further in the hope of better times ahead. It has always been interesting to hear companies state that they do not use price cutting as a matter of policy and that their sales representatives have to get permission from their home office superiors to reduce a price in order to snare a new account or to snatch a customer away from a competitor. Regardless, price competition is endemic during these periods. In some cases it can be rationalized when a back-integrated company uses so-called chain economics to achieve a lower selling price. (*Chain economics* is a term used when a back-integrated company transfers upstream raw materials or intermediates, or both, at a low profit or even at cash cost from plant to plant to end up with a selling price that has little or no profit built into it but generally covers all costs.) However, this practice tends to become the rule when margins shrink to very low levels.

Companies do recognize that there are better ways to compete and to retain or gain customers. Specialty chemicals producers have traditionally been able to do this for some of their products by providing various types of services, typically some form of technical service for which they might or might not charge, or by branding some of their products. Commodities, however, are generally not sold with technical service, making it easier for customers to switch suppliers. Recognition of this unfortunate fact had already led many chemical companies in the mid-1980s to look for other ways to please their customers, to differentiate themselves from their competitors, and to become their number-one supplier of choice. To help make this happen, companies enthusiastically (or pragmatically) signed on to the so-called quality movement, which had already gained advocates in other industries in the late 1970s and soon became the new mantra. Total quality management, or TQM as it became known, focuses on the total process used to produce and deliver the product to the customer, not necessarily on simple product quality alone. With TQM came the principle of continuous improvement, the concept that operational processes can be incrementally improved over time if the firm can motivate its employees to take ownership in the process of production and sales, to become involved, to think about the operation, and to keep suggesting ways to do their work more efficiently. This idea was not new, of course; it stemmed from much earlier "time-and-motion" methods and "work study" groups, which had been used by such firms as ICI decades earlier. TQM had been adopted for some time in Japan, based on the teachings of such individuals as W. Edwards Deming and Philip Crosby. In the 1980s the TQM movement took the chemical industry by storm.

In 1983 President Ronald Reagan appointed a commission to make recommendations for improving U.S. business performance, recognizing the fact that the inability of U.S. industry to compete had become a national issue.[3] One result was the establishment of the Malcolm Baldridge National Quality Award, established by public law and signed by Reagan in 1987. Applicants were to compete in three groups (small business, service, and manufacturing), with two awards made in each group each year.[4] Motorola was one of the first winners.

The ascendance of Japan as a manufacturing power and its success in penetrating foreign markets while protecting its own market has been amply discussed elsewhere. U.S. manufacturers in the 1970s and early 1980s had become aware that they had fallen behind in productivity gains and in manufacturing excellence and realized that something had to be done. Some firms, notably IBM, Motorola, and General Electric, had recognized this problem earlier and had begun to adopt some of the techniques used by Japanese firms. Now many other large Western firms began to take the quality issue seriously, ushering in a new era.

Soon, the terms *best practices* and *benchmarking* were also being heard. For TQM and later for reengineering projects, comparative measurements were important. Companies became addicted to benchmarking, the practice of looking at the specific operations of competitors and those of companies outside their industry to determine which firms achieved the best performance in various operational areas and business processes, such as inventory turns, working capital requirements, and manufacturing cycle times. Companies attempted to identify best practices that could be measured and emulated. Xerox and others had begun this practice earlier, and benchmarking soon became a mantra in the chemical industry.

Then the word *reengineering* and the phrase *business process reengineering*, or BPR, began to be heard. Reengineering was a quite different approach, more of a "step change" than "continuous improvement"; it also became known as "business transformation." Its early advocates, such as Michael Hammer and James Champy, became convinced that companies' operational models, fashioned after World War II and particularly in the 1950s and 1960s, were no longer suitable. These models were based on managements' belief that manufacturing capacity must be carefully controlled to match demand and supply and that companies could lose market share if too little capacity was built to maintain it. Increasingly complex systems were built for budgeting, planning, and control, and the classic pyramidal organizational structure evolved. Companies had contrived to retain the earlier principle of breaking down the work into small, repeatable tasks, which could in many cases be mechanized or automated. But, as Hammer and Champy pointed out, the number of tasks kept increasing, and the processes for making a product or delivering a service became increasingly complicated, requiring more and more people, particularly in the middle of the organization, who often found it hard to coordinate with each other. As senior managers became increasingly divorced from their customers, managements' understanding of and responsiveness to customers' attitudes and needs kept decreasing.[5] Moreover, most industries, including the chemical industry, found that the world in which they had lived relatively comfortably for decades had now changed in a number of important ways: there were uncertain energy and feedstock costs, new global competitors, a slowdown of demand growth, and other problems and discontinuities.[6]

The reengineering gurus like Champy and Hammer strongly believed that three forces were driving companies deeper and deeper into territory that their managers found frighteningly unfamiliar. They identified these forces as customers, competition, and change.

The application of what was eventually called "reengineering"—basically a method for business transformation that looks at the underlying processes making up the activities of the corporation—had its start considerably earlier and in other industries, for example, at General Motors in the Saturn project; at the Topeka, Kansas, pet food factory of General Foods; and in the First Direct subsidiary of Midland Bank. These firms used the "clean sheet of paper" design of business processes (that is, they looked at all the elements that formed part of these processes) to reengineer certain operations. Other firms that used reengineering concepts at an early stage included Cigna, Xerox, and Bell Atlantic.[7] An oft-cited case, which became a sort of poster child for reengineering, involved IBM Credit Corporation, which dramatically reengineered its process for financing the purchase of computers and related software in the 1980s. The process was shortened from as long as two weeks—with an average time of one week—to a few hours. This was the result of having two senior managers take a typical financing request and walk it through all the steps necessary to get it approved: logging of the financing request, credit checking, creating the loan covenant, pricing, and a quotation letter. They found that the actual work required only *four hours* and that none of it took a real specialist. IBM replaced the entire system with one generalist, supported by a sophisticated computer system. These changes amounted to a 90 percent reduction in cycle time and a hundredfold improvement in productivity.[8]

Few chemical company managers were familiar with reengineering when it hit the industry in the early 1990s, but most were quick to understand and embrace the basic concepts. The departments under which various business functions were carried out—order taking, operations planning, manufacturing, inventory management, logistics, and invoicing—often operated like individual, somewhat isolated "silos" that communicated imperfectly with each other, part of a system that required excessive amounts of working capital and slowed down the process of getting product to the customer. The idea that this process could be carried out by crossfunctional teams was immediately appealing, particularly since it was essentially an extension of TQM.

A wave of reengineering projects soon swept the chemical industry, and because the more efficient ways to conduct business required considerably more computer and software support, information technology (IT) investments were also significantly stepped up. But it was soon found that the available software was unable to provide companies with the comprehensive transaction systems that would help implement the results of reengineering. Soon a new type of system, known as enterprise integration software, came on the scene, and the installations were called enterprise resource planning (ERP) systems. SAP AG in Germany shortly became the leading vendor of such software, although in early reengineering work, companies often stayed with earlier software packages (later called "legacy software"), which eventually had to be replaced. Chapter 6 explains the increasingly pervasive role of IT in the chemical industry in the 1980s and 1990s and the emergence of powerful chief information officers (CIOs), who would make expensive purchasing decisions after recommending them to top management.

It also became clear that reengineering and IT investments had to be closely linked to corporate and business strategy to be effective. Companies needed to decide which businesses to keep and which to sell, to decide who and where their most important customers would be, to trim their product offerings by eliminating losers, to examine their new product development process, and to protect and, if possible, increase market share not by price cutting but by getting closer to their customers, who would in turn want to continue doing business with them. In examining a company's various business processes, it was critical to decide whether to fit the reengineering work to the relatively rigidly constructed software or to modify the software to fit the results of reengineering. Neither tradeoff turned out to be completely successful, as many companies later discovered.

Reengineering introduced a new way of thinking about business processes and seemed to be the right methodology at the right time. Companies embraced it enthusiastically because it made a great deal of sense. Since the concept was new and initially not well understood, companies used consultants extensively on these projects. Reengineering often involved massive assistance from such large consulting firms as Accenture, CSC, and Gemini, for which the new methodology became a bonanza as it also led to a lot of systems integration work. It became the right thing to do, even though the early results were not always what companies had expected and paid for. But once reengineering was undertaken, there was no turning back.

Embracing Quality Management

Speaking at a Chemical Marketing Research conference in May 1986, Ernest Drew, then the group vice president of Celanese Corporation, discussed how his company had managed to survive the difficult conditions of the earlier part of the decade. While sales had declined from $3.7 billion to $3.1 billion between 1981 and 1985 and net income had turned negative in 1982 and was negligible in 1983, the firm's income had reached $161 million by 1985, exceeding the figure for 1981. The ROE rose from 13.1 to 18.0 percent between 1981 and 1985. During the same period the salaried head count had been reduced from 11,700 to 7,400 so that sales and administrative costs had dropped from $316 million to $261 million. To explain how these results were achieved, he went on to say that the dramatic improvements were primarily owing to implementation of a quality management program: "In 1981, responding to requests from a key customer to improve the consistency of our fiber products, we initiated a Quality Management process based on Philip Crosby's 'Quality is Free' concept." Citing an example of the results of applying this process, Drew explained that the firm's Palmetto polyester staple plant had improved its output per employee by 54 percent from 1981 to 1985. He went on to say that the savings for the total corporation in 1985 over those in 1981, which were attributable in the firm's opinion to the quality program, were $70 million in operating income.[9]

The strong advocacy of quality management was symptomatic of the 1980s, when most firms were implementing such programs. However, in 1992, *Chemical Week* carried a cover story titled "Quality: Evolving through the Backlash." It began:

> Total Quality Management (TQM), an orphaned U.S. business discipline that grew up in Japan, became a business obsession when it returned home in the 1980s. Adopting the Japanese model, some companies showcased their commitment by hiring quality managers—sometimes called gurus—and signing up consultants with a packaged program for sale. . . . Few organizations that went that route managed to achieve a fundamental positive change. Instead, many established quality as a discipline unto itself—one that cost a great deal of money, misdirected employees' efforts, and never came close to a return on investment . . . [and] . . . while many have succeeded in establishing a quality process, so many have run into trouble . . . that TQM is now viewed by many as another excess of the 1980s.[10]

This viewpoint is in sharp contrast to the successes that Ernest Drew recounted to the CMRA in 1986. Was the Celanese situation relatively unique, or did many other firms see the same beneficial bottom-line results from quality management initiatives? Was there, for example, a lot more slack at Celanese that made it easier to reduce overheads through head count reduction? What was really achieved by companies that adopted TQM, particularly by the ones that did it correctly? In the same issue of *Chemical Week*, two CEOs from the chemical industry gave some cautious opinions on their TQM initiatives. Earnest Deavenport, CEO of Eastman Chemical, stated: "Since we began in 1986, customer satisfaction has been the keystone of our (Quality) program. A second major effort . . . has been employee empowerment . . . and the use of self-directed work teams. . . . The process of creating value through Quality remains an ongoing concern." And Robert Gower, CEO of Lyondell Petrochemical, said: "There are many success stories offsetting the tales of discouragement." At Lyondell, over a six-year period 4,000 employee ideas were cycled through Lyondell's quality program: "On average, you have a $10,000 improvement per usable idea."[11]

In trade journal articles published in the 1980s, few stories can be found that attributed large sales, general, and administrative (SG&A) savings to installation of TQM alone. Most companies that successfully implemented quality programs instead cited improvements in production processes, much greater involvement of their workers in day-to-day activities, higher-quality products,

and more satisfied customers. To obtain these outcomes was after all the original reason that companies focused on TQM.

But how new was TQM? And was it really a Japanese invention? The quality management concept was largely conceived and taught by several non-
Japanese individuals whose names are now well known and who became gurus
to Japanese industry: Deming, Crosby, and Joseph Juran are usually considered the fathers of the quality movement, with such disciples as Tom Peters,
Genichi Taguchi, and Armand Feigenbaum.[12]

W. Edwards Deming, a Ph.D. in mathematical physics, went to Japan to
lecture top business leaders on statistical quality control and told them they
could become world-class quality leaders if they followed his advice. Eventually, the Deming Prize became the most coveted business award in Japan, while
Deming was called the founder of the third wave of industrial revolution. He
preached that productivity improves as variability decreases and that worker
participation in decision making is essential. His "Fourteen Points" and "Seven
Deadly Diseases" are showcased in his landmark book *Out of the Crisis*.[13]

Philip Crosby worked for the Martin Corporation on the Pershing missile
project and later became director of quality for ITT. In 1979 he founded Philip
Crosby Associates. He is associated with the concept of "zero defects," which
equates quality management with prevention. He also proposed a series of fourteen steps to quality improvement and four absolutes of quality management
and stated that a zero-defect strategy (that is, producing a product with no
defects), which he believed was successfully implemented in Japan, required
total management commitment. Crosby also made the initial link between high
quality and reduced costs with his "Quality Is Free" focus and a recognition
that defects increase costs across the whole organization.

Figure 1. Quality management pyramid. *Source:* Bruce Brocka and M. Suzanne
Brocka, *Quality Management: Implementing the Best Ideas of the Masters* (Homewood,
Ill.: Business One Irwin, 1992), 23, by permission of the McGraw-Hill Companies.

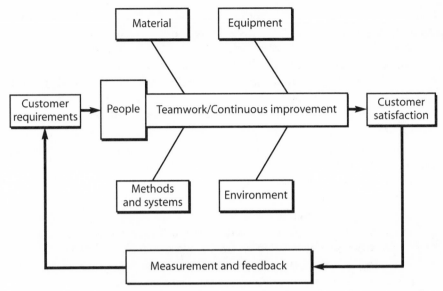

Figure 2. The General Motors quality network process model. *Source:* Bruce Brocka and M. Suzanne Brocka, *Quality Management: Implementing the Best Ideas of the Masters* (Homewood, Ill.: Business One Irwin, 1992), 24, by permission of the McGraw-Hill Companies.

Joseph Juran was quality manager at Western Electric and then a professor of engineering at New York University before becoming a consultant in 1950. He lectured frequently in Japan and is regarded as one of the architects of the TQM revolution there. In 1979 he founded the Juran Institute, which conducts TQM training seminars and publishes related books. Juran lists the following ingredients underlying the Japanese quality revolution:

- The upper management took charge.
- They trained the entire hierarchy.
- They improved quality at a revolutionary rate.
- They provided workforce participation.
- They added quality goals to the business plan.

There was considerable commonality among the precepts taught by these masters. Brocka and Brocka identify the eight "Pillars of TQM" as follows: organizational vision, barrier removal, communication, continuous evaluation, continuous improvement, customer and vendor relationships, worker empowerment, and training.[14] The authors present four key components—top-down strategic vision, continuous analysis, employee empowerment, and listening to customers and vendors—from which the other concepts flow. These key components make up the strategic portion of the quality pyramid shown in Figure 1. A quality management network process model is shown in Figure 2.[15]

By the mid-1980s many large and small chemical firms had embraced the

TQM concept in its various manifestations. Corporate leaders were telling staff that they wanted their company to be recognized as the number-one supplier of choice in all their marketplaces.

Companies wanted better performance, improved efficiency and effectiveness, and satisfied customers. Companies usually initiated a series of projects in various areas of business, such as communications, customer satisfaction, human resources, training, and waste generation. Team building and problem solving, which often involved several-day workshops were common. An important corollary of quality control was benchmarking, as described above. BP Chemicals, for example, at its Antwerp ethylene oxide and derivatives plant compared various aspects of its operations with similar ones of another nearby chemical company. Teams from both companies established that each firm was better in certain functional parts of their operations, thus generating a lot of ideas on both sides for improvement. BP Chemicals reduced staffing at the site by 20 percent, which gave the company a substantial gain in productivity.[16]

Another important step for many firms was to develop supplier partnerships to improve product consistency and overall manufacturing throughput. Dow concluded that "purchased goods and services are the greatest opportunity to improve [the] quality and cost structure of the business."[17] By 1991 Dow had forty-seven qualified suppliers. Cryovac, Grace Specialty Chemicals' packaging subsidiary, began a vendor certification program in 1987 to identify qualified suppliers to be "partners," thus replacing what was often an antagonistic relationship based primarily on pricing. Vendor certification was instituted at many firms, such as GE Silicones, which started a Partners-in-Quality program. Of twelve original suppliers for raw silicon metal, GE Silicones found that only two furnished consistent quality product, and only one of them, Elchem, had a statistical quality control program. GE and Elchem set up joint teams that met once a quarter to set standards and monitor results. Savings of several million dollars were initially realized and continued at a somewhat lower rate thereafter.

European firms led their U.S. counterparts in one respect by establishing a registration system for companies that met the requirements of the International Standards Organization. ISO 9000 became an accepted international hallmark of quality, first in England and then among European companies generally. In the late 1980s and early 1990s chemical firms on both sides of the Atlantic were striving to have their plants and businesses registered by the ISO. This procedure required an audit by an accredited organization. For U.S. companies such registration became almost a prerequisite for doing business in Europe.[18] At first, U.S. firms, including Eastman Chemical and Ethyl Corporation, actually used European auditors to certify plants and processes.

The Chemical Manufacturers Association (CMA) gave broad support to the move for ISO registration. CMA's Quality Council developed a so-called process safety–quality matrix that covered the relationship between quality,

ISO 9000, the Malcolm Baldridge National Quality award, OSHA's (Occupational Safety and Health Association) process safety management code, and Responsible Care's process safety code. Many domestic chemical firms applied and vied for the Malcolm Baldridge award: Eastman Chemical, which had first applied in 1988, finally won the prestigious award in 1993.

Looking back, some general conclusions can be drawn with respect to the benefits chemical firms gained from TQM and similar initiatives during this period. Unquestionably, many firms became more efficient, gained closer relationships with suppliers and customers, improved their production processes, and empowered their workers to a degree not often seen before that time. Companies also became significantly more aware of the need to instill in their staff a desire for quality and good practices. They recognized that even such little things as having a friendly voice for order takers and switchboard operators was important and that customer complaints needed to be handled quickly by trained and friendly staff. The improvements they made in various parts of their operation allowed these firms to work more efficiently with less staff and less overhead. But TQM programs were not created or necessarily intended to achieve the kind of savings that Drew showcased in his CMRA presentation. TQM programs were geared primarily toward achieving small continuous improvements and high-quality, effective means to conduct existing operations rather than toward dramatically changing business processes. Some companies had hundreds of such programs at any given time, most of which were quite small and would not lead to large savings.

The shortcomings of TQM gave rise to the concept of "reengineering," a much more fundamental and dramatic approach to business management. In the early 1990s reengineering caught the attention of management still involved in various aspects of TQM. According to Michael Hammer and Steven Stanton, "TQM assumes the underlying process is sound and looks to improve it; reengineering assumes it is not and seeks to replace it. . . . Success with TQM can position an organization to take that next step."[19]

Nevertheless, the quality management movement continues and recently has been more broadly embraced by a number of firms in the form of the "Six Sigma" methodology. This initiative's success first with Motorola and then with others, most notably General Electric, impressed many companies. Six Sigma is a statistical term that describes how close a product comes to meeting its quality goal or, in other words, how far it deviates from perfection. One sigma means that 68 percent of products are acceptable; three sigma, 99.7 percent; and six sigma, 99.9999997 percent, or 3.4 defects per million. Motorola was one of the first to use this technique and was said to be well above the five-sigma level by the early 1990s.[20] Chemical firms that have adopted Six Sigma include Allied Signal (then headed by Lawrence Bossidy, a former GE executive), Dow Chemical, DuPont, Great Lakes Chemical, Air Products and Chemicals, W. R. Grace, Albemarle Chemical, Celanese, and Cytec.

A *Fortune* article in 2001 explained the technique as training employees to become "black belts" who will then employ drastic steps to "chop, kick and block until errors are virtually nonexistent."[21] Six Sigma techniques also apply to productivity improvements. G.E.'s CEO Jack Welch claimed that Six Sigma initiatives saved the company $2 billion in 1999, three years after implementing the technique. The *Fortune* article is, however, quite skeptical of Six Sigma as a guaranteed route to success, having found that the share price of their sample of companies that instituted Six Sigma did not necessarily show an upward trend compared with a sample of others that did not use this technique. Perhaps with a little tongue in cheek, *Fortune* went on to say that "defects don't matter much if you're making a product nobody wants to buy," referring to Motorola's originally disastrous foray into satellite-linked mobile phones.[22] But the company was merely ahead of its time.

Dow, DuPont, Allied Signal, and other chemical firms committed to Six Sigma are convinced that this initiative is paying off handsomely. Dow has a savings target of $250,000 per project and $150 million in profits for the year 2000, as reported in a quotation by Kathleen Bader, then vice president of quality and business excellence.[23] In addition to using Six Sigma as a tool for boosting productivity, Dow has implemented a new set of order management processes called the "Perfect Order," which is also based on Six Sigma methodology. For this Dow received the 2000 CIO-100 award for achieving "the highest level of operational and strategic excellence for innovation and improved business performance through the use of customer-oriented practices."[24]

Reengineering: The Sequel to TQM

Although experts continue to argue about the relationship of TQM initiatives to reengineering, including the question of whether the latter is an extension of the former, this much is clear: TQM is rooted in the philosophy of "continuous improvement" of existing processes, while reengineering, in its proper application, involves dramatic restructuring of business processes, usually resulting in crossfunctional organizational change. J. Robb Dixon and his colleagues have given the following useful description.

> Reengineering is not a fundamentally new approach to performance improvement, but the combination of emerging technological capabilities and evolving market demands [that] enable and encourage organizations to take the risks associated with radical customer-oriented change. . . . [It is] a radical or breakthrough change in a business process. Reengineering process designs seek dramatic orders of magnitude, as distinguished from incremental, improvement in business value. Key value creation processes involving manufacturing operations include order fulfillment (the customer supply chain process), product development, order creation (selling and configuration) and customer service (post product delivery processes).[25]

The concept of improving business processes certainly comes from continuous process improvement (now often called quality) approaches, to some extent dating back to industrial engineering concepts many decades earlier. Such work was generally aimed at processes small enough for single work groups. The concept of working on broad crossfunctional processes is more recent and is the underlying principle on which reengineering is based.[26]

Michael Hammer and James Champy offer this definition in their book on reengineering: "Reengineering properly is the fundamental rethinking and radical redesign of business processes to achieve dramatic improvements in critical, contemporary measures of performance, such as cost, quality, service and speed."[27] In the authors' view, *fundamental* means asking "why do we do what we do and why do we do it the way we do?" *Radical* means disregarding all existing business structures and procedures and inventing completely new ways of accomplishing work. Dramatic improvements must be much greater than those set as targets for TQM. They seek change amounting to 30 percent to 50 percent or more in the process being reengineered.

Duane Dickson, who had in the 1990s headed up the reengineering effort at Union Carbide as a partner at Gemini Consulting, said that reengineering came about through the merging of several ideas and methodologies. Process mapping came largely from Kepner-Tregoe, a respected smaller consulting firm. Rummler-Brache, founded by ex-Tregoe consultants, furthered this concept and became a boutique reengineering consulting firm hired by a number of chemical companies as trainers of the concept. Gemini (ultimately acquired by Cap Sogeti in the 1990s) was created through the merger of United Research Company, a consulting firm with a strong background in process-driven efficiency and quality consulting, and the Mac Group, which specialized in corporate strategy assignments, with close ties to the Harvard Business School and other academic institutions.[28] Champy and the Index Group—an early IT development and consulting firm—joined with Computer Sciences Corporation to form CSC Index, which quickly became a leading reengineering consultant.

Accenture, which became one of the other leaders in the reengineering movement, placed business process reengineering (BPR) in the middle of a continuum of change, with streamlining a single function as the least dramatic initiative and business invention as the most dramatic (Figure 3).[29] BPR matches or creates best practices and is carried out for business processes across an entire manufacturing unit. Most BPR efforts initially focused on the *supply chain*, defined as the sequence of operations that leads from sales to order fulfillment. It can be argued that streamlining a single function is TQM, while business invention occurs when reengineering uncovers an entirely different way of serving customers. Reengineering could then also lead to "changing the rules of the game," as when Dell Computer in the 1990s completely reengineered its business model to eliminate inventories by having a number of "just-in-time" suppliers. Competitors, such as IBM and Compaq, have not yet caught up to Dell.

Robb Dixon noted that while most reengineering projects do not result in "business invention," they do tend to have one factor in common that involves

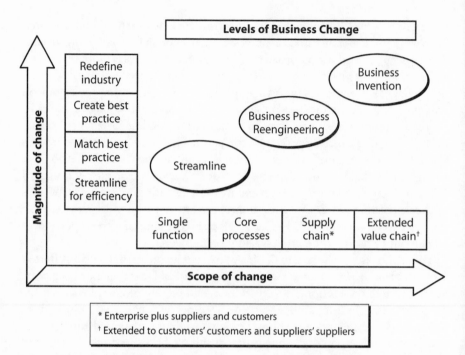

Figure 3. Levels of business change. *Source:* Accenture, by permission.

"changing direction." While major improvement was obviously sought, the direction of the goal had frequently changed. He went on to say that "competitively successful firms learn how to improve in the direction their environments reward, . . . knowledge accumulates and translates into superior performance and [becomes] an effective shield against existing competitors. . . . We believe it is this demand for improvement along new dimensions that makes reengineering both so difficult and so necessary."[30]

John Aalbregtse, a partner at Accenture who heads up the firm's chemical strategy practice, recently discussed the development of reengineering in the 1990s.

> Many companies have used these initiatives to implement strategic shifts in their businesses. Reengineering in the early 1990s was heavily about cost reduction and restructuring; then it was how to move to a global structure and supporting global processes; then it was about growth and new business models. Depending on the chosen strategy there might be more focus on "flexibility" over "cost reduction" or "time to market" over "product performance," but the shift in emphasis was company specific, not for reengineering in general.[31]

Reengineering projects almost always involved the use of information technology to enable the new ways of carrying out the worker-engineered processes. Thus, IT-oriented management consultants had an edge in the reengineering movement and quickly became leaders. It was seen that IT, on which companies were spending a lot of money, could not alone achieve the desired results. When CSC Index and Michael Hammer collaborated on a research project covering crossfunctional systems, they found that such firms as Ford and Hewlett-Packard had adopted many of the components of reengineering much earlier, with substantial use of IT to achieve dramatic improvements.[32]

Hammer and Champy devote an entire chapter of their book on reengineering to "The Enabling Role of Information Technology." In the chapter's introduction the authors propose that "most executives and managers think *deductively*, while applying information technology to business reengineering demands *inductive* thinking—the ability to first recognize a powerful solution and then seek the problems it might solve, probably doesn't even know it has." They deem that IT has disruptive power in a positive sense (for the company), because it can change the way work is done and thereby confer competitive advantage.[33]

The authors go on to say that companies need to make technology exploitation one of their core competencies if they are to succeed in a period of ongoing technological change. Unfortunately, some firms that lagged behind the leaders in understanding the power as well as the limitations and pitfalls of installing complex ERP systems incurred the high costs of implementation without seeing all the benefits.

The rush by many chemical firms into reengineering initiatives can be partly blamed on the hype that this concept created when first proposed by advocates like Hammer and Champy. But the management of these firms also recognized that TQM in its various manifestations was not enough to allow their firms to succeed in the quest for higher market share and greatly enhanced profitability. In particular, these leaders began to see very clearly that the old ways of doing business were not good enough and that their firm's survival might be at stake if they did not take dramatic action. In a strange way corporations owed a debt of gratitude to raiders like Boone Pickens, Carl Icahn, and Hanson Trust, who dissected businesses and made the concept of "breakup value" real.[34] These raiders ironically helped to popularize the concept of shareholder value analysis and economic value-added analysis, which forced managers to question the traditional methods of asset allocation and to look for new ways to examine the source of cost and profitability.

The specific elements and characteristics of reengineering are spelled out by Hammer and Champy. They list several recurring themes that are frequently encountered in reengineered business processes:

- Several jobs are combined into one.
- Workers make decisions.
- The steps in the process are performed in a natural order.
- Processes have multiple versions.

- Work is done where it makes the most sense.
- Checks and controls are reduced.
- A case manager provides a single point of contact.
- Jobs change—from simple tasks to multidimensional work.[35]

There are obvious common elements as well as differences between reengineering and TQM initiatives, according to Duane Dickson. Changing work processes can be done without TQM, but it becomes more successful when TQM is combined with reengineering. Both methodologies need heavy involvement of staff and measurements. However, reengineering can be mandated on a top-down basis, whereas quality management projects, which are usually much smaller, generally originate with staff.[36]

Dixon and coworkers believe that reengineering that generates radical change is usually brought about by "crisis," which can drive organizational change and dramatically improve a firm's competitive, technological, and administrative capabilities. Crisis can be caused by outside forces or generated by top management to stimulate innovative response. Management may create a crisis when it perceives capability gaps that pose a threat to the firm's viability. However, a survey carried out by the authors indicated that only three out of fifteen reengineering projects they reviewed were driven by a crisis. The others were initiated because management had developed a new vision for the firm that could not be supported by existing operating capabilities. They further concluded that reengineering initiatives often came about as a result of corporate strategy making. Managers saw reengineering not as a change driven by crisis, but as a proactive step toward the future.[37]

Another view of reengineering, one that many CEOs of chemical companies probably adopted when reengineering became the rage, was expressed by Roger Hirl, CEO of Oxychem, when he said: "My definition of reengineering is that after you've gone through downsizing and rightsizing and all the classic kind of top-down kind of things, you take a look at all of the processes you have and say, 'What's the most efficient way to get the job done?' " He went on to say that reengineering is particularly useful when amalgamating acquisitions, where disjointed systems and redundant functions can be rationalized. This thought was echoed by Lawrence Farmer, senior finance director at Elf Atochem, which had made a number of acquisitions and decided that reengineering would be the proper response to integrating the various functions. The firm concluded that quality teams, effective when within departments, broke down when they were asked to work crossfunctionally.[38]

A survey of North American and European companies carried out by CSC Index in 1994 showed that reengineering activity was quite prevalent, with 69 percent of North American companies and 75 percent of European companies having at least one project under way. Many of these firms had three or more reengineering initiatives completed. Not unexpectedly, the companies cited demanding customers, increased competition, and rising costs as the main drivers.[39]

In typical projects, companies' business processes were fundamentally "reengineered." Manufacturing processes might be completely changed and logistics systems for serving customers dramatically altered, using ideas generated by a team created within the manufacturing part of the organization, which studied all aspects of order fulfillment to achieve breakthrough results. An important part of the reengineering work—the definition of customer needs and wants—drove much of the reengineering effort.

Reengineering teams would, for example, model the order fulfillment process to understand it completely before seeing how it could be made better. They would usually do this by first modeling the "as is" process and then defining the "to be" model, using benchmarking and other goals. The common procedure was to use what is often called "brown paper modeling." On large sheets of paper tacked to the wall, the existing process is mapped, showing the material flow laid out in detail with all the handoffs, time delays, manufacturing processes, inventories, warehousing, packaging, and logistics indicated where they occur. When the "as is" process is completely mapped, the reengineering team is almost always astounded by its complexities.

Some of the changes and improvements are often obvious; these are frequently termed "low hanging fruit" or "quick hits." These changes can be effected rapidly and the resulting savings used to support and partly finance the reengineering effort. As the team studies the current process, many ways to simplify or combine become obvious. In brown paper modeling, Post-it notes are affixed to various places on the paper to identify areas of greatest need for improvement or to recommend a specific change in the current process. The team does not take anything for granted and is prompted to challenge every existing way that things are done.

Brown paper modeling is followed by workshops in which the reengineering team, often split into several groups of company people and in many cases consultants, redesigns different parts of the process to develop the "to be" model. These workshops also serve to estimate the financial benefits of operating in accordance with the reengineered process. Such estimates usually cover both one-time savings—for example, in lower required working capital and warehouse closures—as well as continuing savings, including staff reductions, better logistics, and elimination of activities that create no value.

Quantum Chemical: Polyethylene

One of Chem Systems' reengineering projects carried out jointly in the early 1990s with Accenture was for a polyethylene producer, with order fulfillment the main target. The client was Quantum Chemical (formerly National Distillers and Chemical Corporation), which had over the previous several years added a number of acquisitions to its polyethylene business to become the largest domestic producer of this polymer. Quantum's high-density, low-density, and linear low-density polyethylene plants were in numerous locations, including several in Texas, Illinois, and Iowa. Warehouses were located all over the country, and a large number of distributors and shippers were used. The number of grades for high- and low-density polyethylene was by far the largest in the industry: even before the acquisitions, Quantum had many more grades than most of its competitors. Ronald Yokum, a former Dow executive, was CEO of Quantum and the sponsor of the reengineering effort, which was headed up by Peter Hanik, formerly with one of the acquired businesses. Accenture and Chem Systems were selected because the former was very strong in IT, organization, and change management, and Chem Systems knew polyethylene, which promised a quick start.

Quantum had done a customer survey, which showed that many customers were unhappy with the firm's order-fulfillment process, including dissatisfaction with delivery time and handling of complaints. The company realized the enormous problem it faced in amalgamating the many different reactor technologies and in coming up with a manageable product line that would serve a very large number of customers with requirements for both normal and special grades.

The reengineering project began with several days of intensive interviews with thirty or more Quantum executives and staff members at the Cincinnati headquarters and in the plants. After analyzing the results of these interviews, the consultant team recommended several areas for reengineering, including sales and operation planning, product line management, manufacturing, and freight and logistics. A generic view of a company's typical business processes as they relate to the supply chain (sales order to delivery to customer) is shown in Figure 4. The selected areas were approved by management, which then appointed a number of Quantum employees to head up and serve in the five teams, along with Accenture and in some cases Chem Systems personnel. To get the teams to work together, Accenture ran a one-day workshop where team members were given problem-solving assignments having nothing to do with polyethylene or for that matter chemicals. In one exercise the teams were told that they had been on an airplane trip and had made a forced landing in a desert with no sign of civilization. They were given a list of fifteen items available on the plane, such as a map, a mirror, drinking water, and blankets, and told they could only take three items with them. The teams competed to determine which would arrive at the best (predetermined) answer for "survival."

With team spirit thus established, the analysis of the selected areas for reengineering began. Chem Systems' contributions were recommending strategy, developing benchmarks for different parts of the supply chain, and helping the Accenture consultants with their understanding of the polyethylene business. In seeking good benchmark information in such areas as cycle time, invento-

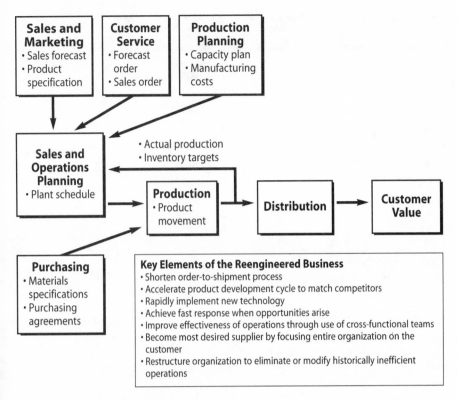

Figure 4. Business processes subject to reengineering. *Source:* Accenture, by permission.

ries, and product grades per reactor, Chem Systems contacted several other polyethylene producers, most of which were willing to participate in this survey. In return for providing information on their business (which remained confidential), they would at the end of the study receive information on where they stood in the range of results representing all the producers. A fictitious example of such a benchmark exercise is shown in Figure 5.

Product line management involved identification of customers, the grades they ordered, and the profitability of specific pieces of business. Using elements of activity-based costing, customers could be separated into three categories, with key customers identified. The results of this phase of the work would allow Quantum to consider dropping some grades and customers, while providing better service to top customers.

Another phase of the program analyzed the so-called reactor wheels, where operators would move from making one grade to another, with off-specification material produced in between. It was found that reactors could be operated much more efficiently, making fewer grades with fewer changeovers and some products made only in one plant. It was also found that Quantum's

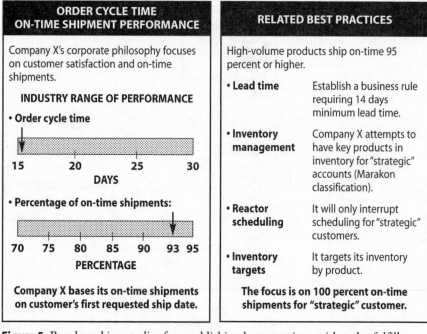

Figure 5. Benchmarking studies for establishing best practices, with order fulfillment used as the example. *Source:* Nexant, Inc./Chem Systems, by permission.

salespeople were too much in control, mandating large inventories of many grades simply because some customers might order them. In the future fewer grades would be made, and some would be made to order. As a result inventories and working capital savings could be achieved.

Reengineering at Quantum proceeded over the next several months. Estimates were developed for one-time and continuing savings that would result when different parts of the polyethylene business were reengineered. An important part of the program was the installation by Accenture of new software packages that would help in planning and scheduling. Chem Systems did not participate in this part of the work but remained close to Quantum, which was apparently very satisfied with the first phase of the project. However, much more work remained to be done. In the meantime the heavy debt load incurred by Quantum in connection with the 1989 recapitalization, along with poor business conditions, caused the firm to be acquired by Hanson Trust, where reengineering would eventually continue. Today these businesses form part of Equistar Chemical, owned by Lyondell and Millennium Chemicals.

Union Carbide: Reengineering the Entire Firm

Union Carbide, which had survived the tragedy at Bhopal (see Chapter 7) as well as an unsuccessful raid by GAF, had started a companywide reengineering program several years earlier, in 1989, under the leadership of CEO Robert Kennedy and later President William Joyce.[40] Early on, the firm had engaged the United Research Company and the Mac Group (later joined to become Gemini Consultants) as its reengineering consultants to work with the Union Carbide teams. The first comprehensive reengineering initiative was in the Specialty Chemicals Division (SCD), headed by Joseph Soviero. The pilot project would focus on sales effectiveness, managing the business portfolio at the divisional level, and work simplification, involving primarily the Kanawha Valley plants in West Virginia. The program was considered a major step beyond the existing corporatewide quality program known as EQ (Excellence through Quality). The project turned out to be hugely successful, reducing workforce by 13 percent, cutting manufacturing costs by 20 percent to 25 percent, and achieving savings of $20 million.[41] Owing to the success of the first project, Soviero recommended extending it to the entire division, which was one of five divisions in the company, headed at that time by William Lichtenberger, the president of Union Carbide. Noting the success of reengineering experienced in the SCD, Lichtenberger and Joyce decided to go ahead with reengineering on a companywide basis. Each division was to take one key process through reengineering, and if this was successful, then that division would work with the other divisions on the same process. SCD took on the development of a new production management process and a worldwide business management process. The Industrial Chemical Division would pioneer the development of a new maintenance process. Polyolefins took on logistics, while the Solvents and Coatings Division took outbound logistics.

Management initially set a goal of $200 million in savings. With a strong push from the top the reengineering effort proceeded across the entire corporation, including virtually every part of the firm's value chains. Division heads were asked to sign on to their share of the expected savings without knowing whether they could actually be achieved.

Early in the program the divisions were quite skeptical about whether the ambitious cost-cutting goals could be reached. Joyce instilled an internal competitive spirit by challenging the divisions with the results in specific areas achieved by other divisions. "If Specialty Chemicals can reduce its manufacturing costs by $10 million, why can't you?" But at the same time he strongly encouraged, or forced, divisions to share knowledge. Plant operation teams were composed of both operators and mechanics. Crossfunctional reengineering teams were composed of members of several divisions to encourage cooperation and knowledge sharing. These teams developed a series of business cases with specific cost-cutting targets that had to be approved for funding by division heads who needed to trust the process without being directly involved in it.

When the goal of $200 million in savings was in sight, Joyce raised the figure to $400 million and later to $575 million. This number was reached in

1994. According to Duane Dickson, only 30 percent to 40 percent of the total savings was a direct result of head count reduction. The balance was a result of simplification of work processes and savings *associated* with head count reductions. Taking out layers of management was a big factor. For example, it was possible to take out one layer of supervision in Carbide's operating plants, empowering workers to make more decisions.[42] Carbide also streamlined its management to cut layers, replacing its four-division reporting structure with nine strategic business units reporting to Joyce, which eliminated the divisional vice presidents entirely.[43]

As a result of earlier decisions and programs, including considerable portfolio trimming, a new Union Carbide emerged, now a highly focused company. It was able to concentrate on making sure that its commodity businesses (olefins, glycol, and polyethylene) would not lose money at the bottom of future petrochemical cycles, while much of the rest of the smaller firm, such as performance chemicals and other specialties, including Unipol licensing, was a big contributor to earnings. As the 1990s progressed, with the usual ups and downs for the cyclical petrochemical industry, Carbide was able to show that much of its business would have relatively stable margins over the cycle, thus attenuating the swings in its commodity businesses. Reengineering had succeeded in reducing its average fixed cost per pound from 15 cents to 10 cents, and many administrative functions had been dramatically improved. The company's record on worker safety and environmental compliance was substantially better. Further, from an investor standpoint, Carbide's return on equity was now at the top of the industry, when it had been near the bottom years earlier.[44]

A significant consequence of the program was the 1993 spinoff to the public of the company's Linde Industrial Gas Division, renamed Praxair, with William Lichtenberger as CEO. The stock market validated this move when the combined market value of the two stocks soon reached a figure of about double that of the original Carbide shares.

While Union Carbide's reengineering effort was "crisis driven," most other companies carried out such efforts for many of the general reasons discussed earlier in this chapter: to become more focused and more efficient competitors. Dow Chemical undertook a comprehensive reengineering program in 1993. Using 1992 as a base year, the company set as its goal to cut in half its unit conversion costs by 2000 for all its businesses on a companywide (global) basis. Unit conversion costs include maintenance, labor and salaries, benefits, plant overheads, and other such fixed, or period, costs. Dow had insufficient information on competitors to determine "best practices" in a number of areas, and so it relied on such outside sources as the Solomon report for ethylene plants and Phillips Townsend Associates for some other benchmarks. By 2000 the company reported that it had slashed conversion costs by 48 percent.

Specifically, maintenance costs, which according to Dow, usually run about a third of total conversion costs, were reduced from $1 billion to $650 million.[45]

In 1997 William Stavropoulos, then CEO of Dow, presented to Security Analysts the results of Dow's 1992–96 reengineering effort over a period when volume growth increased by 27 percent: sales per employee were $310,000 in 1992 and $539,000 in 1996; cost reduction from 1992 to 1996 was $1.7 billion; and fixed cost productivity improvement from 1992 to 1997 was 46 percent.

Engelhard: Cut Costs and Increase Sales

Even though Engelhard had restructured its operations in 1990 and reduced costs via head count reductions, Donald Torre, its chief operating officer, recognized that the firm had not fundamentally changed and that it now needed to change its business processes. This realization led to a major reengineering effort. Torre's goal was to reduce manufacturing costs by 20 to 33 percent or 10 percent below best in class, whichever was better. Moreover, he mandated that the effort be self-funded.[46]

Two high-level teams were established: a sponsor team (ST) consisting of Torre and three other high-level executives and a reengineering task force (RTF) headed by Stephen Pook, who was director of information systems. As a first step six other teams were created, each to focus on one manufacturing site in a different business unit. In workshops held with the ST and RTF, many participants were shocked to discover that they did not have a really good understanding of how the business worked as a whole. Beginning from this point, the six reengineering teams spent the next several months creating process models of the six units, each consisting of six to twelve processes. They identified "leverage points" where small improvements would have a major impact on overall performance. One example was production scheduling, since small improvements in accuracy can have a major impact on inventory levels. After deciding on action items, a number of changes were implemented. By the end of the year all six teams had achieved at least 20 percent lower costs, with some reaching 33 percent. As an important part of the new operation the worker-to-foreman ratio increased from eight-to-one to fifteen-to-one or even eighteen-to-one.

The next development was poignant. Reengineering of the petroleum catalyst business had increased its potential capacity by 45 percent, and the team wanted to receive credit for this achievement since the company could now cancel a planned multimillion-dollar expansion of capacity. However, the ST told the RTF that additional capacity was not a bankable benefit: only the additional output, when sold, could be counted. This led to the next step, namely, reengineering the customer acquisition process. Engelhard was much more successful in selling to small, independent refiners than to the major refiners, which required more relationship building, team selling, and the involvement of many departments, from purchasing to R&D. The reengineering team accordingly created a new selling process in which crossfunctional teams were paired with new, large target accounts. These potential new customers proved more demanding: they wanted to receive catalyst in specific ways, including packaging delivery schedules.

Within six months the new capacity had been sold out, and the petroleum catalyst unit had increased its market share by about one-third while achieving a major improvement in profits. After this success the RTF turned its attention to foreign operations, as well as to enabling processes, including information systems and finance.

The results at Engelhard validated the concept that effective reengineering involves a brand-new look at existing processes and operations. By gaining a true understanding of the entire sales, manufacturing, and order-fulfillment process, the teams were able to achieve major benefits that would for the most part not have been identified by the incremental TQM approach.

The Geon Company

The Geon Company, previously the Geon Vinyls Division of BFGoodrich, had been spun off from its parent in April 1993 when the managements of BFGoodrich and the subsidiary agreed that the needs of both entities would best be served if they were separate public entities. Chem Systems had been working with management at Geon—John Lauer, William Patient, and Thomas Waltermire—helping the division reach this conclusion. The main reasons for the split related to the cyclical performance of the PVC business and the division's need for back-integration, presaging a capital investment program for Geon that Goodrich's management was unwilling to undertake, given its ambitious plans for expanding its aerospace business.

A top priority for the new firm was to reduce the cost of PVC manufacture and to reengineer its PVC compound operation, which had too many products and high operating costs. Chem Systems benchmarked Geon against competitors to develop a basis for action. One of the first steps was to identify Geon's best PVC plants as a pilot, work on removing bottlenecks to increase capacity, and then assess all the other plants against that plant. As a result three regional PVC plants were shut down, while the remaining plants were soon able to produce more than the original total PVC capacity. In fact, while Geon employment decreased by 35 percent, Geon increased its total PVC output by 5 percent (110 million to 125 million pounds) with three fewer plants at a 34 percent increase in productivity.[47] Geon also pruned a number of low-volume specialty resins that had been added to the product line as a result of customer requests. Eventually seventy of the hundred or so resins were eliminated without losing market share. Accenture assisted Geon in this effort and also installed software used to provide information on manufacturing formulas for all grades and reactors. Reengineering was later extended to order entry, delivery, and payment as well as to the firm's R&D activity.

To address the problems in compounds, Chem Systems carried out cost analyses on a number of Geon's PVC compounds. It found that in a number of grades, particularly in extruded compounds, the firm made no profit and in many cases incurred a loss: the company would have made more money just selling the resin used to make the compound. Pruning the compounds product line substantially simplified the manufacturing cycles, eliminated some raw materials, and increased productivity.

Geon later decided to install SAP R/3—a later version of the company's software—as its transaction system and became one of the relatively few chemical firms to achieve rapid success with it. William Patient, the CEO of the new firm, later said he attributed the success of Geon's adaptation of SAP to his issuance of a mandate that everybody had to use the new software beginning on day one, regardless of possible early glitches. This worked well, even though it was traumatic for some of the staff. Patient related that one of the people taking orders for customers called him and screamed that the SAP system was not working, that it would not let her take an order from a good customer. He decided to check into this and found that there was insufficient inventory for the grade requested and that the customer would have been very unhappy if the order and delivery date had been accepted. Another early finding concerned requests from Geon's compound plants when they ran out of specific color pigments. Before installing SAP, the central warehouse shipped out a number of drums of the pigment whenever the plant signaled that it was running low. After installing SAP, the warehouse shipped out one drum or a few pounds, as specified by the formula and based on information in the system on total back orders.[48]

Under BFGoodrich the Geon Vinyls Division had conducted business as usual, since it had no particular incentive to become the most efficient producer. As a new public entity, with initial poor profitability and with financial analysts breathing down its neck, Geon reengineered its entire operation to become the most efficient domestic PVC and compounds producer. Later it merged with MA Hanna to become, as PolyOne, the largest domestic producer of PVC compounds and an important supplier of other compounds and resins.

Role of Management Consultants in Reengineering

The establishment of reengineering practices became a major driver of growth for such firms as McKinsey, Accenture, PriceWaterhouse, and KPMG Peat Marwick. Between 1980 and 1994, estimated North American consulting revenues shot up from around $2 billion to $15 billion per year.[49] Consulting became the career of choice for top MBAs from Harvard and other leading business schools, with 32 percent of Harvard Business School graduates in 1994 joining consulting firms. Such jobs attracted the highest starting salaries, which had jumped from $30,000 to $35,000 to $80,000 or more, plus a substantial signing bonus. Many of these graduates stayed with consulting firms for perhaps five years before assuming key management roles in client organizations they had worked for in their consulting capacities.

The use of consultants to aid in or take charge of reengineering activities in major firms seemed fairly logical: the reengineering "vogue" had created an excellent market for selling highly specialized skills; given the fact that reengineering involves substantial change, operating companies found it easier to bring in consultants to lead this effort and to absorb some of the inevitable negative fallout resulting from head count reductions; and the intellectual resources gained by consulting firms from various reengineering assignments provided a valuable source of ideas for other clients.

Many of the top MBAs hired by consulting firms, who had turned down lower-priced job offers from operating firms, had now gained experience and were working as management consultants for firms that had wanted to hire them in the first place. This worked well for both sides. The company and the consultant grew to know each other well before entering into a permanent working arrangement. Moreover, the consultant, having gained a great deal of experience working with a number of other firms, could come into a higher-level job than he or she could have aspired to after a few years of entering the same firm's employment with an MBA.[50]

The role of consultants in reengineering projects varied considerably. Consultants were primarily used to advise and provide technical expertise. Sometimes they were involved in the design but usually not in the management of the project. They were used to benchmark project progress and often played the role of scorekeeper, establishing a baseline before the project and monitoring progress. As an observer, the consultant could be objective about progress. A key part of the consultant's contribution was training the staff in reengineering methodologies. Previous TQM training was considered helpful in undertaking reengineering projects, even though the focus tended to be quite different.

Companies hiring reengineering consultants would first decide whether to use a boutique firm (for example, Rummler-Brache) primarily for education and training of staff and to deliver experience along the way or to hire a large management consulting firm with a reengineering practice (for example, Accenture or Gemini). Such a firm would provide a partner to lead the effort, joined by a large number of consultants with varying skills and, later, another group of consultants to select and help install the IT software inevitably required for implementation. Each of these consultant companies had developed its own approach to reengineering, differing in some details from each other but with a common goal. Accenture defined business reengineering as the process of aligning people, processes, and technologies with strategies to achieve step changes in business performance.

Accenture's methodology was called "value-driven reengineering." It consisted of five phases, covering the following steps abbreviated below:

- *Develop shared vision:* Understand the values of stakeholders in relation to the enterprise's competencies and capabilities. Deliverables include an operational vision and a customer value analysis.
- *Assess/align:* Identify the gap between the operational vision and the current processes ("as is") and identify initiatives that will bridge the gap.
- *Create master plan:* Develop a view of the "to be" and an implementation road map, along with performance measures and cost-benefit projections.
- *Design, pilot, implement:* Design the business reengineering approach, put it into practice, and prepare to operate.
- *Operate:* Deploy the business reengineering approach and set the stage for continuous improvement.

As part of this approach Accenture would assess the client's current IT systems and look for ways to improve them or use its systems development methodology to assist with the design and development of a new or modified IT system.

At the start of the reengineering vogue in the chemical process industries, there was a considerable amount of resistance to a full-blown commitment to IT as an enabler for business productivity. Senior management was reluctant to embrace IT because of unfulfilled promises made by their information systems staff several years before, when large capital outlays did not deliver large savings. However, firms like Dow and DuPont, as well as the large German companies, ultimately put IT at the top of the agenda for such requirements as globalizing their business structure and integrating their various processes. Dow began using SAP R/2—a somewhat earlier version of the company's software—at an early stage for customer services and financial accounting on a global basis. IT was soon broadly accepted as an enabler for reengineering companies' business processes, such as customer service, manufacturing and operations, order processing, delivery and logistics, and production planning and scheduling. In fact, chemical companies played a major role in the development of application software that could back up comprehensive reengineering activities.

Reengineering (as a term or discipline) lost favor or was deemphasized in the late 1990s, as firms became more preoccupied with ERP system installations and, soon thereafter, e-commerce strategies and implementation projects. Nevertheless, the concepts embodied in reengineering have continued, although in most cases under different terms and methodologies. As mentioned earlier, General Electric, Dow Chemical, and others use the Six Sigma technique not only for continuous improvement but also for process redesign. According to Arnold Allemang, executive vice president of operations for Dow, the very successful Six Sigma program at Dow consists of two initiatives. The first (measure, analyze, improve, control, or MAIC) is an extension of Dow's quality program but is now supervised by "black belts." A main objective of this program is to work on facilitating and improving relations with customers so that they want to do more and more business with Dow. To accomplish this, small and medium-sized projects are launched that aim to bring the operations toward Six Sigma in terms of "defects," perhaps starting from two or four sigma. The second initiative (design for Six Sigma, or DFSS) involves looking at processes "from scratch" and changing these dramatically to achieve Six Sigma at the time the new process is implemented. Dow's "perfect order" process, which completely reengineered Dow's order-taking system was developed under the DFSS regime.[51]

Still, at this writing, it is fair to say that reengineering as a discipline probably failed to live up to its promise and in some chemical and other companies it eventually became a bad word.

While in most firms dramatic head count reductions lowered fixed costs, the firms incurred correspondingly high costs to install expensive ERP systems. These systems took a long time to work as effectively as originally promised and mainly improved internal processes rather than allowing the companies to work more closely with suppliers and customers. John Aalbregtse of Accenture looked at it this way:

> There was a lot of value generated in the industry from reengineering initiatives. However, not all companies were successful, just as not all companies will be successful with Six Sigma. It takes a strong commitment from the top. It takes rigorous monitoring and tracking of the savings. It takes changes in IT and organization structure to support the new processes. And it takes ongoing continuous improvement of the processes to keep them relevant. . . . I believe we are on the cusp of the next reengineering step change: e-commerce. E-enablement will fundamentally change many of the processes that were reengineered in the early 1990s. And capability hubs serving multiple companies will fundamentally change the industry cost structure.[52]

Aalbregtse is one of the best consultants around, and he strongly believes in what he says. But consultants are prone to believe that there is always a new "breakthrough" methodology that can fix the problems of older approaches while creating new opportunities for the consulting profession. Although each new concept adds considerable value and e-enablement may be among the most important new initiatives, "silver bullets" are hard to find, and new concepts may not deliver their original promise.

Endnotes

1. *U.S. Chemical Industry Statistical Handbook* (Arlington, Va.: Chemical Manufacturers Association, 1995), 3.
2. Emily Plishner, "Third Quarter 1991: Exports Buoy Volume but Margins Shrink On," *Chemical Week* 149 (4 Dec. 1991), 32–38.
3. Clyde V. Prestowitz, Jr., *Trading Places: How We Allowed Japan to Take the Lead* (New York: Basic Books, 1988), 188.

4. Bill Creech, *The Five Pillars of TQM: How to Make Total Quality Management Work for You* (New York: Truman Talley Books, 1994), 213.

5. Michael Hammer and James A. Champy, *Reengineering the Corporation* (New York: HarperBusiness, 1993), 16–17.

6. Peter H. Spitz, *Petrochemicals: The Rise of an Industry* (New York: John Wiley & Sons, 1988), 462–505.

7. Thomas Davenport, "Business Process Engineering: Its Past, Present and Possible Future," Harvard Business School case no. 9-196-082, Nov. 1995.

8. Hammer and Champy, *Reengineering the Corporation* (cit. note 5).

9. Ernest H. Drew, "Managing Assets in a Changing Environment," Chemical Marketing Research Association conference, New York, 5–7 May 1986.

10. Rick Mullin, "Quality: Evolving through the Backlash," *Chemical Week* 151 (30 Sept. 1992), 41.

11. Rick Mullin, "CEOs on the Long March to Quality: Extracting Values along the Way," *Chemical Week* 151 (30 Sept. 1992), 58, 60.

12. Bruce Brocka and M. Suzanne Brocka, *Quality Management: Implementing the Best Ideas of the Masters* (Homewood, Ill.: Business One Irwin, 1992), 55–97.

13. W. Edwards Deming, *Out of the Crisis* (Cambridge, Mass.: MIT Press, 2000).

14. Brocka and Brocka, *Quality Management* (cit. note 12), 22.

15. Ibid., 24.

16. Rick Mullin, with Michael Roberts, "A Focus on Customer Satisfaction," *Chemical Week* 151 (30 Sept. 1992), 41, 44.

17. Ibid., 49.

18. Karen Heller, with Rick Mullin, "ISO 9000: A Framework for Continuous Improvement," *Chemical Week* 153 (22 Sept. 1993), 30–31.

19. Michael Hammer and Steven A. Stanton, *The Reengineering Revolution* (New York: HarperBusiness, 1995), 97.

20. Bill Creech, *Five Pillars of TQM* (cit. note 4), 250.

21. Lee Clifford, "Why You Can Safely Ignore Six Sigma: The Management Fad Gets Raves from Jack Welch, but It Hasn't Boosted the Stocks of Other Devotees," *Fortune* 143 (22 Jan. 2001), 140.

22. Ibid.

23. Bill Schmitt, "Moving Ahead with Six Sigma," *Chemical Week* 162 (26 April 2000), 64.

24. "The Dow Chemical Company Receives CIO-100 Award for Excellence in Customer Service," *Business Wire* (15 Aug. 2000), 2192.

25. J. Robb Dixon et al., "Business Process Reengineering: Improving in New Strategic Directions," *California Management Review* 36:4 (Summer 1994), 94–95.

26. Davenport, "Business Process Reengineering" (cit. note 7).

27. Hammer and Champy, *Reengineering the Corporation* (cit. note 5), 32.

28. Duane Dickson, personal communication.

29. Accenture, personal communication.

30. Dixon et al., "Improving in New Strategic Directions" (cit. note 25), 97.

31. John Aalbregtse, personal communication.

32. Davenport, "Business Process Reengineering" (cit. note 7), 3.

33. Hammer and Champy, *Reengineering the Corporation* (cit. note 5), 91.

34. Francis J. Gouillart and James N. Kelly, *Transforming the Organization* (New York: McGraw-Hill, 1995), 100.

35. Hammer and Champy, *Reengineering the Corporation* (cit. note 5), 50–68.

36. Duane Dickson, personal communication.
37. Dixon et al., "Improving in New Strategic Directions" (cit. note 25), 98.
38. Rick Mullin, "Manufacturers Determine the Scope of Change," *Chemical Week* 153 (8 June 1994), 25.
39. Davenport, "Business Process Reengineering" (cit. note 7), 4.
40. Gouillart and Kelly, *Transforming the Organization* (cit. note 34), 155–161.
41. Allison Lucas, "Union Carbide Changes Its Mindset," *Chemical Week* 153 (24 Nov. 1993), 39.
42. Duane Dickson, personal communication.
43. Lucas, "Union Carbide" (cit. note 41), 39.
44. William Joyce, personal communication.
45. Arnold A. Allemang, Dow Chemical Company, personal communication.
46. Hammer and Stanton, *Reengineering Revolution* (cit. note 19), 204–210.
47. Ira Bleskin, "Reengineering Critical to Turnaround at Geon," *Chemical Week* 153 (8 June 1994), 48.
48. William Patient, personal communication.
49. Richard L. Nolan, "Role of Management Consulting in Reengineering," Harvard Business School case no. 9-195-200, 8 Feb. 1995, p. 1.
50. Ibid., 6.
51. Arnold A. Allemang, personal communication.
52. John Aalbregtse, personal communication.

Chapter 6

Riding the Waves

The Changing Role of Information Technology
in the Chemical Industry

David A. Crow

How will the chemical industry use information technology (IT) in the twenty-first century? What lies ahead, and how is the industry to prepare for it? If the past is a reliable guide to the future, the story of IT in the chemical industry can best be portrayed as waves of change working through the industry, carrying companies forward to new levels of efficiency and profitability.

This process has been and remains inherently messy; one wave does not run its course to be followed neatly by the next. Waves overlap one another, at times magnifying the power of previous waves and at other times negating their force. To a swimmer adrift in strong currents, surrounded by these tumbling, overlapping waves, the scene can seem daunting; no less disorienting is the experience of a corporation being tossed about by this trend or that and trying to understand what is about to hit next. Distance becomes the prerequisite for perspective.

The chemical industry has ridden at least six waves over the past three decades, with several new waves building on the horizon. From early process control leadership, through the advent of packaged software, to outsourced IT solutions, and more recently through the highly anticipated e-commerce wave, the chemical industry has weathered it all well. Some companies have sensed certain waves earlier than other companies and have capitalized on their foresight.

Analyzing IT as waves of change has the added benefit of casting in sharp relief the responses corporations make as executives confront successive waves of change. The technology may change and the complexity continues to increase, but the fundamental decision remains the same. Each company must decide how it is going to respond to the waves of change that press upon it. Each company must make its choice: ride the wave and perhaps run the risk of getting ahead of it, or follow the wave and risk being either left behind or outperformed by superior competition.

The First Wave: Process Control Leadership

The process industries have long seen themselves as leaders in the use of process control technologies (Figure 1). As early as 1948 there was speculation about running a computer-controlled plant. In November 1952 two scientists from Shell Development talked about computer applications to petroleum and chemical processes. According to Thomas M. Stout and Theodore J. Williams, the Shell scientists described "supervisory control," in which the computer changes the set points of analog controllers, as well as "direct digital control," where analog controllers are eliminated.[1] By 1956 measurements taken at a DuPont plant in Niagara Falls were being transmitted by telephone wire to Philadelphia for processing by a Burroughs digital computer. Yields and material balance data were sent back to Niagara Falls for use by the plant operators.

The first industrial control computer system, according to Stout and Williams, was that of the Texaco Company at its Port Arthur, Texas, refinery, using an RW-300 manufactured by the Ramo-Wooldridge Company, which led the way in the development of industrial computer control.[2] Closed-loop control was achieved at the Port Arthur facility on 15 March 1959. (This type of system senses its output and makes corrections to the process. The feedback to the control system regulates the process.) Early installations were of the supervisory control type, and in addition to Texaco, such companies as Monsanto, BFGoodrich, and Union Carbide Chemical Company boasted similar systems.

The first serious proposal for direct digital control of full-sized chemical plants—that is, control without using an intermediate analog, electronic, or pneumatic control system—was made by Imperial Chemical Industries (ICI) in the early 1960s. A system using the Ferranti Argus 200 computer was operating at ICI's soda ash plant in Lancashire, England, by November 1962. A so-called third generation of direct digital control computers, including the GE 4020, the CDC 1700, the IBM 1800, and the SDS Sigma 2, appeared in the

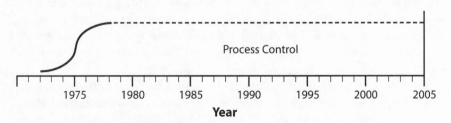

Figure 1. The first wave—process control.

mid-1960s. This early phase of technology culminated in the advent of Honeywell's TDC 2000, which was unveiled in 1975 and marked the close of a pioneering period for industrial computer control.

In the first wave of technological implementation the emphasis was squarely on process control, the focus was the plant, and the cost was typically measured in hundreds of thousands of dollars.

Although these process control technologies were leveraged by many chemical companies, with corresponding benefits at the plant level, the early "business systems" were still in their infancy. Most were focused on basic financial control and reporting processes and were rarely integrated with their process control counterparts. This dichotomy was also reinforced through separate departments within a company (that is, engineering and IT) with somewhat different objectives and technologies.

The Second Wave: Process Manufacturing Planning Tools

The rapid proliferation of computers throughout the world of business and commerce inaugurated a new wave of technological innovation around process manufacturing planning tools (Figure 2). These systems went well beyond control of production processes to the planning of the volume and variety of outputs from those processes. Computer manufacturers such as Digital Equipment Corporation (DEC) and IBM introduced integrated information and control systems architectures for manufacturing automation, addressing a wide range of process, batch, and discrete manufacturing processes. These early software solutions were designed largely with the "widget maker" in mind, and not surprisingly, manufacturers using discrete production processes were among the first to seize on the power of material requirements planning, and later the closed-loop manufacturing resource planning (MRP II) technologies. The MRP mantra heard from such gurus as Ollie Wight resonated widely, but such process industries as chemicals did not show much interest. The market drivers behind this activity were fairly clear: markets demanded increasing levels of specialization, and manufacturing plants were responding.

From the perspective of process industries batch processors, such as pharmaceutical giants and food and beverage companies, were among the first to pick up on the innovations of MRP. Software companies, which had focused solely on the discrete industries up to this point, soon recognized the unique needs of process industries, as well as the untapped market potential, and moved to modify their systems accordingly.

> MRP II for process industries . . . must take into account the special characteristics of both batch and continuous-flow. Rather than having its products assembled or formed, as in discrete manufacture, in the process industries products are blended, or substances may be broken down into constituent ingredients eventually used as products. Co-products and by-products come into play. Processed materials can be packaged in many different ways.[3]

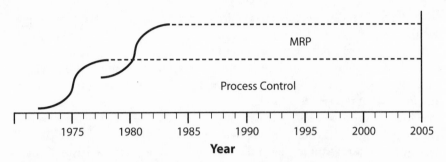

Figure 2. The second wave—process manufacturing tools.

Despite the fact that MRP software had deep penetration in discrete manufacturing, several MRP vendors, including Datalogix, Marcam, SSA, and ASK, targeted the process industries. Chemical firms, however, were comparatively slow to accept the commercial software programs; the chemical industry strongly believed that its needs were unique and, according to a *Chemical Week* article in 1991, was just as strongly biased against packages developed in-house. Gartner Group's Jon Borelli put North American MRP sales to chemical firms at only about $60 million and estimated that such commercial packages had achieved only a 15 percent penetration of the potential market.[4] Some executives viewed IT as a necessary evil, something you had to have, but never as an integral or strategic part of the business. Others, even proponents of IT, were far more inclined in the early years to develop their programming in-house. The largely unspoken belief that the unique operating requirements of the chemical industry necessitated the custom development of software had the unintended but very tangible consequence of making early software development programs costly to initiate and to maintain.

An order-entry system being developed at Rohm and Haas in the early 1990s reflected the hybrid nature of systems in use. Dave Stitely, chief information officer, described a 1993 initiative to upgrade order processing for a global client-server environment as a blend of rewritten in-house software, a planning system purchase from Datalogix, and a Marcam software module for MRP control at the manufacturing level.[5] At that time Rohm and Haas had order entry online across North America and in one-third of its European sites, with the Asia-Pacific area and Latin America to be included in 1994, at a total cost of $35 million to $40 million, including in-house development time.

Geon—Linking Process Control and Enterprise Applications

The first focus of process control may have been the in-plant processes, but that simple, pristine state of affairs did not last for long. An excellent example is the experience of the Geon Vinyl Division of BFGoodrich, now PolyOne. Geon has consistently chosen to ride the forward edge of the waves of change sweeping the industry.

In the mid-1980s the Geon Vinyl Division of BFGoodrich Company was running thirteen plants and earning $1 billion in revenues by pumping out billions of brightly colored polyvinyl chloride (PVC) plastic pellets. But the market had become flooded, leading Geon to switch strategies, abandon its heritage as a commodity producer, and become a custom manufacturer of specialty plastic compounds.

"[Geon] had made a fundamental decision to go from a commodity producer to a specialty manufacturer, and that kind of change requires a change in business procedures," said Accenture partner Karl Newkirk, who was working with the company at the time. "You're talking about changing all the processes in the company when you go from doing one massive order a week to 200 orders a day for different customers."[6] To accomplish this, Geon had to reengineer many of its processes, rebuild its software systems, and switch from traditional long production runs to a more complex batch-oriented schedule. John Menyes, vice president of information technology at Geon Vinyl Division, and Bruce Gordon, director of advanced systems operations and manager of the reengineering effort, scoured the market for a system that could handle the new complexity, linking control systems with their business counterparts.

"We had this notion that we could buy some magic software off the shelf, slam-dunk it into the computer, and it would be the answer to all our woes," said Gordon.[7] But of several systems considered, only one had functions geared to the industry. To make matters worse, the project team discovered that no software was available to integrate plant-floor computers with the planning systems used at headquarters. With no acceptable software in sight Geon commissioned the development of new software, dubbed Process/1, which had the useful ability to link a company's process control system and distributed plant operations with the enterprise system at headquarters.

Process/1 was an early example of a manufacturing execution system, which attempted to take integrated computing to the next level by implementing a comprehensive approach to enterprise integration. For all its success Geon's experience with Process/1 was also a good reflection of an industry that was still getting its IT feet wet, an industry in which many managers still were ambivalent about the role of IT.

Geon's experiment also pointed toward the next wave of technological change. Industry observers at the time noted that the chemical industry would eventually have to address the problem of integrating across business units and downward to process controls. And commercial software was certain to play a key role. Accenture's Karl Newkirk predicted as early as 1993 that most major chemical companies had essentially stopped writing their own software because of prohibitive development costs and maintenance and support costs.[8] "Our strategy is to purchase applications whenever we can, and develop applications when necessary," said Tom Carpenito, then global information utility manager at DuPont.[9] Industry standardized software was now a clear direction for the chemical industry, but few knew how important one small German software company would become.

The Third Wave: ERP and the Emergence of SAP

Despite the small, but growing, penetration of "process" MRP tools in the chemical industry, there was an increasing and often quite vocal dissatisfaction with the limitations of these early technologies. A January 1992 *Datamation* magazine survey of manufacturing software complained: "Global manufacturers are pushing into the 21st century with the latest manufacturing technologies—but are stuck using yesterday's manufacturing software. Some manufacturing resource-planning (MRP II) packages have changed little since their introduction in the mid-1960s."[10] Others were blunt in their criticism: "The traditional MRP system never worked with the chemical industry," said Advanced Manufacturing Research's Ted Rybeck.[11] Few MRP systems were designed to meet the needs of a global multinational company.

Around this time industry watchers began to speak of a new level of software emerging in the process manufacturing arena with the advent of global computing. They cited a break from MRP systems that evolved as a standard in the 1980s and a move toward architectures that united different departments of a manufacturer, its suppliers, and its customers. In November 1991 Datalogix, a vendor of software packages for such process manufacturers as chemical and pharmaceutical makers, launched its second-generation global enterprise manufacturing management system (GEMMS). About this time the term *enterprise resource planning* (ERP) began appearing with greater frequency in the industry trade journals and research reports (Figure 3). In truth, the distinction between MRP II and ERP was not always crystal clear. ERP was still concerned with manufacturing, and many of the modules are the same in both systems, but more emphasis was placed on integrating financial and distribution capabilities. What was really new and different in the ERP systems were such qualities as flexibility, capability, and scalability. To achieve these ends, the newer ERP systems used client-server platforms rather than mainframes, relationship database systems, the Unix operating system, and work-flow methods for modeling organizations that were less obsessed with rigid functional pigeonholes. As the name indicated, the software embraced more than just a single manufacturing plant; conceptually, it extended across the entire enterprise. And as more chemical executives began talking about ERP, more discovered an interesting but not-so-little German company known as SAP.

The Emergence of SAP and Its Impact on the Chemical Industry

In 1972 four former IBM engineers founded SAP GmbH to market an application software system, and during the first ten years of the company's existence it concentrated on selling mainframe software exclusively to the German-language market. By 1988 SAP, headquartered in Walldorf Baden in

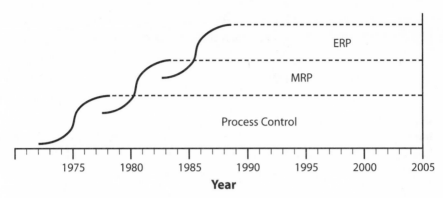

Figure 3. The third wave—enterprise resource planning.

what was then West Germany, realized the highest revenues within the German domestic software market.

SAP began its international expansion effort in earnest in the mid-1980s by establishing small subsidiaries around Europe. These sales offices were initially also research projects. By using various European locations as a set of multicultural test laboratories for its global efforts, SAP learned how to adapt to unfamiliar styles of business, customers with different priorities, and different language requirements. Profits from these early foreign subsidiaries were used to fuel further international efforts.

In 1988 SAP made the United States the centerpiece of its international strategy. That year the company also earmarked more than 10 percent of its total yearly sales and 22 percent of its staff for R&D, establishing overseas laboratories in France, India, and the United States that would concentrate on creating new products for local markets.

Given SAP's heritage, the chemical sector has been one of the company's main target markets. The chemical industry itself has strong roots in German industry; it is therefore not surprising that one of the most innovative IT solutions for the chemical industry should come out of Germany. In addition to its global capabilities many functions unique to the chemical industry were gradually supported by SAP software, including regulatory compliance, tolls and exchanges, and process environment modeling with powders, liquids, and gases. Surprisingly, by 1995 the chemical industry had the largest percentage of SAP installations and accounted for 40 percent to 50 percent of SAP's annual revenues, according to Advanced Manufacturing Research.[12]

For the chemical industry the era of ERP can easily be seen as the era of SAP, which opened the door to the use of IT on a broader scale. SAP allowed companies to analyze their customer and product profitability as well as to do activity-based costing. SAP implementations grew rapidly throughout Europe, with ICI being its model global chemical company. In fact, ICI became SAP's first North American customer in Canada, followed closely by manufacturers like Dow, DuPont, and Mobil, which dedicated time and money to provide a consistent global solution.

SAP Dominates the Chemical Industry

Soon *Chemical Week* was reporting, "when chemical companies start selecting ERP vendors, the 'how' question seemed to have one answer—SAP."[13] According to the Gartner Group, the worldwide ERP market, which was worth $3.5 billion in 1995, had by 1998 ballooned to exceed $10 billion. SAP held a 35 percent share of the market that year, followed by Oracle at 16.7 percent, J. D. Edwards at 8.7 percent, and Baan at 8 percent.[14] The *Chemical Market Reporter* called SAP "the ERP standard among large chemical companies, rewarding companies that adopt it with rich advantages over their less wired competitors."[15]

Rarely has a software solution taken an entire industry by storm the way SAP came to dominate the top tier of the chemical industry. In fact, comparisons were soon being made between SAP and a certain company based in Redmond, Washington. SAP's R/2 release for mainframes was supplemented in 1993 by R/3, which only accelerated the company's growth. Some 70 percent of the chemical companies recognized as global players were implementing R/3 by 1999, including fifteen of the seventeen largest chemical companies in the world.

The reasons for this dominance were not hard to find. Many companies were able to reduce inventory levels and system life-cycle costs after implementing SAP across their financial, materials management, controlling, sales, and distribution functions. More important, SAP's global capabilities and its highly integrated architecture were in stark contrast to its primarily U.S.-based competitors. And SAP also realized better than most software companies that an ERP sale had to be focused toward the senior executive level, reflecting the increasing sophistication and importance of IT. Stories of SAP's success still continue, even as they are overshadowed by the newer waves of IT.

Studies in SAP Implementation

One of SAP's largest clients in the chemical industry, Eastman Chemical, brought its first pilot customer service center online in 1993. Jack Spurgeon, chief information officer for Eastman, reported that year that the company purchased SAP's R/2 system "to integrate the entire company worldwide, so that it can act locally within a culture and a currency."[16] Eastman's SAP system managed order processing, financial management, inventory management, purchasing, manufacturing, and maintenance functions—tying in all the company's customer service centers. The company installed its R/2 system using in-place mainframe computers, which were kept as data warehouses when the installation was upgraded to R/3 software on client-server technology.

Dow Chemical was another big SAP user. "We run a single, large SAP complex," reported David Kepler, then director of information applications. By 1996 Dow was running key business applications on a mainframe, and engineering and manufacturing systems mainly on Digital VAX (virtual address extension) computers. Dow's long-range goal was coexistence of R/2 and R/3. "We believe we are going to continue to leverage our investment in the R/2 mainframe environment," Kepler said at the time. "What we are really trying to do is integrate that with client-server."[17]

Not without Weaknesses: SAP Competitors

It was perhaps inevitable that the SAP juggernaut would also garner its share of naysayers. Critics began to comment that SAP software was too complicated and too costly to implement, that for every dollar of software purchased, you had to budget ten dollars for implementation. "No one has ever [explored all the options] of SAP's decision-making tree and returned alive," said Kirtland C. Mean, president of the consulting firm CSC Index, in Cambridge, Massachusetts, in 1996. "SAP is like everything German. It's over-engineered, with more options and flexibility than ever seen before."[18]

Serious commentators acknowledged that the processes SAP sought to control were exceedingly complex from the start.

> Chemical production is a very complicated business. The complications of balancing preferred sequencing in production against the customer demands (which never matches that sequence!) against the need to keep this high fixed cost facility running to drive unit costs down is just one example of pressures that will be placed upon an ERP system. . . . The selection of the right tool is not easy. . . . The power of an ERP system is the easy access to, and interchange of, information through all parts of the organization. If a group has to play tricks on the system to get it to work, if they are making mistakes because they don't understand how to use the system, or—worst case—they develop their own "side systems" which work better and then half heartedly fill in the blanks for the accountants in the ERP system, your hard work and investment in your ERP system will be a wasted effort.[19]

Realists in the executive ranks were sympathetic. "If you think SAP is complicated, look at your legacy systems," said Robert E. Barrett, Monsanto's director of worldwide operations and financial programs, toward the end of the decade. "Now *that's* complicated. I can't tell you what I'm selling when I'm on 16 different computer systems."[20]

But competitors were more than happy to capitalize on the rising tide of criticism. "Although SAP AG has long dominated the burgeoning enterprise resource planning (ERP) software market, some analysts said the German company's phenomenal growth would be slowed by increasing competition from 'second generation' contenders such as PeopleSoft Inc., Oracle Corp., Baan Co., and J. D. Edwards," wrote one observer.[21] In addition to these companies there were also Marcam's new Protean software and Datalogix's new GEMMS product.

Although the SAP blitzkrieg was barely slowed by these competitors, they were not doomed to failure. Boston-based AMR Research in 1998 described the ERP market as pyramid shaped, with a few large users on top and an expanding number of mid- and low-range businesses below.[22] When some of the largest companies sought out an ERP application, only SAP and one or two others had the range and scope of applications to satisfy demands. But when medium and smaller users went shopping for an ERP, the arena opened to a larger number of competing bids.

If SAP did have a significant weakness, it was that the platform was weak in a number of areas, such as forecasting and network optimization—a void that drove the growth of the "bolt ons" from such third-party vendors as Manugistics and AspenTech. Manugistics, for example, focused on integrating its transportation, distribution, and manufacturing planning applications with MRP-ERP systems from SAP, Oracle, and Datalogix. MIMI, originally provided by Chesapeake and later acquired by AspenTech, was for many chemical companies the standard for planning and optimization.

Throughout the 1990s SAP had no serious competition in the tier-one core ERP space. PeopleSoft focused on the human resources segment, J. D. Edwards targeted the smaller chemical companies, and Baan soon abandoned direct competition with SAP before collapsing altogether in 2000. Even a competitor such as Oracle, with deep pockets and a mature solution like the Datalogix platform it acquired, found the going tough. "They [SAP] have a leg up on the competition," said Victor Muschiano of the ARC Advisory Group in Dedham, Massachusetts. "I don't see the Oracle initiative running at the same level of sophistication as the SAP model."[23]

The Continuing Evolution of ERP Solutions

In an era when corporate funding for capital investments was comparatively easy to come by, the 1990s were dominated by massive SAP implementation projects involving hundreds and even thousands of people. Once this high tide receded, the issue facing chief information officers everywhere was mining the value from SAP systems they had in place.

The benefits of R/3 were well established. Many companies reported returns on their implementation investment of at least 15 percent, and many saw much greater rewards, with the savings often coming from inventory reductions in the range of 20 percent to 30 percent.

With most major companies having already implemented ERP systems, ERP software vendors began to extend core ERP products by adding functionality in areas from customer relationship management to supply chain management.

According to AMR Research, traditional ERP software generated 91 percent ($16.8 billion) of ERP vendors' revenues, while 9 percent ($1.6 billion) came from the sale and installation of strategic extensions, such as supply chain management (SCM), e-commerce, and e-business relationship management

(ERM).[24] Those strategic extensions grew by 92 percent in 1999 and contributed significantly to the overall ERP market's 11 percent growth that year. This trend is expected to continue; platform extensions should generate five-year compound annual growth rates of 36 percent for ERM, 40 percent for SCM, and 56 percent for e-commerce. By 2004 strategic extensions are expected to account for 43 percent of traditional ERP vendors' revenues.

The Fourth Wave: Outsourcing Changes the Competitive Landscape

The wave of change that moved industry leaders from custom-developed systems to industry-accepted ERP software was so large and so powerful that it in turn helped trigger still another wave: the widespread use of outsourcing as a technique for cost-effectively implementing and maintaining major IT investments in the chemicals industry (Figure 4).

As more companies turned to commercially available packages, outsourcing of IT tasks won favor as companies recognized that the implementation of large systems such as SAP's was beyond the scope of their internal IT resources. Moreover, as newer software was being implemented, the costs to support legacy systems continued to grow. In those early days outsourcing arrangements were effectively cost-reduction plays, many times involving guarantees of cost savings along with up-front asset transfers.

With the complexity of IT infrastructures mounting and the number of software applications used by any given company climbing from the tens into the hundreds, a new, more expansive, and more permanent brand of outsourcing appeared on the scene in the mid-1990s. Challenged by an intensifying shortage of IT professionals, internal IT functions were finding it more difficult to support applications, man the help desks, and still respond in a timely fashion to management requirements for reports and changes to existing systems.

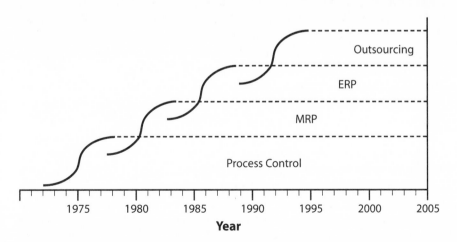

Figure 4. The fourth wave—outsourcing.

Senior executives were becoming concerned that the IT infrastructure on which their company's operation depended seemed to require ever-greater investments of people and assets. Their question was a fairly simple one: when is this going to end?

For many the solution was to ask the external programmers to hang around, which is precisely what many corporations effectively did. By outsourcing applications maintenance and other technology tasks to external specialists, large corporations accomplished several objectives at once. They solved pressing manpower issues. They were able to manage the expense of their IT infrastructure by contractually locking in attractive long-term rates with clear incentives to reduce costs. And by actually transferring internal IT professionals to the external service provider, these companies enabled large numbers of their IT employees to be better managed and motivated. Further, working with the external service providers gave companies access to state-of-the-art technology and innovative solutions.

Outsourcing enabled companies to focus on maximizing their competitive advantages in the marketplace and improving the way business is managed. With more focus on business challenges and less on managing technology's "plumbing," companies found that they could be more competitive. The experiences of two global giants reflect the different approaches to outsourcing taken in the chemical industry.

Dow Chemical

Since 1995 Dow Chemical has outsourced IT resources to such firms as Accenture for large implementation and support activities. Outsourcing has given Dow's IT unit flexibility in managing its resources and superior performance, including improved time to market, increased productivity, and more efficient delivery of IT solutions.

However, Dow's approach to outsourcing is anything but standard. Rather than simply hand over IT and e-business development and management to an external services provider, Dow has placed 450 IT employees under an organization called the "Alliance," managed jointly by it and Accenture. Under this arrangement hundreds of Accenture employees are dedicated full time to Dow. Based on workloads and the cyclical nature of the chemicals industry, that contingent can vary by up to several hundred people.

Dow staff members are responsible mainly for managing projects and selecting architectures; Accenture staff charts the methodologies and operating disciplines and brings new technologies and concepts to the table. Accenture is compensated on productivity and time-to-market improvements as well as on output. According to corporate vice president and CIO David Kepler, Dow spends more than $400 million annually on IT resources, and Accenture has helped Dow reduce those costs by $70 million a year, while improving time to market by 10 percent.[25]

During the Alliance's existence, Dow has completed more than eight hundred projects, including a Web procurement system, a global management reporting system that ties Dow's SAP systems to its Oracle data warehouse, and its MyAccount@Dow customer extranet. In 1999 the Alliance was focused largely on year 2000 remediation. A key project in 2000, in addition to Dow's e-commerce activities, was IT integration of acquired companies such as Union Carbide.

DuPont

In 1997 DuPont took a different path when it entered into a leading-edge IT outsourcing arrangement with the transfer of its data centers and IT personnel to Accenture and Computer Sciences Corporation (CSC). With IT staff inherited from DuPont in the outsourcing deal, both companies established new chemical industry centers in Delaware. Accenture reestablished a dedicated IT outsourcing center in Wilmington to design, build, and run IT solutions for DuPont. This center was later expanded to include similar outsourcing arrangements with other chemical industry clients. Thirty minutes away, in Newark, Delaware, the CSC chemicals center has pursued a similar business as well as consulting work, based on the DuPont contract.

The new chemical centers illustrate how consultants are pooling skills from multiple clients to develop products and services that will be sold to other chemical companies. The leading-edge structure in the DuPont arrangement, where two competing service firms form a relationship, was a fairly new phenomenon at its inception in 1997, but it has since been replicated elsewhere. Intellectual capital is pooled, but carefully, with complex nondisclosure agreements and proprietary know-how still held close to the chest. A great deal of systems experience that was once considered proprietary, however, is now treated as commodity know-how. Some industry watchers looking at the DuPont arrangement speculate that such outsourcing companies as Accenture and CSC may be the corporate "utilities" in the twenty-first century, providing a kind of power that is different from electricity but just as essential for effective operations.

Management and governance issues are important to address when two competitors are working side by side for the same company. G. Frank Conway, former director of global alliance management at DuPont, reported he visited the Accenture and CSC facilities "every few weeks to go over what's going right and what's going wrong." He said: "The power of ERP is integration. It requires cooperation from people with parochial points of view. Once the tools are in, we have to get people to use them. Our challenge is to educate users as to what the system can and can't do. We're looking to Accenture and CSC to see how we can get the most out of IT."[26]

Y2K: The Fifth Wave or a Major Discontinuity?

For a few years running up to 31 December 1999, the chemical industry and indeed the entire corporate and institutional world worked through what will long be remembered as the "Y2K" era, when everyone worried about the millennium bug and struggled through billions of lines of programming code to find and fix sources of potential problems. Many companies were still operating legacy systems with custom programming, and these systems potentially posed vast risks in terms of Y2K exposure. Concern for business interruption grew at the board level as the new millennium appeared on the distant horizon, and aggressive plans were set for Y2K remediation programs, particularly among the global enterprises, which had the most to lose from Y2K-related disruptions (Figure 5).

Of course, the bedeviling thing about the Y2K bug was that no one could ever be really sure how bad the disruptions would be, if there were disruptions at all. This uncertainty forced many companies to take a conservative, no-tolerance approach. Even so, the hysteria rose as the calendar advanced. In October 1999 the U.S. Senate's Y2K committee cited a survey showing an alarmingly high number of small chemical firms that had yet to complete their Y2K repair work. The survey found that 86 percent of the small and medium-sized chemical handlers and manufacturers were not prepared for Y2K and had no coordinated contingency plans with local emergency officials.

Midnight of 31 December 1999 came and went, as we all know, and virtually nothing went wrong. Either the business world vastly overrated the danger, or companies collectively did their work exceedingly well: the number of actual problems caused by the arrival of a new year, century, and millennium turned out to be embarrassingly small. "The global chemical industry has crossed over into 2000 unfazed by the so-called millennium bug, or Y2K problem," according to the *Chemical Market Reporter* on 17 January 2000.[27] The hundred-member Texas Chemical Council heard of no problems. The Environmental Protection Agency reported no disruptions in chemical manufacturing. None of the three hundred members of the Synthetic Organic Chemical Manufacturers Association reported any problems. The American Petroleum Institute reported that essentially all the world's refinery capacity and oil and natural gas production made the transition successfully. After all the hype it seemed almost inconceivable not to have some mighty miscue, and yet that is precisely what happened.

Inevitably, the question arose as to whether all the work was worth the money. Beyond this, the potential threat of Y2K disruptions drove many enterprises to shelve major IT initiatives until the new millennium had arrived; the last thing IT executives wanted was to deal with Y2K issues in the midst of major new implementation problems. Deferral of major projects combined with diversion of IT budgets to Y2K remediation had a major effect on most corporate IT budgets in 1998–99. According to the Giga Information Group, Y2K budgets as a percentage of overall IT budgets hovered in the 25 percent to 30 percent range across industries for about two years running up to the new millennium. Financial and insurance sectors attacked the problem early, and often incurred higher costs owing to the lack of remediation tools in 1996–98 and the heavy

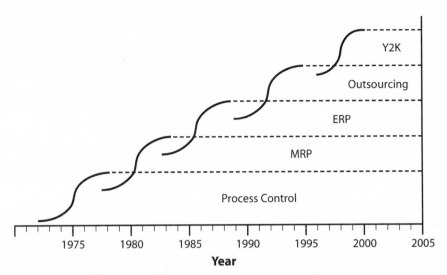

Figure 5. The fifth wave—Y2K.

use of in-house staff. Giga Information estimated that given an industry statistic of 2.8 percent for IT spending as a percentage of revenues, the average Y2K project consumed between 0.70 percent and 0.84 percent of gross revenues. These industrywide estimates were confirmed by a review of U.S. Securities and Exchange Commission filings of public companies concerning Y2K spending. A survey of 125 firms found that the average Y2K cost was $31 million, with the median at $8.5 million. The average Y2K cost as a percentage of total revenues was 0.55 percent, and the median was 0.69 percent. Projecting these averages to the gross national product suggests that the U.S. economy spent $51 billion on Y2K remediation. A Department of Commerce report put overall Y2K spending at $100 billion. The Gartner Group at one time estimated worldwide IT costs of $300 billion to $600 billion.[28]

So the Y2K craze was indeed a major wave, but it was in many ways a rogue wave, aberrant and moving at cross-purposes to the normal course of business. Y2K had two interesting effects on IT in the chemicals industry. It brought major implementation efforts to a virtual halt, as scarce programming resources were diverted to deal with the perceived Y2K threat, and it to some degree served to obscure another, even larger and more significant wave that had been building for some time—e-commerce.

The Sixth Wave: e-Commerce Holds Great Potential

Whether owing to new technologies, pent-up business need, or other factors, the next major wave of technological innovation, e-commerce, took the chemical industry by storm (Figure 6). Economic conditions were ripe; chemicals was a large, fragmented industry that made an ideal target for e-commerce initiatives. Although the first rush of enthusiasm, media attention, and stock market interest was centered on Internet-enabled business-to-consumer, or so-called B2C, sites, it soon become obvious that the big money was to be made in the business-to-business, or B2B, sector. And no B2B sector seemed to offer more promise than the commodities-dominated chemical industry.

The Early Promise

The industry trade publications started tracking the story in 1998. *Chemical Market Reporter* wrote in September of that year that "electronic commerce is making its way into the chemical industry. With external assistance from chemical distributors, logistics firms, and consultants, a growing number of chemical companies are beginning to adopt Internet technology through broad e-initiatives, contained e-commerce pilot programs or information-oriented web sites."[29] *Global Energy Business* wrote enthusiastically about the potential for e-commerce in the chemicals industry:

> At first glance, the chemical industry appears to be a textbook example of a business that could benefit tremendously from the use of e-Commerce. Its vastly dispersed production, hundreds of thousands of consumers, and expensive and relatively slow distribution system are exactly the sort of business inefficiencies that e-Commerce was designed to address.[30]

In 1999 five vertical online exchanges were established for the buying and selling of chemicals, including such leaders as CheMatch and ChemConnect; a year later the competitive set had ballooned to over forty-five exchanges. Despite the questionable economic logic of building so many versions of what is essentially a common digital marketplace, the new ventures were attracted by the immense forecasts being made for B2B commerce:

- Forrester Research predicted that B2B e-commerce will reach $1.3 trillion by 2003 as compared with the market's $8 billion in 1997.
- By 2003 Forrester estimated that B2B sales online will account for 9.4 percent of corporate purchasing in the United States.
- The Aberdeen Group estimated that the B2B e-commerce market may now be ten or more times larger than the online consumer area and that it will hold at that level.
- In 1998 the Organization for Economic Cooperation and Development (OECD) predicted that the B2B e-commerce market would account for about 80 percent of overall e-commerce for the next five years.[31]

Working on the leading edge of the wave, two Accenture partners issued a projected timeline for adoption of Internet technology in the chemical industry. In their version of the adoption curve the "true innovators" undertook e-commerce in 1996 and the "early adopters" between 1997 and 1998; both had a strategic objective and "added value" in mind. The "early majority" joined between 1999 and 2000, attracted by the potential for added value and cost

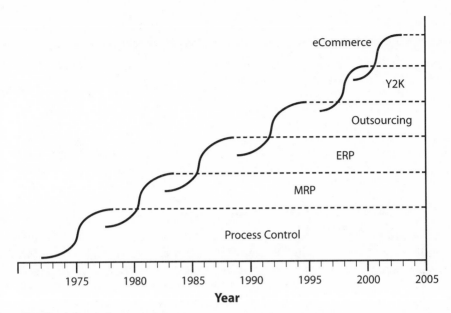

Figure 6. The sixth wave—e-commerce.

savings. The "late majority" and "laggards" entered in 2000 and after, out of technical necessity and the need to catch up with projected cost savings.[32]

Roger K. Mowen, Jr., vice president and chief information officer of Eastman Chemical Company, joined the prediction game in his keynote address at the e-Commerce in Chemicals Conference on 21 February 2000. "So here's my prediction: when it comes to e-commerce, the winners and losers in the chemical industry will be determined within the next three years. Some people in my group think it's closer to one year. But certainly within three years, everyone in this room will know who the big winners and losers are. Our challenge— certainly my challenge at Eastman—is to get there first."[33]

Eastman Chemical was determined to ride the crest of the e-commerce wave and get to the future first. It was interesting to watch the path of the latest IT wave sweep across the chemical industry, a wave that is arguably still moving today, and see how right or wrong these projections turn out to be and whether it was smarter to be an "innovator" or part of the "late majority" when it comes to e-commerce.

Climbing the Learning Curve

Despite the clear application of e-commerce efficiencies to the needs of the chemical industry, the learning curve was steep and the climb at times painful. Early e-commerce enthusiasts all came from the B2C side of the revolution, and chemical executives were rightly skeptical of B2C-style flash and hype.

Initial emphasis was on the corporate Web site and on the broadcasting of information to the Internet audience. One publication put it this way:

> The strategy within B2C environments is to simulate the cultural experience that consumers derive when visiting a retail store. The consumer expects to browse through a catalogue, drag and drop items to a shopping cart icon, and complete the transaction with a credit card. However, the B2C format lacks the speed and efficiency demanded by business buyers to ensure that tasks are completed promptly and accurately. In a B2B environment, time is of the essence and a solution must be designed around this reality. "Real-time" inventory availability, accurate pricing, and immediate delivery of information are necessary for the business buyer to react instantly to their own customers' demands, and to enable increased efficiency and profitability. In addition, the successful B2B solution must match existing business processes.[34]

John Mulholland, vice president of strategy at e-Chemicals, one of the first third-party online e-commerce providers, remained optimistic, commenting: "The chemical industry, like some other forms of heavy industry, is seeing an evolution of technological offerings with regard to electronic commerce. Look back 18 months and we have seen a progression from companies developing Web sites to showcase their products and services, toward actually carrying out online transactions."[35]

An early competitor of e-Chemicals, ChemConnect, was also optimistic. Michael Eckstut, then senior vice president at ChemConnect, said:

> By 2004, the chemical industry is forecast to account for a quarter of all B2B e-commerce, making it both the biggest industry and the biggest e-industry in the world. . . . B2B space has come into its own as the biggest opportunity online, attracting serious capital investment with a potential four times greater than that of business to consumer (B2C). Within this burgeoning B2B space, the chemical industry has evolved into a dynamic, dominating force that is redirecting sales away from traditional channels and pushing into new pathways.[36]

Yet a survey carried out by Arthur D. Little showed that chemical companies were slow to adopt e-commerce as a fundamental part of doing business. The survey was carried out among leaders from thirty major chemical companies and included these findings:

- 80 percent of respondents had dedicated e-commerce teams, but only 57 percent were selling products online.
- 80 percent of respondents had made no e-commerce alliances of any sort.
- Less than 10 percent of respondents expect the impact of e-commerce to be big enough to make them restructure their basic processes.[37]

Clearly, strong skepticism was rampant throughout the chemical industry. The explosive euphoria surrounding the emergence of B2B exchanges did little to comfort the skeptics.

The Euphoria of Exchanges

Between 1998 and 2000 several distinct types of online markets emerged, all battling for attention in the super-heated dot-com environment of the time. There were catalogue-inventory sales sites, usually offering laboratory or research chemicals, especially for the specialty, fine organics, biochemical, and biotechnology sectors. Examples of such sites included Chemdex (Ventro Life Sciences), e-chemicals, fobchemicals.com, and Sciquest.com. In the United States, Ventro's life sciences marketplace reportedly contained 1.3 million products and 2,200 suppliers, accessed by more than 24,000 users in June 2000. There were also auction–reverse auction sites, where deals were done through price bidding. Examples included chemicalbid.com and sourcerer.co.uk. Last but certainly not least were the trading exchanges themselves, such as Chem-Connect and CheMatch, which focused on direct materials and offered public dealings and private transactions in password-protected trading "rooms."

Specialty Chemicals provided a snapshot of operations at one exchange, i2i Chemicals.

> Buyers and sellers using i2i choose the transaction model best suited to their needs: classifieds, auctions and exchanges. i2i's Chemical Exchange was launched in September of 1999, and initially transactions revolved around a select group of about 70 high-activity chemicals (acetic acid to vinyl chloride monomer) and plastics (ABS to PET) that are typically bought and sold in large quantities. i2i formed a strategic partnership with SAP, designed to enable members to integrate Internet-based trading with back-end ERP solutions for such functions as inventory management and accounting.[38]

At the same time ChemConnect boasted the largest e-commerce marketplace for the industry—serving buyers and sellers of petrochemicals, plastics, basic industrial chemicals, pharmaceuticals, agricultural chemicals, and fine and specialty chemicals—globally, online, and in real time. The site brought together suppliers of chemicals and plastics, buyers that use chemicals or plastics in manufacturing operations, and intermediaries that assist clients in buying and selling these products. These participants used ChemConnect's World Chemical Exchange to find trading partners, negotiate pricing, and complete transactions.

The ChemConnect Story

John Beasley and Jay Hall founded ChemConnect in 1995 in Atlanta, Georgia. Their goal was to build an online trading community of chemical buyers and sellers; the first offering was an online suppliers directory. In 1997 Beasley and Hall joined forces with Patrick van der Valk, founder of BetaCyte, and expanded the service to enable online transactions between buyers and sellers. In December 1998 the company raised its first round of venture funding and relocated to San Francisco. In the summer of 1999 a second round of capital funding gave ChemConnect the financial support and industry backing to launch the World Chemical Exchange, which grew to include more than 5,000 members from more than 4,000 companies in 105 countries. At one time or another such companies as ICI, BP Amoco, BASF, and DSM have chosen the Chem-Connect World Chemical Exchange as their platform for e-trading in chemicals.[39]

John Beasley, then chairman of ChemConnect, set forth the value proposition for online trading.

> In the past, the global and fragmented nature of the chemical industry meant that buyers and sellers faced significant and costly obstacles in their attempts to find one another and to complete transactions. Today, ChemConnect supplies chemicals and plastics industry professionals with instant desktop access to real-time supply and demand information about the chemical products that most interest them.[40]

The Consortia Fights Back

Beasley's argument is persuasive, perhaps too persuasive. As the global chemical companies considered the clear efficiencies of the early exchanges, they also wondered why they should not be capturing the value created from the marketplaces they themselves were making.

In time this reasoning led to the creation of consortia, shaped and led by industry leaders. Elemica, a neutral intercompany commerce enabler for the global chemical industry, was designed and founded by twenty-two of the chemical industry's leading players, including BASF, Bayer, Dow, DuPont, BP, and DSM, to function as the hub for the chemical industry, serving both buyers and sellers. The company offers an integrated, end-to-end solution, enabling buyers and sellers to streamline their business processes for contract sales and improve order fulfillment and management. Elemica's comprehensive solution promises powerful efficiencies and offers support that extends from pre-sales through the transaction, supply chain, and after-sales process. At about the same time a second similar trading hub, Envera, began operating.

Another major consortium-based selling-side exchange, Omnexus, aimed to provide similar services to the plastics marketplace. Dow, DuPont, Bayer, BASF, and Ticona are the founding companies. The new marketplace was intended to simplify business transactions greatly, allowing customers to choose from a very wide range of products and services from participating resin manufacturers. Customers would also be able to purchase other plastic-related materials, molding equipment, tooling, maintenance supplies, packaging materials, and other related services.

Early e-commerce initiatives may have been concentrated in the United States, but it is an increasingly global phenomenon. A series of online chemical exchanges such as cc:Markets and Chemployer opened in Europe, and heavyweight producers like BASF (in Ludwigshafen) and Shell Chemicals (in London) have rolled out major initiatives. Some forty-nine Asian producers also backed a new online exchange, ChemRound, which debuted in July 2001, and companies like Reliance Industries (in Mumbai)—India's largest petrochemical group—have made substantial commitments to e-commerce.

In the very early years of the millennium the plethora of chemical exchanges has consolidated rapidly. Many of the smaller players were based on unsound business models, and with the drying up of venture capital, many simply closed their doors. Other more robust exchanges are trying to build alliances with other survivors, and in some cases they are acquiring or merging with the hope of continued existence. One good example was the acquisition of CheMatch and Envera by ChemConnect, creating a "two-horse race" between Chem-Connect and Elemica. Owing to the strange course of consolidation, both companies have the same major investors. We will see what the future holds.

Place Your Bets

Many chemical companies are still refining their e-commerce strategies, but an increasing number have completed implementation of major e-commerce projects.

BASF AG invested 75 million euros in e-commerce in 2000–2002 and aimed to conduct 40 percent of its business online by 2002 and 50 percent by 2005. The company also planned to have about one-fifth of its 5 billion euros per year of raw material spending online by 2002 and has a number of other internal and external e-initiatives under way.[41]

DuPont is a founding member of many of the consortia and had a host of internal e-initiatives under way. The company initially teamed with Chemdex and others to launch IndustriaSolutions, a fluid processing procurement site, and with agricultural groups Cargill and Cenex Harvest States Cooperatives to launch Rooster.com, an e-market for the agricultural supply chain. Unfortunately, these sites recently folded under the weight of consolidation.

Eastman Chemical, one of the early e-commerce leaders also announced investments and participation in various e-ventures, such as the plastics e-procurement group, PlasticsNet.com; e-supply chain enabler e-Chemicals; online credit management company eCredit.com; and e-logistics provider Cendian. As with other e-venture forays only a few have survived and prospered.

Two companies that have clearly decided to ride the wave, and perhaps even get out in front of it, are PolyOne, which was discussed earlier as it implemented an MRP system, and Dow Chemical.

PolyOne and Dow: Two Companies in the e-Commerce Curl

Geon, in Avon Lake, Ohio, had always been on the leading edge of the changes sweeping the chemical industry. This company has shifted the way other companies in the industry use information technology. Shareholders approved the merger of Geon with M. A. Hanna, a Cleveland manufacturer of chemical products for making plastics and rubber, to form PolyOne Corporation, which is now a leading supplier of polymer chemicals. PolyOne is a $3 billion company competing in an industry dominated by such companies as BASF, Dow Chemical, and DuPont. Its edge comes from e-commerce, and to that end PolyOne has used technology to establish Internet links between its ERP systems and the ERP systems of its largest suppliers, from which it buys 80 percent of its raw materials. Customers can go to the GetPolyOne Web site to place orders, check order status, and review invoice data. The chemical manufacturer also moved into online marketplaces; it is a member of ChemConnect's World Chemical Exchange and the intercompany commerce enabler Elemica.

Since the launch of its Web site, www.dow.com, in 1994 Dow has built a diverse array of e-business initiatives on an IT foundation integrated across its many global businesses. "Our goal has been to move beyond the standard Web-based, buy-sell transactions that typify consumer e-commerce. . . . Today's technologies should not only make us more productive, they should also make it easier for our customers to do business with us," said David Kepler.[42] Specific e-business projects are geared toward improving Dow's customer interface, supplier interactions, electronic sales and distribution channels, and future e-business and e-commerce opportunities.

To maximize electronic transparency among its many employees and business units, Dow has unified its companywide IT foundation with an ERP system from SAP for global operations. Kepler reports that the SAP system allows the company to integrate every aspect of its organization—from procurement, to manufacturing and logistics, to sales, customer service, accounting, and human resources. Its "sizeable investment" in implementing and maintaining SAP integration "is now returning over $200 million/year," claims Kepler.[43] To promote the sharing of data and best practices, Dow has also used its SAP foundation to create a shared data network. This central data repository gives employees from all over the world read-only access to current information on financial performance, sales, global inventories, and the supply chain from Dow's various business units.

To provide customers with a faster, more convenient interface with the company, Dow launched MyAccount@Dow. This private, customer-specific extranet gives registered customers online access to transactional functions and customer-specific information. Customers receive via e-mail a password to enter MyAccount@Dow and the URL for the extranet. From then on customers had immediate access to such information as the status of and changes to new and repeat orders, account history, and payment information. Customer-service representatives provide added support by phone and e-mail. MyAccount@Dow was initially piloted for two hundred key customers, in a wide array of industries, across many of Dow's fifteen global businesses; more than three hundred companies were registered by mid-2001, with significant potential for expansion to the tens of thousands of companies Dow sells to in seventy-five countries.

Dow has also implemented an advanced customer relationship management system from Siebel Systems, Inc. The Siebel system provides a centralized repository for all customer information collected by customer-service call centers across six of Dow's global businesses and three internal functions.

Dow Chemical's 123 manufacturing sites, involving 600 plants and warehouses in 32 countries, currently produce more than 3,500 products. With such a diverse slate of manufacturing activities, it is no surprise that the company is counting on Web-enabled technologies to improve its procurement initiatives of everything from raw materials to equipment to office and lab supplies. To help over fifty thousand employees automate the purchase of non–raw material items, Dow established eMart. This internal Web site catalogs more than a hundred thousand items from over a hundred suppliers. For its Internet-based global purchases of chemicals and plastics raw materials, Dow selected ChemConnect as its preferred third-party exchange. Dow has also joined two industry consortia for e-commerce referenced earlier: Elemica, the intercompany commerce enabler for contract chemicals, and Omnexus, the plastics exchange for the injection molding and thermoplastics industries.

Moments of Truth

The level of e-commerce activity in the chemical industry was extraordinary, reflecting both the potential size of the prize and the uncertainty felt by all the major players about which models and methods will prove successful in the long run. However, gone are the heady dot-com days; everyone has seen too many start-up businesses crash and burn to want to repeat the experience. We have entered a period of consolidation and integration. Today there is just as much focus on execution as on concept and an attempt to assess realistically where e-commerce adds value and where it does not. Increasingly, e-commerce is recognized as one buy-and-sell option among several, as an enabler, not a panacea, and certainly not the exclusive channel of the future. Leading chemical companies are thoughtfully backing those exchanges that seem to hold the greatest promise, while also understanding that private marketplaces will inevitably exist side by side with public marketplaces.

The Waves Continue: The Future of IT in the Chemical Industry

Looking back over more than three decades of change and innovation, there is a temptation to think that certain tasks are completed and can be checked off the to-do list. But even in such traditional areas as process control, interest continues today.

Every chemical industry executive understands that change is constant, that competition never ceases, and that the waves just keep rolling on. When looking toward the future and assessing the likely impact of IT on the chemical industry, a smart first step would be to track the same waves that have driven the industry's progress in the recent past. Continuing expansion around core ERP platforms, Internet-enabled netsourcing, and the maturation of e-commerce are all clear extensions of important trends we have already studied. Alongside these, entirely new advances such as mobile commerce will become rising waves in their own time. The final section of this chapter postulates an agenda for the chemical industry in the first decade of the new century.

Mining Value from ERP Systems

With most companies having already implemented ERP systems, the focus in the future will be to continually leverage existing infrastructure for maximum value. Many companies are planning to do this by adding functionality in such areas as e-commerce, supply chain management, manufacturing, customer relationship management, and human resources.

SAP has taken dramatic steps to shatter its image as a monolithic platform with mySAP, which encourages users to shape the solution's functionality to fit their needs. "Leading ERP vendor SAP AG has worked to catch up in the e-commerce market through significant investments in its mySAP.com Internet-enabled product suite."[44] The company has also focused on complementary solutions, such as SAP Business One, which is tailored to the needs of smaller sales offices, as well as the development of various cross-application software packages that leverage their core ERP system and data.

J. D. Edwards is another ERP vendor emphasizing extended applications. In fact, the company has gone so far down the path of expansion that it no longer regards the term *ERP* as an adequate generic description of its offering. Their most recent suite of applications includes not only an ERP suite but also customer relationship management, supply chain, supplier relationship, business intelligence, and collaboration and integration products.

Still other companies are gearing up to challenge SAP's domination in key portions of the ERP area. AspenTech has recently announced plans to expand its enterprise offerings in the supply chain and manufacturing areas. In a similar thrust Invensys PLC, i2, and Honeywell have all recently announced plans to expand their product suites in the production and supply chain–planning areas, all targeted at the chemical process industries.

The conceptual power of enterprisewide platforms remains tremendous, and there are still large areas of corporate operations where solution providers and their corporate clients have only scratched the surface, suggesting that these are definitely areas to watch in coming years. For example, enterprisewide platforms for the preservation and sharing of knowledge capital would appear to be a logical next step in the development of ERP. Solutions that seek to capture new knowledge as it is created and then disseminate and leverage that knowledge capital within the organization have significant potential. Systems that facilitate codevelopment of products with customers through information sharing take the same principle across organization borders. Another major industry need is "middleware," which helps facilitate the integration of different ERP and legacy systems of newly merged or acquired companies.

Outsourcing: Beyond the Traditional

Outsourcing is another area likely to elicit a "been there, done that" reaction. Yet the power of the Internet and its growing acceptance in industry is enabling new variations on the outsourcing theme.

In the conventional "one-to-one" outsourcing scheme, teams of IT professionals outsourced to an external services provider continue to handle applications management, run help desks, and execute core IT tasks for their former employer. Under new models now emerging in the chemical industry and in other sectors, alternative approaches to service delivery are appearing. In one scenario the same outsourced IT professionals who are supporting their former employer may also deliver shared services to other companies with comparable needs.

In another scenario a company may decide it is more efficient to "netsource" a solution platform that is maintained by a third party, and an entire subindustry of application service providers has emerged to meet this need. The same principle applied on a far broader scale points to the provision of entire business processes via the Internet. One working example of this approach was e-peopleserve, an Internet-enabled U.K.-based human resources service provider, which delivered a comprehensive spectrum of human resources services to corporate clients. By handling everything from payroll and benefits administration to pension accounting, these netsourced solutions free internal human resources professionals to focus their time and energies on higher-value strategic services.

The most promising view is that of "business transformation outsourcing." Business transformation is not a new concept; companies have been reinventing themselves for a long time and will continue to do so. What is new and different about business transformation outsourcing is that it offers executives a way to achieve transformational outcomes quickly. CEOs who use this approach start with a bold strategic agenda, sharing risks and gains with outsourcing partners well beyond the IT area. Whether the strategic agenda involves growth, market repositioning, or rapid diversification, the results of such an alliance can be stunning. More important, the scope of outsourcing has recently expanded in the chemical industry to include finance and accounting, human resources, procurement, customer service, and other "core" and "non-core" business processes, along with their underlying IT applications.

e-Commerce: Consolidation in a Multimodal Market

Despite the spectacular fall of the dot.coms, high on the action agenda is the development of Internet and Web-based technologies in order to enhance communications internally and with customers and suppliers. After the implosions of 2000–2001 no one is under any illusions about the status of e-commerce today. Tremendous upheaval lies ahead, as mergers, acquisitions, and consolidations sort out the survivors. The focus is on performance rather than promises, and everyone has adopted a "show me" approach.

Leading Wall Street investment firms expressed concern about the long-term viability of large chemicals e-commerce consortia, stating that governance issues in the groups may make them too slow to react quickly in the e-business environment. The speed, clear strategy, and ultimate competitiveness of such large online business-to-business (B2B) trading groups as the Elemica e-commerce trading site, the ChemConnect-Envera e-commerce hub, and the Omnexus trading site for plastics resins and moulding equipment were the issues of concern. Despite these pessimistic perspectives, all three exchanges were still in business as of this writing.

The predicted boom in which analysts forecast a rapid swing to plastics buying on the Web that would rival the invention of electricity in scope has not yet materialized, but the proselytizers remain optimistic. Peter McCullagh, former CEO of e-Chemicals, said:

> The industry has taken advantage of the Internet and e-commerce to improve information flows and basic transactions, but it is just starting to dive into Internet-enabled supply chain optimization. We're on the verge of the third and most dramatic phase of the e-commerce revolution—the ability to reach and understand customers more effectively and realize greater efficiencies through the development of an electronic supply chain.[45]

The potential for value creation is immense, but moving from traditional to e-supply chain management is not a linear progression. There will be many twists, turns, and failures before the marketplace sorts itself out. Traditional channels of doing business are not going away, but e-business will profoundly affect these channels. Alongside traditional channels will be the large consortia and private networks. E-commerce will coexist with conventional forms of buying and selling, while positively influencing the way in which the industry operates, enabling companies to deepen existing relationships with major customers. Enterprises can use e-business to exchange planning information and inventory over the Internet, and they can use it as a one-to-one communication tool. E-commerce will also allow suppliers to expand their business to new customers without adding staff, using a Web site to sell and its transaction side to take orders, lower the cost of fulfillment, reduce mistakes, and increase customer satisfaction.

New Industry Standards

Regardless of how the e-commerce side of the business shakes out, the emergence of the Internet has already had a significant impact on an existing industry standard: electronic data interchange (EDI). EDI is a standards-based protocol for conducting electronic transactions between trading partners and, until the advent of the Internet, had dominated the electronic processing of orders be-tween chemical suppliers and customers. Commonly used by larger companies with large trading partners, EDI typically occurs over private networks and is considered to be effective in reducing transaction costs. E-commerce, which relies on the use of Internet-enabled technologies, obviously sets forth a differ-ent path.

Peter Lyon, global director of customer care at Dow Chemical, views both EDI and the Internet as tools for enterprise customer management. He em-phasizes that EDI has always been used with a specific customer base and that the Internet is likely to find its own niche. "There are a number of our cus-tomer segments where EDI may not be the most feasible tool, and instead the Internet may be a more feasible tool to automate the order, payment and sell-ing process."[46]

One industry group that is addressing the issue of electronic commerce is the Chemical Industry Data Exchange (CIDX), a standards body and trade as-sociation of nearly eighty member chemical companies, their trading partners, and e-commerce vendors. The mission of CIDX is to provide the chemical industry with cost-effective e-commerce solutions that either lower the cost of existing processes or enable new commercial business processes to be used in the chemical industry. A transaction from a trading partner's e-procurement system can differ from an identical e-marketplace order or one passed between marketplaces. To help address this issue, the CIDX membership, including Accenture, defined and published industry conventions for EDI. Known as Chem eStandards, these conventions consist of a series of free, open transaction mes-sages based on the extensible mark-up language, or XML, e-commerce com-munication standard. There are two things a company must have to implement Chem eStandards: data mapped to Chem eStandards and its successors and an infrastructure designed to support the flow of these standards. The numerous messages published by CIDX should substantially reduce the number of unique data mappings required among systems, thereby facilitating buy-and-sell inte-gration of today's customers, suppliers, and e-marketplaces.

Mobile Commerce: A Fast-Moving Wave

Imagine the impact on supply chain management if you were able to track an individual pallet or container, continuously, automatically, passively, with the use of a low-power radiofrequency-emitting tag, attached at the plant, which sends a steady signal to a satellite in low Earth orbit. Among all the technologi-cal innovations on the near horizon, one of the most disruptive, yet potentially attractive for chemicals companies is mobile commerce.

Datamonitor is projecting stunning increases in the m-commerce market-place, with Internet spending via mobile phones growing fivefold between 2001 and 2004, with the wireless market being valued at $416 billion by 2005.[47] As wireless infrastructure such as the Intelligent Network Foundation (INF) spreads, it brings with it the technology to provide continual monitoring of equipment and supply without requiring human intervention. Wireless devices installed in every product tank within a manufacturing facility can alert an electronic supply chain when replenishment is required. Another application would be the response of a supply chain manager to an exception alert, where a railcar of product is delayed to a facility. M-commerce technology will be able to track the transportation of the product and notify the appropriate people in the supply chain immediately so they can adjust production schedules. Dow plans to implement INF globally across the Dow enterprise.

The Yankee Group forecasts that worldwide cellular carrier revenue from value-added services will increase from a low number to $25 billion in 2003. Many other new technologies could ride atop the m-commerce wave. Take speech recognition, for example. In a study called "Voice Recognition Systems," Allied Business Intelligence concluded there would be 17 million by 2005.[48] In the m-commerce world, speech recognition will play a vital role as new tools are used to access wireline and wireless Web-based applications. If the predictions hold true, IT departments will be at the center of a massive enterprisewide business process reengineering effort.

Transforming these technological advances into practical business applications will take time. Largely on the strength of m-commerce and broadband service potential, global telecommunications rushed to wire the world with fiber optics, only to watch their market valuations crash as excess capacity flooded the market. Third-generation communications services, which were supposed to be around the corner, are not projected to be introduced in Europe and the Americas until much later in this decade. Despite these delays and disruptions, the power of m-commerce is real, and the most valuable immediate applications can be found in the industrial rather than the consumer marketplace.

Conclusion

At the turn of the century, industry surveys of chemical executives revealed e-commerce, customer relationship management, knowledge management, simulation and modeling, and better leveraging of ERP as key priorities in chemical companies' IT strategies. New IT investments are more like rifle shots than enterprisewide changes; business units are focused on returns from specific IT-enabled programs in such defined areas as procurement or customer relationship management.

These priorities include a long list of action items for the years ahead:

- Sizable investments in the development of customer relationship management systems;
- Improved integration through front-end and back-end systems;
- Increased focus on human resource issues, including knowledge management, collaboration, productivity measurement, and improved leverage of data;

- Improved information organization and analysis capabilities;
- Increased focus on simulation and modeling;
- Advanced logistics management and supply chain optimization;
- Collapsing data and voice telecommunication systems;
- Data mining applied to R&D, marketing, and quality issues;
- Collaborative business-to-business links;
- Embedded processors using telemetry to communicate with the network; and
- Multimedia capabilities for training and collaborative needs.

To this list, a final to-do must be added: the development of effective measures for the return on investment in IT. In the past the case for IT investments was often so compelling that executives did not even bother to construct a business case. When attempts were made to build a business case for major ERP investments, for example, the results were impossible to isolate and difficult to quantify. Moving forward, the industry is becoming increasingly focused on value creation and will demand more rigorous methods for calculating the return on capital investments. When chemical companies are spending an estimated 1.2 percent to 1.5 percent of total revenues for total IT expenses (capital and expense), with some corporations reaching 2.5 percent during such enterprisewide efforts as ERP, senior executives clearly need to know what they are buying with those resources. Some companies, such as Dow Chemical, already do. As Dow looks forward over the first decade of the twenty-first century, it expects to see $2 billion of business value out of its IT investments, while at the same time reducing the cost of IT annually. Every global chemical company needs to have equally clear and compelling targets for growth, performance, and profitability.

Endnotes

1. Thomas M. Stout and Theodore J. Williams, "Pioneering Work in the Field of Computer Process Control," *IEEE Annals of the History of Computing* 17:1 (Spring 1995), 6–18.
2. Ibid.
3. Kevin Parker, "Batch Is Unmatched in Automation Gains," *Manufacturing Systems* 10 (Oct. 1992), 30.
4. David Rotman, "New Wave of Software Entices the Industry," *Chemical Week* 148 (27 Mar. 1991), 37.
5. Rick Mullin, "Software's New Promise: The Global Network—Decision Makers Key-in on Application Programs," *Chemical Week* 152 (19 May 1993), 24–27.
6. Doug Bartholomew, "Vinyl Victory—How One Company Found That Big Problems Require Big Solutions," *InformationWeek* (4 May 1992), 32.
7. Ibid.
8. Mullin, "Software's New Promise" (cit. note 6).
9. Monua Janah, "A Search for the Right Formula," *InformationWeek* (9 Sept. 1996), 92.

10. Mike Ricciuti, "Connect Manufacturing to the Enterprise (Manufacturing Software Vendors Offer Enterprise Resource Planning Packages)," *Datamation* 38 (15 Jan. 1992), 42.

11. Mullin, "Software's New Promise" (cit. note 5), 25.

12. Kara Sissell, "Competing Software Muscles in on SAP: New Uses Tackle Internet Concerns," *Chemical Week* 158 (21 Aug. 1996), 26.

13. Rick Mullin, "IT Integration: Programmed for Global Operation," *Chemical Week* 159 (12 Feb. 1997), 21.

14. Cynthia Challener, "Leveraging ERP Investments in the 21st Century," *Chemical Market Reporter* 257 (24 Jan. 2000), 24.

15. Dan Scheraga, "SAP Leads the Way as Chemical Companies Focus on ERP," *Chemical Market Reporter* 255 (25 Jan. 1999), 10.

16. Mullin, "Software's New Promise" (cit. note 5), 26.

17. Janah, "Right Formula" (cit. note 9), 92.

18. Sissell, "Competing Software" (cit. note 12), 26.

19. Susan Connor, "Data Watch—Fatal Flaws (4344)," *Data Quest* (FT Asia Intelligence Wire) (31 Jan. 1999), 1.

20. Mullin, "IT Integration" (cit. note 13), 26.

21. Martin Stone, "ERP Davids Nipping at Goliath SAP's Heels: Study," *Computing Canada* 24 (2 Nov. 1998), 23.

22. Ibid.

23. Howard Solomon, "SAP AG Has Leg Up on Competition: ARC," *Computing Canada* 25 (26 Nov. 1999), 4.

24. Cynthia Challener, "Extensions in e-Business, SCM and CRM Are Key for ERP Vendors," *Chemical Market Reporter* 258 (25 Sept. 2000), 16.

25. Robert Preston, "Andersen Delivers for Dow," *InternetWeek* 822 (24 July 2000), 55.

26. Rick Mullin, "IT Centers Launched by Consultants DuPont, Monsanto Are Springboards," *Chemical Week* 160 (19 Aug. 1998), s38.

27. Don Richards, "Chemical Industry Unvexed by Y2K Computer Problems," *Chemical Market Reporter* 257 (17 Jan. 2000), 5.

28. Kazim Isfahani, "Evaluating Y2K Expenditures," Giga Information Group—Ideabyte, www.rightnow.com/news/giga.html (7 Feb. 2000); U.S. Department of Commerce, Economics and Statistics Administration, Office of the Chief Economist, *The Economics of Y2K and the Impact on the United States*, by William B. Brown and Laurence S. Campbell, Nov. 1999; Bruce Caldwell, "Y2K Under Control—By Spending Tens of Billions of Dollars, IT Organizations Think They Have the Year 2000 Issue in Hand," *InformationWeek* (10 May 1999), 52.

29. Teresa Ortega and Patricia Van Arnum, "Untangling a Web of Promise," *Chemical Market Reporter* 254 (14 Sept. 1998), FR16.

30. Alessandro Vitelli, "What's Up with Energy e-Commerce?" *Global Energy Business* 2 (1 Aug. 2000), 45.

31. Tom Mulligan, "e-Commerce Comes to Chemicals," *Specialty Chemicals* 20 (1 June 2000), 190.

32. Ortega and Van Arnum, "Web of Promise" (cit. note 29).

33. Roger K. Mowen, Jr., keynote address, e-Commerce in Chemicals Conference, Amsterdam, 21 Feb. 2000.

34. "The Challenge of e-Commerce (in the Chemical Industry)," *Specialty Chemicals* 20 (1 Oct. 2000), 319.

35. Christopher Reilly, "Providers Expand Services as Technology Evolves," *Purchasing* 129 (2 Nov. 2000), 48C30.

36. Hilfra Tandy, "Chemicals Industry—On-line Sales May Slash Prices and Profits," *Financial Times*, 3 July 2000, p. 4.

37. Mulligan, "e-Commerce Comes to Chemicals" (cit. note 31).

38. Ibid.

39. Ibid.

40. Ibid.

41. Michael Roberts, "Chemical Industry e-Commerce Has Landed: Hype Becomes Reality," *Chemical Week* (Special Supplement: Internet Focus 2000) (26 July 2000), s5.

42. Suzzane Shelley, "Launching a Battery of Dot.com Offerings," *Chemical Specialties* 2 (1 Sept. 2000), 49.

43. Ibid.

44. Challener, "Extensions in e-Business" (cit. note 24).

45. Roberts, "e-Commerce Has Landed" (cit. note 41).

46. Ortega and Van Arnum, "Web of Promise" (cit. note 29).

47. Jake Sorofman, "European Chemical News: m-Commerce Expected to Boom," *Chemical Business Newsbase* (online service), 4 July 2000, 14.

48. Deborah Mendez-Wilson, "Voice Portal Din Ups Intensity," *Wireless Week* (13 Nov. 2000), 30.

Chapter 7

The Chemical Industry and the Environment

Meeting the Challenge

Peter H. Spitz

The eventful history of the chemical industry over the past three decades was arguably dominated by three overarching influences, each of which has strongly shaped the industry's recent characteristics. First came the two oil shocks in 1973 and 1978, which initiated the shift of petrochemical production to low feedstock cost regions, the increasing domination of petrochemical manufacture by large downstream integrated oil companies, and pervasive cyclicality for the global petrochemical business (although the continued cyclical characteristics of the industry can no longer be ascribed primarily to feedstock-related issues). Second was the rapidly growing influence of the financial community, which led and is continuing to lead to restructuring, consolidation, demergers, and other structural changes, as firms attempt to define core businesses and increase earnings to meet financial community expectations for higher shareholder value. The third influence was the strong increase in regulatory controls and community pressures to reduce various types of emissions, to treat waste streams, and to protect consumers and plant workers from exposure to toxic chemicals, including restrictions on the production of chemicals identified as harmful, or potentially harmful, to humans or to the environment.

Other chapters in this book deal with the first two items, while this chapter presents a view of the changes that environmental controls and policies and public pressures wrought on the industry, especially in the way it conducts its day-to-day business. The chapter also discusses the often dismal public perceptions of the chemical industry. At this writing, polls still identify the chemical industry as having a poorer image than most other industries, despite the strides it has made in "becoming a good citizen." While most people realize that chemicals and plastics are essential to modern civilization, the companies making these chemicals—including chemicals that are transformed into drugs, whose manufacturers have a better image (even though pharmaceuticals are as bad as petrochemicals on the Resource Conservation and Recovery Act inventory [see appendix at end of chapter])—are still considered by many as major polluters and as dispensers of dangerous products: recalcitrant entities that let the profit motive overshadow their environmental and public responsibilities. The American Chemistry Council (ACC, formerly the Chemical Manufacturers Association, or CMA) has for many years led campaigns, most recently Responsible Care, to promote various progressive actions by its member companies beyond compliance with regulatory statutes and to inform the public of the progress the industry keeps making to act responsibly. It appears that promotion of a better understanding of the industry in people's minds has been somewhat successful. Thus, a campaign by another industry association, the American Plastics Council (now merged with the American Chemistry Council), has greatly helped in convincing the public of the indispensable contribution plastics make to modern living. But there is still a long way to go, and every time a major spill or explosion occurs or another chemical is newly identified as a conceivable source of cancer or birth defects, there is another setback, as the media and some of the environmental groups seize on the publicity to paint the industry in a poor light once again.

The chemical industry has now succeeded in modifying most of its environmentally harmful manufacturing and waste-disposal techniques and in understanding and dealing with the toxic nature of many of the chemicals that are indispensable for our way of life. It now also recognizes the need to protect its workers and to be completely aboveboard in providing information to its employees and surrounding communities about potential hazards inherent in its manufacturing processes. But it had to come a long way and left a history of incidents that many people find hard to forgive and forget.

The rapid, often painful transformation of an industry primarily concerned with technological innovation, large-scale manufacturing, and selling into an industry that would strongly embrace corporate environmental management is a cautionary tale. Initial rejection of Rachel Carson's book *Silent Spring* as alarmist and scientifically unsound, denying the environmental effects of the Santa Barbara channel oil spill and initially downplaying the problems with lead in gasoline, showed an industry about to face dramatic societal change. Andrew Hoffman, in a groundbreaking book, states in his introduction that

> these industries were composed of proud people who believed that they were taming nature for the good of humankind: they were a remnant of . . . the "Golden Age of Engineering." At first, they regarded as a personal affront the aggressive claims that their actions were misguided. . . . However, over time, as societal perspectives changed, so did perspectives within the industry [and] environmentalism, at first seen as a threat from the fringes of society, became a central component of businesses' "competitive strategy."[1]

Perhaps most important, the chemical industry began to recognize that it had to submit itself to public accountability, with its leaders saying that "we have to have a public license to operate."[2] Among other things, this eventually led to over three hundred community advisory panels, internal advisory boards, and many other actions that had never been contemplated by an industry mainly bent on pushing out its products and selling them to the world.

The case can easily be made that issues related to the environment and toxicity have in many respects had a greater effect on the industry than almost all the other traumatic events it had to face in the declining years of the twentieth century. These issues can be divided into two categories: reaction to the specific regulations, including the steps taken for compliance, and changes in company operations that were directly or indirectly brought about by environmental issues.

The breakthrough legislative acts passed in the 1970s brought about a sea change for the ways that chemical companies had previously conducted most of their operations. These acts are described later and will be referred to at various points in this chapter. In the United States the Environmental Protection Agency (EPA) was given birth in 1970, when President Richard Nixon created the National Environmental Policy Act (NEPA) by executive order. Along with analogous state and local agencies, the EPA became the official watchdog and almost immediately a body that industry considered its adversary. This situation came about largely because the EPA's actions were not always governed by purely scientific or legal considerations. The organization was also a political body, with key executives appointed by the party in power and with its own agenda on environmental issues. How key pieces of legislation affected various aspects of industry operation will be discussed later.

Companies also had to review many of their traditional ways of doing business. Between 1970 and 1999 companies discontinued the production of certain chemicals as a result of legislation, global treaties, or voluntary withdrawal. Siting for new plants became much more difficult. New grassroots installations for large plants designed to make a variety of high-production-volume chemicals became very difficult in many of the countries belonging to the Organization for Economic Cooperation and Development (OECD). Expanding the existing plants also became more difficult. In both cases local concern about emissions and potential hazards were the main cause. Some companies decided to locate new plants where environmental regulations were less strict or public attitudes less hostile, in some cases losing proximity to markets and creating fewer new jobs in the home country. BASF's decision to build some of its new heavy chemicals manufacturing operations in Antwerp, Belgium, instead of Ludwigshafen, Germany, is a case in point. In an effort to reduce high-volume emissions, companies became very active in recovering and in many cases recycling chemicals previously sent to land disposal, to bodies of water, or into the atmosphere. In many cases these actions had a positive effect on the balance sheet. However, substantial additional expenses of various kinds, attributed to environmental control, increased the capital investment and operating cost of new and existing plants. A number of companies added environmental remediation services to their product offerings, including scrubbing equipment, water-treating chemicals, and engineering services. Alternative technologies were developed to make processes safer and less polluting or to develop and commercialize new processes and in some cases new chemicals substituting for those being phased out. In considering acquisitions of companies, divisions, or plants, chemical firms hired environmental consultants to study what liabilities the firms might have to deal with owing to "past sins" incurred at the site they would be acquiring. Such concerns often led the acquiring firm to drop the acquisition target or in any case to put responsibility on the seller. Many companies incurred substantial charges to pay claims to persons harmed by exposure to toxic chemicals or to other chemical-related incidents. In extreme cases—for example, asbestos litigation—this has led to a number of bankruptcies. A great deal of management time was devoted to communication with various company stakeholders to outline the companies' policies and actions designed to produce chemicals in an environmentally friendly manner. So-called life-cycle analyses (LCAs), also known as "cradle-to-grave" studies, came into vogue to assess the total environmental impact of the manufacture and use of specific chemicals. These were at times used by firms in an attempt to prove that a certain chemical or other product was environmentally less harmful than a competitive product.

Looking back, it is hard to realize how far the industry had to go to be able to keep operating in a manner acceptable to its stakeholders and the government. The changes were so pervasive that it is difficult to understand how and why the industry was able to operate under such truly laissez-faire conditions before deciding, or being forced, to change its ways. As an example of the result of benign neglect for disposal of wastes, by the mid-1980s the organic chemical industry had become the single largest source of hazardous wastes in the United States,[3] and it remains so today. It has been estimated that during the rapid expansion of the industry after World War II, the leading producers had deposited 762 million tons of chemical wastes in more than 3,300 locations around the country.[4]

To understand how this came about and how industry participants began to deal with pollution and waste disposal, it is necessary to go back several decades.

Growing Concerns

In the early twentieth century chemical plants generally disposed of their wastes in on-site impoundments, landfills, rivers, and lakes and in the ocean. Dilution found some support among public health officials, who believed that water bodies had natural purification powers and that some toxic discharges were therefore considered beneficial because of their alleged germicidal effects ("the solution is dilution"). However, uninhibited use of waterways for chemical discharges eventually led local and state agencies to oppose wide-scale use of riverine environments for this purpose, and some pollution edicts were issued. Much early attention was directed toward operators of the particularly malodorous coke oven chemicals installations, forcing some of these producers to install equipment for treating phenolic and tar wastes. In Illinois public outcry over drinking-water contamination caused a steel works with coke ovens to divert its waste from the Calumet River to the Illinois River drainage basin, relying on dilution in another, larger waterway.[5]

As manufacturers increasingly recognized the problems inherent in dumping waste streams into sources of drinking water, including contamination of groundwater through leakage of chemicals into aquifers, plant siting became an issue. So plants began to be located away from urban locations. But absence of treating equipment, even in sparsely populated areas, gave chemical manufacturers a poor reputation. Treating methods were generally known, but they were not often used, although neutralization of acidic or caustic wastes was sometimes practiced. In the 1930s the perception about the presumed beneficial effect of toxic chemicals on pathogenic bacteria gave way to recognition that real dangers were being posed to fish and wildlife. Toxicologists identified a number of synthetic chemicals as an increasing menace and along with biologists argued against dilution as a means of chemical waste treatment. So land burial of toxic wastes was adopted as an alternative. From the 1940s on, Monsanto buried some of its waste on its property to avoid dumping into the Mississippi River. Hooker Chemical began burying toxic chemical wastes in a landfill it owned at Love Canal, located close to Niagara Falls, New York.[6]

The Clean Air Act (CAA) became law in 1955 and the Solid Waste Disposal Act in 1965. By 1969 the federal government was beginning to work with the states to regulate the major forms of pollution.[7] These earlier pieces of legislation were important starting points, but only later, when amendments were passed, did the regulations become really effective.

In 1948 Congress had passed the Water Pollution Control Act, but its effects were limited. It preserved the states' authority to legislate abatement measures and provided funds for technology research on waste treatment. Companies generally continued existing practices. However, in 1950 the Manufacturing Chemists Association (MCA, later CMA, and now ACC) established a task force to coordinate pollution-control strategies. Its basic objectives were "to stimulate interest of member companies in controlling their own wastes, emphasize the importance of clean water resources to chemical producers, underscore public relations benefits of proper waste disposal, foster uniform state pollution control legislation, and encourage the exchange of technical and regulatory information."[8] The response by member companies was mixed. Use of advanced treatment methods began to be used when severe pollution problems from previous years were recognized, with an accompanying threat of litigation or regulatory enforcement. Dow and DuPont were among the more enlightened industry participants, with Dow, for example, installing a three-phase treatment for 50 million gallons a day of organic wastes. DuPont hired a biologist to conduct a "biodynamic survey" of a river into which a Texas plant was discharging wastes. After this survey, treatment and disposal techniques were installed to prevent toxic materials from entering the river, while solar evaporation was selected to eliminate the discharge of dilute aqueous solutions.[9]

Pressures from several sources began to mount. The U.S. Fish and Wildlife Service had earlier expressed concern about discharge of toxic wastes into waterways, and toxic waste disposal finally emerged as a major public health issue in the 1950s. Stung by criticism, the MCA proclaimed that chemical producers devoted between 2.5 percent and 4 percent of construction costs to pollution control and spent $40 million annually to abate pollution. Nevertheless, the U.S. Department of Health, Education and Welfare, which had surveyed waste-treatment facilities, reported serious shortcomings of industrial waste-treatment facilities. Twenty-four percent of 5,967 plants surveyed had no treatment at all, and 34 percent used methods normally reserved for domestic sewage treatment. Many toxic or acidic wastes were harmful to the municipal treating facilities receiving them, destroying the bacterial flora or causing severe corrosion.

Waste management policies employed by chemical firms often included transferring potential legal responsibilities to third parties. Insurers offered such coverage in many cases, often not realizing the extent of their exposure in future damage suits. This became the subject of much litigation between insurers and the insured in later years.

By the 1960s waste management was receiving much attention at the corporate level. High-level executives were appointed to oversee pollution abatement programs rather than having these handled largely at the plant level. While internal expertise on toxicology was rapidly increasing and technologically advanced treatment methods were becoming widely available, companies nevertheless still largely used dilution and isolation to dispose of waste streams or opted for land burial for solid wastes. When public officials claimed that continuing chemical pollution posed serious health and environmental risks, the industry responded that it was dealing with the problem. The lag time between recognizing the serious and growing environmental problem and implementing solutions was significant.

The rise of environmentalism in the 1960s sharply increased awareness of the potential and actual hazards of chemicals and chemical wastes. Rachel Carson's 1962 book *Silent Spring* alerted Americans to the possible dangers of the broad-based use of certain pesticides. Soon thereafter the world learned about the poisoning of fishing grounds in Minimata Bay in Japan that had resulted from Chisso Corporation's longtime dumping of a mercury compound. The death of close to a thousand inhabitants was reported, as well as numerous chronic diseases and abnormalities of the population in a village next to the bay, including blindness and birth defects.

At the end of the decade Earth Day gave a specific rallying point to the environmental movement, whose agenda significantly included pollution and toxic threat from chemicals.

In late 1970 representatives of industry, government, and conservation groups met at a three-day Coast Guard symposium in New Orleans to figure out how to prevent environmental pollution from spillage of hazardous materials—mainly chemicals. The group, including representatives of Union Carbide, DuPont, and Dow, reached the following conclusions, as reported by *Chemical Week*:

• The best way to control pollution is to prevent spills.
• More federal regulations will be needed to encourage industry to use the best controls possible, technologically and economically, to prevent spills.
• Reporting and response centers must be set up to provide round-the-clock advice and help if spills occur.
• Heavy fines and civil actions should be taken against individuals and companies that fail to report spills immediately.

The *Chemical Week* article's subhead read "Nixon may ask for more controls over manufacture, storage, and shipping of hazardous chemicals and spill-reporting network."[10] And it was during the Nixon administration that some of the most important environmental regulations were passed. (These and other regulations passed in the 1970s and 1980s are described in the appendix to this chapter, using various sources of information, including the very useful Web site maintained by the Battelle Institute [www.chemalliance.org]).

The negative image that the chemical industry had already developed was further exacerbated by periodic events that received wide publicity and, along with growing pressures from government and public sources, created the background for the broad legislative responses that occurred in the 1970s, which related to pollution as well as to the harm caused by toxic chemicals. Thus, as news came out about the effects of worker or general public exposure to certain chemicals, reaction by lawmakers was not just inevitable; it was also badly needed, regardless of whether the companies involved had prior knowledge of the dangers to which their workers were exposed.

During this period it became known that workers in Italy and in the United States who were exposed for long periods to plant emissions of vinyl chloride developed angiosarcoma of the liver. In 1974 BFGoodrich announced that several workers in its polyvinyl chloride (PVC) plants had developed liver cancer, for which there was no cure. Short-term exposure to high levels of vinyl chloride was also shown to affect the central nervous system, causing such symptoms as dizziness, headaches, and giddiness. The EPA eventually classified vinyl chloride as a "group A" human carcinogen of medium carcinogenic hazard.[11]

In Hopewell, Virginia, in the early 1970s, Allied Chemical, and later a contractor working for Allied, produced a pesticide named Kepone, which caused various severe medical problems for workers as a result of massive exposure in a poorly maintained plant. Kepone wastes were also dumped into the James River, which was later closed to fishing, and into the municipal sewage systems, killing the beneficial bacteria being used to treat residential sewage. Lawsuits by various injured parties claimed damages of up to $8 billion.[12]

In Bristol, Pennsylvania, a number of workers in the Rohm and Haas ion-exchange production unit developed a fatal type of lung cancer because of long-term inhalation of bis-chloromethyl ether, an intermediate in the manufacturing process. In 1975, CBS devoted a fifteen-minute segment of the nationally televised news program *CBS Magazine* to this story, and the *Philadelphia Inquirer* ran a story on the same subject under the title "54 Who Died."[13]

In Kyushu, Japan, in 1968, a polychlorinated biphenyl (PCB) mixture leaked out of a factory pipe and contaminated some rice oil. People who ate the oil developed a number of serious symptoms, including deformities of joints and bones. It was shown that PCBs could be transferred from mother to fetus, and some babies were born dead. Two U.S. producers of PCBs, Monsanto and General Electric, discontinued making the chemicals in the 1970s.[14]

Unfavorable publicity involving toxic waste dumps or toxic emissions in some cases received intense media coverage even when there eventually was reasonable doubt about the level of harm inflicted on people living close to these sites. Three well-known examples follow.

In 1976 residents along Love Canal near Niagara Falls began complaining about chemical odors from a landfill and, according to surveys, suffered various medical disorders. Hooker Chemical had from 1942 to 1953 disposed of hazardous waste on this managed landfill. Hooker later ceased using this site for waste disposal and deeded the property to the town with a specific written warning regarding the chemicals buried below ground. Nevertheless, the town proceeded to build a school over the site, and homes had eventually also been built on the land, which understandably represented inexpensive real estate in the growing community. When the situation at Love Canal became a media event, President Jimmy Carter declared a state of emergency. About 2,500 residents were temporarily relocated. Eventually a number of homes were sold, with sales managed by a state agency.

While there is no question that exposure to toxic chemicals can be extremely harmful, there is continuing disagreement about whether the incidents of cancer found at Love Canal were greater than would be expected from normal statistics. In 1981 a *New York Times* editorial on the subject noted that "Love Canal, perhaps the nation's most prominent symbol of chemical assaults on the environment has had no detectable effect on the incidence of cancer. . . . It may well turn out that the public suffered less from the chemicals there than from the hysteria generated."[15]

Also in 1976 an explosion in a hexachlorophene plant in Seveso, Italy, released a vaporous cloud of chemicals, including tetrachlorodibenzo-p-dioxin (TCDD), which settled over a populated area. Affected residents soon developed a series of reactions, including a skin disorder, chloracne. Since dioxin levels found in the blood of residents were several magnitudes higher than normal, dire consequences were predicted, and for some time every incidence of a birth defect or cancer was attributed to dioxin poisoning. Eventually, however, studies by the *Journal of the American Medical Association* of fifteen hundred Seveso children concluded that only very slight abnormalities were found and that these disappeared with time.[16]

Possible dioxin exposure became a much bigger media event in the Times Beach, Missouri, incident several years later. Thus, in 1982 the chemical industry received another jolt when flood waters containing dioxin spread throughout the town of Times Beach. It was discovered that years earlier a contractor had sprayed a nearby road with asphalt that happened to contain the dioxins. Soil levels measured hundred of times higher than what was to date considered as "safe" levels of the chemical. It was decided to evacuate the town at a cost of $33 million. Years later, in 1991, it was concluded as a result of studies that no residents had in any way been affected by the dioxins and that there had been no health threat from the dioxins bound to the ground.[17] But after the Times Beach incident dioxins became a "poster child" for toxic chemicals as seen by environmentalists, who even came out against incineration of chlorinated materials because of the likely formation and emission of dioxins under the conditions in the furnace.

Evidence of worker exposure to toxic chemicals provided much of the background for passage of the Toxic Substances Control Act (TSCA) and the Resource Conservation and Recovery Act (RCRA) passed later in the decade. It is probably fair to say that in most cases in which companies became convinced that its workers were exposed to harmful chemicals, they took the appropriate action to eliminate such exposure or reduce it to safe limits. Unfortunately, there were some egregious exceptions, where companies waited too long or acted in an adversarial manner, which resulted in devastating publicity.

Then in 1980 another very important piece of legislation, known as the Comprehensive Environmental Response, Compensation and Liability Act (CERCLA), was passed by Congress. As more and more concern was raised about largely abandoned hazardous waste sites, pressure was building about the need to clean up these sites to avoid other "Love Canals" and similar discoveries. This legislation was followed in 1986 by another act, known as the Emergency Planning and Community Right to Know Act (EPCRA), which revised some of the rules and added further provisions. This act was passed shortly after the accidental release of methyl isocyanate from a Union Carbide plant in Bhopal, India, which killed thousands of residents living close to the plant. The legislation recognized the fact that communities near a chemical plant have as much right to know about toxic chemical releases (normal or accidental) as workers at the plant.

The tragedy at Bhopal in December 1984 remains the most devastating accident ever to occur at a chemical facility.[18] The official death count was 1,800—although local doctors say it could have been 500 to 1,000 higher—with 14,000 others "seriously affected." Union Carbide was excoriated worldwide. The company became a major target of environmentalists and other adversarial factions, causing it to remove the directions to its headquarters in Danbury, Connecticut, from the exit ramp on Interstate Highway 84. The unfavorable publicity generated by the Bhopal accident resulted in a sharp drop in the firm's share price. Sam Heyman, a financial entrepreneur who had previously taken control of GAF, started acquiring Union Carbide stock. In defending itself, the firm incurred massive debt for stock buybacks, later requiring it to sell a number of its key downstream assets (for example, Prestone antifreeze and Glad Bag plastic bags). Further pressure by the financial community later resulted in the sale of UCAR, its carbon graphite subsidiary, and of its silicones business. To create more shareholder value, the company also spun off its industrial gas division, Linde, as Praxair and thereby became a small, "focused" company. Union Carbide then became a much easier target for Dow Chemical in 2001, which acquired the firm. This resulted in the demise of one of the best-known chemical companies, which, it might be argued, never really recovered from the effects of the Bhopal incident.

In 1990 the Pollution Prevention Act was passed by Congress to essentially complete the list of regulations under which the industry would henceforth operate.

New Operating Rules for the Industry

The body of environmental laws passed between 1970 and 1990 had a much broader purpose than singling out the chemical industry for restrictive regulations. Chemical companies were hardly alone as historical polluters, with steel mills, paper mills, cement plants, refineries, and many other process industries equally culpable. Toxic waste had been dumped into land sites by many types of industrial operators, waste transporters, and waste disposal firms. And air pollution, including smog, formed by release of gases covered by the CAA, was to a much greater extent caused by motorists and utility plants, particularly those burning high sulfur coal. This is clearly shown by statistics about the six major atmospheric pollutants identified in the CAA—sulfur dioxide, nitrogen oxide, carbon monoxide, total suspended particles, volatile organic compounds

(VOCs), and lead. According to statistics for 1970, chemicals and allied products were responsible for the release of only 2.5 percent or less of sulfur and nitrogen oxides, carbon monoxide, and particulates, and 5.7 percent of VOCs.[19]

Nevertheless, much of this legislation was aimed primarily at the process industries and particularly at the chemical industry, which had drawn unfavorable attention because of historical emissions into the air, rivers, and lakes, in some cases with serious effects on workers and the surrounding population. Moreover, the public and regulatory authorities recognized the imperative need to do something about the thousands of toxic waste dumps all over the country, particularly in such states as New Jersey, Texas, and Louisiana, where a preponderance of chemical plants and refineries were located.

As chemical firms began to take measures to comply with the new rules, their operating costs mounted rapidly. In 1971 Monsanto stated that it now had ten times as many people working on pollution as they had just two years earlier. Hercules's pollution abatement expenditures in 1971 were double those in 1970. DuPont's nearly $10 million increase in pollution control costs came to 9 cents per share after taxes, according to Edwin Gee, chairman of the firm's environmental quality committee.[20] Cyanamid in 1971 spent the equivalent of 12 cents per share for control equipment, while Hooker Chemical's costs rose from $3.5 million to $11 million.[21]

There was considerable concern whether processes and equipment were available to comply with the new laws. Thus, Union Carbide chairman Perry Wilson called for a proven means of abating sulfur oxides emission from coal-burning power plants, such as the company that operated at its Kanawha Valley in West Virginia site.[22] The need for new technologies and for various types of equipment and services to limit plant emissions was, of course, also seen as an opportunity for firms to market such services. Many of the companies were already marketing or starting to market pollution control equipment, as well as water-treatment and other suitable chemicals, either through existing divisions or with newly established branches, such as Monsanto Enviro-Chem. Among the various types of equipment now finding a much expanded market were fabric filters, electrostatic precipitators, cyclones, scrubbers, fixed-bed adsorption units, high-temperature incinerators, centrifuges, flotation equipment, pumps, chemical feeders, sedimentation and aeration tanks, electrodialysis equipment, and various types of control equipment.

Compliance with the new rules governing air pollution required the installation of much new equipment. Sulfuric acid plants, for example, dealing with now unacceptable levels of emission of particulates and sulfur oxides, needed to install additional absorption towers, demisters, and gas cleaning equipment, such as cyclones or filters. The phosphate industry, which emits fluoride particulates, needed to improve its collection efficiency from 97 percent to 99 percent to reduce the 13,000 tons then being emitted to 4,000 tons. This required installation of secondary gas-cleaning systems and much other equipment.[23]

Dow Chemical and others saw the new rules as an opportunity to improve processes in order to reduce emissions and waste. Earle Barnes, president of Dow Chemical USA, announced that the company had, as a result of such initiatives, been able to increase the yield in its styrene-butadiene latex operation from 94 percent to 98 percent, reducing waste by 70 percent.[24] Many other companies attempted to offset the higher costs of regulation by improving their processing, thus achieving savings in raw material costs or in waste disposal, or both.

Pressure to do something about hazardous waste had been building rapidly. An EPA survey of waste management practices in 1981 had shown that most waste was managed in surface impoundments, landfills, and underground injection wells, with very small amounts recycled or incinerated. Of the approximately 71 billion gallons of hazardous waste managed in 1981, almost half were buried in surface impoundments, 8.6 billion gallons were disposed in underground injection wells, and 800 million gallons were landfilled, with only 450 million gallons incinerated and only modest amounts recycled. Studies by such bodies as the Office of Technology Assessment and the National Academy of Sciences found that technologies existed to address various types of hazardous wastes but that loopholes or weak regulatory programs provided a disincentive to use these alternatives.[25]

While previous environmental legislation largely involved meeting certain goals over time, the passage of RCRA posed a much greater challenge, since enforcement for noncompliance would result in large fines and potential criminal penalties. Complying with the "cradle-to-grave" provisions of RCRA required tracking of hazardous wastes from generation to ultimate disposal, which could greatly increase operating costs and could cause a loss of customers, while companies were also faced with a shortage of licensed disposal sites. While it had taken a decade to put teeth into the original (1967) Clean Air Act, the tremendous visibility given the waste-disposal problem put pressure on the EPA to move quickly under RCRA.[26] Companies could be liable for $25,000 for each day of violation. Persons responsible for handling hazardous wastes improperly could be held liable for fines up to $250,000 or could be imprisoned for five years.

Recognizing the difficulties faced by small firms needing to comply with RCRA provisions, larger firms would teach customers how to cope with RCRA. Dow Chemical held a number of seminars in various cities and also distributed a booklet on RCRA to about 25,000 of its (mostly smaller) customers.

Passage of RCRA brought about the rapid growth of waste-treatment firms that would take away the hazardous wastes generated by the chemical industry and dispose of them legally—or at least that was what companies hoped and expected, given their potential liability under RCRA's cradle-to-grave provisions. Such companies as Chemical Waste Management and Rollins Environmental Services generally spent $25 million for building a typical regional waste-disposal site, which would have an incinerator and chemical, physical, and biological treating facilities for neutralization, oxidation, and heavy metals separation of chemical waste.[27]

Industry and Government Slowly Come to Terms

Creation of the EPA in 1970 established a new legal basis for corporate environmental management as a matter of regulatory responsibility. A number of functions previously lodged in other departments (Departments of the Interior; Agriculture; and Health, Education and Welfare; and the Executive Office of the President) were transferred to the new agency. An early administrator, William J. Ruckelshaus, had been granted a great deal of power for enforcement. Punishing polluters became the EPA's initial primary goal, with a plethora of enforcement actions justified on political grounds by an administrator who had come out of the Justice Department. These actions, taken over the early and mid-1970s, established a punitive "command-and-control" type of relationship between government and industry. The new rules would force companies to use new pollution-control techniques, and the problem of pollution would presumably disappear. Emphasis was on "end-of-pipe" solutions (that is, dealing with the effects of pollution instead of eliminating the source) rather than on changing processes. But by the end of the 1970s the EPA, now under Douglas Costle, began to feel the weight of unrealistic objectives. Adversarial relations between government and industry had by then become the order of the day.[28]

Executives, who saw the growing burdens of the new regulations as imposing considerable hardship on their firms, felt increasingly under siege as costs rose sharply and as federal and state regulators were often placed in highly adversarial positions. No company likes to face more regulations, including some affecting virtually every aspect of its manufacturing business. Accordingly, many top managers felt they had no choice but to try to influence the course of the new regulatory climate and, in particular, to challenge what they considered nonscientific conclusions about the alleged lethal nature of chemicals they were manufacturing. Paul Orrefice of Dow was one of the most vocal critics of what he considered excessive government regulations in the environmental area as directed against the chemical industry. Dow had over the years come under severe public attack as the manufacturer of DDT (dichloro-diphenyl-trichloroethane) and of Agent Orange (a combination of the pesticides 2,4,5-T and 2,4-D), the defoliating chemical used widely during the Vietnam War, which was blamed for veterans' health problems caused by dioxins. He tried to deflect criticism of chemicals as a source of cancer by pointing out that cigarette smoking was clearly a much greater cause and that nothing was being done by the government about banning cigarettes. Orrefice's strong views about government "overregulation" were well known, as he frequently spoke on the subject, pointing out the high cost. In 1977, with after-tax profits of $556 million, Dow said it spent $186 million to comply with federal regulations, a figure 82 percent higher than two years earlier. Frustrated with the government in a number of ways, Orrefice in a speech to the Detroit Economic Club in 1978 said that the U.S. government was "in such serious decline that it could become a banana republic except that the EPA wouldn't let us produce the pesticides necessary to grow bananas successfully."[29] Several years earlier, in 1970, Ted Doan, then president of Dow, had taken a less confrontational attitude. Speaking to a group around the time of Earth Day at a National Pollution Control Conference in San Francisco, Doan advocated that regulatory agencies "get tough" with companies that pollute the environment. Questioned as to why Dow in Sarnia, Ontario, had allowed mercury from its chlorine plant to flow into the St. Clair River, Doan said "we goofed." He told the conference that Dow had been operating a pollution-control business for twelve years that was just now starting to break even.[30]

Many executives hurried to Washington to complain and try to modify a regulatory apparatus they blamed for various difficulties, from stagnant productivity to higher interest rates. The Business Roundtable, a lobbying group composed of CEOs from a large number of firms from many industries, also assaulted Congress as it considered many of the new regulations. Irving Shapiro, CEO of DuPont, led some of these actions, though his style was to operate quietly with a distaste for public controversy. Orrefice led a CMA faction that opposed anything more than a $600 million fund for hazardous waste cleanup as compared with an original $4.1 billion figure proposed by a Senate committee. Shapiro eventually broke ranks with most of his CMA colleagues and said he would support a $1.2 billion fund, which later led to a compromise figure of $1.6 billion. (The final costs were much, much greater and are still rising.)

Consideration of the passage of TSCA also generated much controversy. Most CMA members strongly opposed this legislation, which they considered much too restrictive. Vincent Gregory, CEO of Rohm and Haas, who had learned his lesson from the bis-chloromethyl ether fiasco (see earlier under "Growing Concerns"), was one of the few chemical executives to testify in Congress in favor of the act.[31]

Pollution abatement costs kept rising rapidly throughout the 1980s. The cost for "all manufacturing" rose from $8.1 billion in 1980 to $17 billion in 1990. For "chemical and allied products," costs rose from $1.8 billion to $3.9 billion, with costs related to solid waste management almost tripling.[32]

The cost would have been considerably greater if the EPA had not promulgated an air-emission trading program consisting of several parts.[33] One of these, the "offset policy," allowed firms located in "nonattainment areas" to build new facilities if they obtained emission reduction credits from other sources that more than offset their new emissions. This allowed trade-offs that often saved firms a great deal of money and also was thought by EPA officials to be helpful in getting air pollution districts into compliance with national ambient air-quality standards by 1982.[34] For example, DuPont at its Chambers Works on the Delaware River slashed by 99.9 percent the airborne emissions of volatile organics from five sources, which obviated the need to reduce by 85 percent the discharges from 205 other sources at the complex and generated estimated savings of $1 million per month. In an example of the "bubble policy," another part of the EPA program, Union Carbide at its Texas City plant would have had to install expensive equipment to cut ethylene emissions from the firm's storage facilities. Instead, the company and the EPA agreed that shutting down an old polyethylene unit would give the same reduction in ethylene emissions.

In another example Southern California Edison (SCE) was able to use the offset policy to great advantage with respect to its emission of combustion gases and particulates covered by the CAA. In the firm's territory there were many furniture makers and other small and medium-sized companies that emitted large quantities of organic solvents also covered by the CAA. Many of these firms were set to move to other states or abroad, or in any case away from Southern California's South Coast Air Quality Management District (SCAQMD), which had particularly severe smog problems that would in the future make it impossible for these firms to continue operating with large quantities of volatile solvents. SCE signed an agreement with SCAQMD that would allow the firm to continue to operate its power plants, with plans to reduce emissions in the future, if SCE could act quickly to bring about a substantial reduction in VOC emissions in its territory. Chem Systems was hired by SCE to make a detailed survey of VOC polluters in the region and to work with these firms to reduce organic solvent emissions significantly. This was done either by installing exhaust vapor collection and recycling systems or through the use of other technologies. SCE in some cases funded some of the required modifications.

Chemical and petroleum industry expenditures for environmental control kept rising steeply, although at different rates. For the chemical industry annual capital and operating costs rose steadily from under $1 billion in 1973 to $4 billion in 1988, rising to almost $7 billion by 1992 (Figure 1). Petroleum industry expenditures stayed under $2 billion until the late 1980s and then rose steeply to almost $7 billion by 1992 because of the passage of CAA amendments.[35]

As interest in solving environmental issues at the source increased, as indicated in public surveys and from the rapidly rising number of environmental articles appearing in the press, the emphasis for attacking pollution problems was shifting dramatically from end-of-pipe solutions to product and process substitutions.[36]

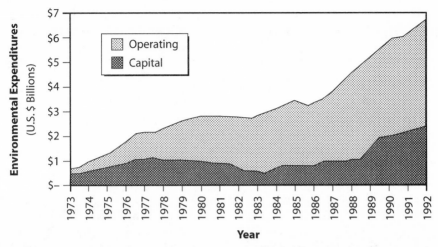

Figure 1. Environmental expenditures for the chemical industry (1973–1992). *Source:* Andrew J. Hoffman, *From Heresy to Dogma* (San Francisco: New Lexington Press, 1997), 146–147, © 2001 by the Board of Trustees of the Leland Stanford Jr. University, used with the permission of Stanford University Press, www.sup.org.

The Industry Adapts Its Product Offering

The challenges inherent in meeting the new regulations were hardly confined to effecting a series of end-of-pipe emission reductions, notifying workers and the surrounding community of the toxic nature of the chemicals being manufactured, and advising the regulatory authorities about new findings on the toxicity of existing and new chemicals. Beyond meeting all the new regulations chemical companies now needed to review their product slate in order to begin to phase out those chemicals that, rightly or wrongly, according to some scientists, had ended up on the list of hazardous and potentially carcinogenic chemicals. Companies also had to decide whether to continue to produce certain highly volatile chemicals that would find their way into the atmosphere at plants and at user locations, causing problems in meeting the VOC-related provisions of the Clean Air Act.

Regulatory actions and pressure by environmental groups had already caused companies to discontinue manufacture of a number of chlorinated chemicals, including herbicides and pesticides (and also a number of drugs, notably Thalidomide, which caused serious birth defects). Now companies began to phase out the production of a number of commodity chemicals. Notable among these were several chlorinated solvents, chlorofluorocarbons (CFCs),[37] and polychlorinated biphenyls (PCBs).

The situation with chlorinated solvents is complicated. Several of these, including carbon tetrachloride, are highly toxic and are now essentially phased out. Others, for example, trichloroethylene, are somewhat toxic and therefore no longer used for such applications as metal degreasing, but they do remain as raw materials for hydrochlorofluorocarbons (HCFCs), now used to replace CFCs. Perchloroethylene, also quite toxic, is now manufactured in much smaller quantities, but it is still used in many dry-cleaning establishments because no good substitutes are available. However, closed systems must be used to avoid emissions.

By 1987 production and use of a number of chemicals had been substantially curtailed as a result of environmental concerns. The number of producers of these chemicals was also sharply reduced.[38]

Production of highly volatile aliphatic and aromatic solvents decreased substantially as paint and adhesives producers began to replace these with other solvents or with different technologies. The need to replace organic solvents in paints and lacquers resulted in the development and rapid adoption of a series of new technologies, such as water-based or high solids paints, powder coatings, and ultraviolet- or infrared-cured radiation coatings. Although large firms applying coatings on substrates, such as sheet steel in production lines, could install systems to collect and recycle volatile solvents, they instead challenged the chemical industry and the paint formulators to develop new technologies that would mostly avoid the use of volatile solvents. Within a relatively short time water-based paints came into use for automobile top coats and for many other industrial applications. Water-based vinyl latex was already widely used for house paints, and it became the coating of choice in architectural applications. Radiation coatings found increasing use to replace the highly volatile nitrocellulose lacquers used in traditional applications and for printing on plastic bags. A number of adhesive formulations were also changed from solvent-based to water-based applications, and flexographic printing inks were changed similarly. These dramatic technological changes transformed entire industries in North America, Western Europe, and Japan. Although traditional producers of chlorinated, hydrocarbon, and oxygenated solvents lost much of a large market, the chemical industry had shown much ingenuity and inventiveness in developing rapid responses to a serious problem—the need to reduce VOC emissions.

The case for CFCs was much the same. By the mid-1970s most countries had agreed on phasing out the production and use of CFCs, as the world became aware of the "ozone hole." These chemicals, developed by Ethyl Corporation in the 1930s, are ideally suited as refrigerants and as propellants for various types of aerosol sprays or plastic foams; no perfect substitutes have been found for many of these applications. CFCs began to be phased out in the late 1980s. After the planned phaseout of CFCs under the approved Montreal protocol, development work by such firms as DuPont, ICI, Elf Atochem, and Allied Chemical, among others, resulted in the relatively rapid commercialization of HCFCs, which have lower ozone depletion potential, although they are somewhat less effective as refrigerants. Here again an entire industry was transformed over a period of several years. Later, with concerns about global warming growing and with the development of the Kyoto protocols in response, HCFCs also came under question.

Detergent builders presented another problem. The widely used phosphates were seen as promoting the rapid growth of algae in lakes and streams, which was unsightly as well as environmentally harmful. In the United States phosphates were largely replaced by other builders, such as the somewhat less effective zeolites, with manufacturers reformulating surfactants and other ingredients for laundry and dishwasher detergents to get around the problem. Low-foaming detergents were developed and substituted in washing machine and dishwater liquids and powders to avoid unsightly foams in waterways where domestic sewage was discharged. With the growing emphasis on the use of "green" products, Henkel Corporation made a big point of the fact that it produced natural rather than synthetic alcohols, namely, alcohols produced from coconut or palm kernel oil. Other firms, such as Procter and Gamble, had long used natural alcohols as well but did not worry about the fact that ethylene-based alcohols, which the company also used, were not "green." Henkel also started producing an "all green" detergent alkyl polyglucoside (APG) that substituted starch for ethylene oxide in formulating a nonionic detergent.

Some companies decided to determine whether one of their products was "greener" than a product marketed by a competitor and would therefore have greater appeal to one segment of the population. As discussed earlier in this chapter, life-cycle analysis was in vogue in the 1970s and 1980s as a technique for calculating the environmental costs of manufacturing specific products and to compare the environmental costs of competing products from cradle to grave. Several studies appeared in public forums that involved comparisons of the environmental cost of using cloth diapers with that of disposable diapers. Both had undesirable aspects, since using cloth diapers involved detergent production and disposal and energy for heating water, while disposable diapers end up as nonbiodegradable material in landfill. The LCA for both products involved a number of obvious and not-so-obvious environmentally harmful steps—for example, the atmospheric pollution caused by burning felled Malaysian or Indonesian palm trees that provide the palm kernel oil subsequently converted to detergent alcohols. Environmental pollution and energy use by the diesel engine trucks used to haul this or that raw material, intermediate, final product, or waste was common to both alternatives. The disposable diapers appeared to be the winners in this contest. But the results of such studies were not always considered conclusive because the study sponsor usually had an ax to grind and was always able to shape the comparison or bury the study if it gave the undesired results. Another competitive LCA looked at the environmental costs of ethylene glycol compared with those of propylene glycol as an antifreeze material. Union Carbide and Arco Chemical were on opposite sides of this issue.

Franklin Associates was a leading consultant hired to perform studies of this type. One of their studies reaffirmed the fact that plastics manufacture involved less energy use than use of natural material, such as wood or paper, in packaging.[39]

The need to meet CAA provisions resulted in the commercialization and subsequent explosive growth of methyl tertiary butyl ether (MTBE), an oxygenate made from methanol and isobutylene. Tests had shown that gasoline containing certain oxygenates burned cleaner, thus generating lower amounts of carbon monoxide and nitrogen oxides in automobile combustion products. While ethyl alcohol was also useful as an additive, MTBE had a higher octane rating and was generally more effective (although in corn belt states grain-based alcohol has been used instead of MTBE owing to a large federal subsidy). Over a few years MTBE, which was either produced directly or as a derivative of a coproduct in some propylene oxide plants, became one of the highest-volume chemicals produced in the United States. Recently, MTBE has come under pressure as a toxic chemical that has found its way into the drinking water supply owing to leakage from a number of gasoline storage tanks. California decided to phase out MTBE by 31 December 2002, and New York and several other states have similar plans for 2002 to 2004. MTBE's future is now unclear, with ethyl alcohol and other oxygenates likely to be increasingly used as gasoline additives to achieve cleaner burning conditions. Much more ethanol capacity is slated to come on stream over the next three years.

In the 1960s the phaseout of tetraethyl lead (TEL) from gasoline had already begun to affect the chemical industry. Lead had been identified as a harmful pollutant, and so the U.S. government, well before establishment of the EPA, had mandated the gradual removal of TEL as a gasoline additive. Under the aegis of the CAA, there arose an additional reason for removing TEL: it would foul the catalytic converters (for carbon monoxide and unburned hydrocarbons) that were being installed on the exhaust systems of new cars coming off the assembly lines. Refiners were able to replace TEL with higher-octane aromatic and aliphatic cuts as well as by adding oxygenates with high-octane blending properties. Ethylene dibromide, which had been used as a carrier for TEL, was phased out in conjunction with the removal of TEL.

Environmental issues also kept certain products from ever coming to market. Monsanto Chemical was involved in two of these cases. The firm had developed a synthetic sweetener, cyclamate, to compete with saccharin, but tests showed that the substitute in massive quantities produced tumors in test animals, and so it never became commercial. Monsanto was also developing a clear bottle resin called Cycle-safe to replace glass in soft drink bottles and in other applications. However, tests showed that the product contained small amounts of residual acrylonitrile, which was on the EPA's list of hazardous chemicals. At the time Cycle-safe was competing with polyethylene terephthalate (PET) for similar applications and had many of the same qualities required for a recyclable, clear bottle resin. When Monsanto could not solve the problem, Cycle-safe was withdrawn from commercial consideration.[40] PET then became the clear winner and went on to become a multibillion-dollar product worldwide.

Plastics were involved in another case. In the early 1990s environmentalists attacked the famous McDonalds' "clamshell"—used to serve hamburgers "to go"—as being environmentally harmful, since they were made of foamed polystyrene and were not biodegradable in the landfills where they ended up. McDonalds, anxious to have a "green" image, quickly gave in, eventually switching to a "quilt-wrap" container composed partly of coated cardboard (which was also largely nonbiodegradable). A study by Franklin Associates showed that the energy consumption inherent in manufacturing the new container was considerably less than that for foamed polystyrene, thus further buttressing McDonalds' position. A number of communities followed by mandating the phaseout of polystyrene for a number of other uses, such as home insulation. Soon polystyrene threatened to become a "non-green" product, and doomsday forecasts were made for future demand of this material and its precursor, styrene. But it became apparent that polystyrene was no different than a number of other plastics and that landfill disposal, recycling, or incineration was always going to be required if the world were to continue using plastics. The McDonalds case became notorious mainly because of the high visibility of the company and its products. A few years later, uses for polystyrene began increasing again. In Europe and Japan, where land is at a premium, incineration is much more widely used than in the United States, where the "NIMBY" ("not in my backyard") syndrome has severely limited the construction of incinerators, which represent an environmentally excellent method for plastics and wastepaper disposal.

While not a new product, sodium chlorate became a major commodity chemical because pulp mills were required to change their bleaching technique, which had resulted in discharging dioxin-containing wastes into adjacent waterways. There were also some issues regarding dioxin residuals remaining in paper products. Instead of using elemental chlorine bleaching, many of these mills switched to sodium chlorate–generated chlorine dioxide as the bleaching agent, at times using hydrogen peroxide as a second bleaching chemical. In Europe some mills totally eliminated all uses of chlorine compounds in any form.

An interesting consequence of the new regulations requiring environmental impact statements for new plants, as well as community notification, was the fact that it became extremely difficult for firms to build new plants for processes that are potentially highly polluting. Potential competitors were therefore unable to consider building such plants at new locations in industrial countries. This created new entry barriers for such processes, in that sense favoring existing producers who had thereby been "grandfathered" at their existing locations. Two processes of this type involved high-temperature thermal chlorinations, such as those used to make an epoxy intermediate (epichlorohydrin), and phosgenations, used in the manufacture of isocyanates and other chemicals.

Furthermore, while the OECD and some of the developing countries imposed serious restrictions on the operations and waste disposal practices of chemical firms, some other countries either did not take environmental responsibilities as seriously or were able to offer sites far away from populated areas for new chemical plant construction. Saudi Arabia, which has rapidly built up a world-scale chemical industry, has not hesitated in offering sites for the construction of certain types of plants that would be difficult to construct on a "green field" basis (that is, on a brand-new site) in compliance with environmental regulations in OECD countries.

Doing the Right Thing: Responsible Care

The chemical industry has for decades now suffered from a poor image compared with that of other industries. A number of negatives are associated with a poor image, such as difficulties in hiring the most competent scientists and engineers, ever-increasing regulatory burdens, and becoming open to attacks by environmental groups that want to ban the production of chemicals even if no reliable scientific evidence supports such a ban. For example, some groups want to eliminate the production of chlorine and chlorine derivatives, many of which are absolutely essential to daily life and often have no good substitutes. The industry finds it difficult to defend itself against the charges of environmental extremists, which tend to be sensational and frequently unsubstantiated by proper science. Leading the fight for a more favorable industry image is the industry association now known as the American Chemistry Council (ACC), which has conducted a number of campaigns to educate the public about the industry and to respond to what it considers irresponsible charges brought by extreme factions. An example of this was the Greenpeace proposal on phaseout of chlorine that was presented to the U.S.–Canadian International Joint Commission considering policies affecting the Great Lakes region. Such an action is patently ridiculous. While a number of chlorine compounds are now banned or being phased out in favor of substitutes, it is considered almost unthinkable to ban the use of chlorine as a drinking water disinfectant, which protects the population from ingesting various harmful pathogens found in untreated surface water. (Even when ozone is used as a primary treating agent, chlorine is added downstream to prevent bacterial regrowth in the pipes.) PVC has been specifically targeted by environmental groups as hazardous owing to alleged dangerous amounts of free vinyl chloride monomer present in the plastic and to the possible formation of dioxins during incineration of discarded

PVC pipe. Here again proper science says that PVC pipe can be used at no risk. The Chlorine Institute, which represents domestic chloralkali producers, spends much of its time dealing with attacks on chlorine in its various forms by certain environmental groups.

The Manufacturing Chemists Association, a predecessor to ACC, was founded in 1872, initially to provide a forum for chemical executives to share their views. It was renamed the Chemical Manufacturers Association (CMA) in 1978, when it had already become an important force for lobbying the federal government on behalf of the association's chemical producers in areas ranging from toxicology and epidemiology to environmental affairs. Consisting of almost two hundred members, it represented about 90 percent of U.S. chemical-producing capacity and also included as members U.S. subsidiaries of many multinational firms.

The ACC holds annual meetings at the Greenbriar Hotel in White Sulphur Springs, West Virginia, every June where executive teams from member companies attend programs to discuss industry issues during the formal program, with discussions often continuing on three excellent golf courses belonging to the hotel. It is thought that what became the Responsible Care program effectively had its origins at the 1983 meeting when Bill Simeral, the executive vice president of DuPont, gave his farewell address. First congratulating the industry for its success in achieving a highly favorable trade balance in chemicals and for its technological achievements, he then said it was time the industry did something about its poor image with the public. He said that the public saw newspaper pictures of leaking drums oozing into dump sites and other unfavorable visual material. Drawing parallels to the steel and auto industries, which had recently lost much public support, he urged members to respond to public concerns and to take a more constructive attitude toward federal regulation.[41]

Shortly thereafter, Superfund came up for authorization. Louis Fernandez of Monsanto, then chairman of the CMA, urged the association to support this legislation—even though chemical companies were only part of the nine thousand companies in thirty-three industries responsible for the identified hazardous waste sites. Fernandez testified in favor of the legislation and offered technical expert support from the CMA.

In late 1984 came the disaster at Bhopal. Ed Holmer, president of Exxon Chemical, who had just become chairman of the CMA, immediately said that all chemical makers had to respond to this tragedy and review their own situation. A survey sponsored by the CMA shortly thereafter revealed that although 95 percent of CMA members had emergency response plans in place, almost all were confined to what went on inside the plants, with little coordination with the outside: at Bhopal almost all the fatalities were people from the surrounding community.

As a result the CMA undertook its first effort to develop a program that would provide safety in local communities in the event of a chemical plant disaster. Called Community Awareness and Emergency Response (CAER), this program was intended to put potentially damaging information before the public before extreme environmental groups could incite public outrage and to persuade the EPA that the chemical industry wanted to be proactive. The program established a mechanism to engage community residents in constructive dialogue on ways to address community issues effectively and improve emergency preparedness. (The EPA undertook a parallel effort called the Chemical Emergency Preparedness Program. When Title III of the Superfund Amendment and Reauthorization Act [SARA] was drafted, legislators adopted whole segments of CAER almost word for word.)

Unfortunately, in August 1985 Union Carbide had another toxic chemical release at its plant near Charleston, West Virginia. Coming just at the time of the CAER launch, this event created a real credibility problem for an industry already severely shaken by Bhopal and now facing another case where the public could point fingers and wonder whether managers were really on top of the situation. The industry "found itself awash in a sea of mistrust and misunderstanding."[42] An editorial in the *Pittsburgh Post Gazette* served to further inflame feelings about the industry. When the CMA claimed that two billion tons of chemicals were hauled around the country with 99.99 percent causing no incidents, the newspaper jumped on this statistic and said that there had therefore been accidents involving 200,000 tons of toxic chemicals, so why should the country be impressed and grateful?[43]

Throughout 1986 the CMA wrestled with how the industry could change the public's negative perception of it and the poor effects this perception was having on its business. But members soon realized that they did not just have a public relations problem. They also had a performance problem, and the industry could not just advertise its way out of the situation. In late 1986 fire destroyed a chemicals storage facility belonging to Sandoz in a town near Basel, Switzerland, resulting in massive spillage of toxic chemicals into the Rhine, which flows up through Germany and the Netherlands into the North Sea. The subsequent investigation discovered that most of the major chemical companies located along the Rhine were dumping into the river, in some cases accidentally and in others as permitted under current regulations. Massive fish kills and other damage to the ecosystem were reported as a result largely of the Sandoz accident. Some assessments predicted that it would take the Rhine ten years to recover.

Around this time the CMA saw the Canadian Chemical Producers Association (CCPA) put into effect an initiative that went beyond CAER to include essentially all the activities of the chemical industry: the association called this initiative "Responsible Care," being an umbrella program to promote new ways of managing the industry's manufacturing, distribution, hazardous waste management, and other functions. An important difference between this initiative and CAER was that Responsible Care was not a voluntary program: members formally accepted its principles and made it a condition of membership.

Because Norman Kissick, then chairman of the CCPA, was also the head of Union Carbide's Canadian operations, Robert Kennedy, then CEO of Union Carbide, became aware of the program and asked the CMA to consider how such a program might fit the U.S. chemical industry. Within a relatively short period the CMA members voted in favor of making participation in the Responsible Care initiative a mandatory part of CMA membership. Over the next several years a number of codes were developed and adopted, including the existing codes for Community Awareness and Emergency Response; Pollution Prevention; Distribution, Process Safety; Employee Health and Safety; and Product Stewardship. Also established were ten guiding principles acknowledging community concerns, need for safety, environmental sensitivity, and participation with government agencies in creating responsible laws and regulations.

Many of the smaller firms had problems with the provisions of some of the codes because of lack of managers and required infrastructure. Nevertheless, CMA members large and small were willing to comply with Responsible Care, given the industry's image problems with the public.

Recognizing the global nature of the issue and that many U.S. firms have subsidiaries in Europe, Asia, and Latin America, the CMA set out to preach the virtues and principles of Responsible Care to foreign countries and companies. In Mexico, which at the time was hoping for rapid implementation of the North American Free Trade Agreement (NAFTA), chemical firms signed on to the equivalent Responsibilidad Integral program in 1991. The Conseil Européen des Fédérations de l'Industrie Chimique (CEFIC), the equivalent of the CMA in Europe, took the lead there, while the European Environmental Council provided incentives to local industry by suggesting that European Union legislation should place more emphasis on voluntary initiatives. CEFIC also developed a product stewardship code, which most members embraced. In Germany the Verband der Chemischen Industrie held its first Responsible Care workshop in November 1993, while in Spain the Federacio Empresarial de la Industria Quimica Española stressed the potential economic benefits of Responsible Care.[44]

It is difficult and certainly beyond the scope of this book to assess the success of the Responsible Care program. Statistics on industry toxic-substance releases have certainly shown considerable improvement, but it is impossible to determine how much of this improvement has been caused by new regulations and how much by the voluntary efforts of companies who joined Responsible Care.

Table 1. Target TRI Chemicals

Organic Chemicals	Inorganic Chemicals
Benzene	Cadmium and cadmium compounds
Carbon tetrachloride	Chromium and chromium compounds
Chloroform	Cyanide compounds
Methyl ethyl ketone	Lead and lead compounds
Methyl isobutyl ketone	Mercury and mercury compounds
Tetrachloroethylene	Nickel and nickel compounds
Toluene	
1,1,1,-Trichloroethylene	
Trichloroethylene	
Xylenes	

Source: Environmental Protection Agency.

A voluntary program, strongly endorsed by the CMA, was the so-called 33-50 release reduction program begun by the EPA in February 1991. It was aimed at the voluntary reduction of a number of the most hazardous chemicals under the EPA's Toxic Release Inventory (TRI), most of which went into the atmosphere (Table 1). The program's goal was a 33 percent reduction by 1992 and a 50 percent reduction by 1995. These figures would be achieved by a number of steps, including material substitution, treatment and reuse, recovery and recycling, and process changes.

Often, the CMA and other industry groups challenged what they considered unwarranted new standards that government agencies, such as the EPA and OSHA (the name derived from the Occupational Safety and Health Act), came up with to provide the safest possible atmosphere for plant workers or the public. Such challenges were generally based on the fact that the agencies had not relied on the "best available science" (a phrase often used by the EPA), but had arbitrarily set such standards, given that in many cases little applicable science was available. While TSCA had a cost-benefit provision, most of the other regulatory acts did not, so that the imposition of extremely low emission or exposure standards would require immense expenditures by industry. Some of these cases went as high as the Supreme Court for a decision, which could go either way. Thus, in 1980, the Supreme Court struck down a proposal by OSHA to reduce the permissible benzene exposure limit for workers from ten parts per million to one part per million on the basis that there was no evidence that exposure to benzene was harmful below a certain threshold. The proposed OSHA ruling had been challenged by the CMA, the American Petroleum Institute, other industry organizations, and such firms as Uniroyal, DuPont, and Exxon.[45]

Statistics covering key pollution indicators and TRI releases show some progress. Between 1985 and 1996 chemical and allied products achieved substantial reductions in emission of carbon monoxide and lesser reductions in nitrogen oxides, VOCs, and sulfur oxides (Figure 2). The fact that the chemical industry is actually a small factor in VOC emissions is shown by the statistic that in 1998 only 2 percent of such emissions was caused by the industry, with most of the balance caused by other industries and users. TRI releases by the chemical industry dropped 55 percent between 1988 and 1996, with the largest reduction achieved in releases to surface water (Table 2).[46]

For the U.S. chemical industry, including pharmaceuticals and fertilizers, TRI emissions since 1988 have been reduced 63 percent, while production has increased 27 percent.[47] An EPA report issued in 2000 indicated that of the 7.3 billion pounds of 1998 TRI releases, metal mining was responsible for 48 percent, electric utilities for 15 percent, and chemicals for 10 percent, with the balance from other sources.[48]

Other chemical industry accomplishments include the fact that process safety has improved by 14 percent since 1996 and chemical transportation safety by 23 percent since 1995. ACC members have established over three hundred local plant community advisory panels throughout the United States.

The Responsible Care program had less success in convincing the public that the industry was taking its responsibilities more seriously. Thus, in 1994 three polls, including one sponsored by the CMA, showed that the chemical industry still had one of the least favorable public ratings—better than the tobacco industry but poorer than nuclear power and much poorer than the oil, automobile, and paper industries. Interestingly, the plastics industry had a

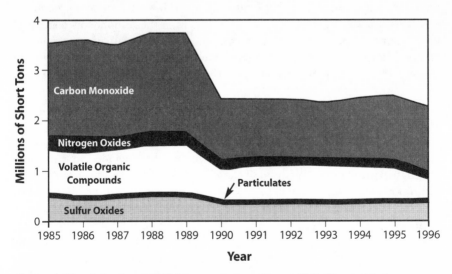

Figure 2. Trends in chemical industry emissions. *Source: U.S. Chemical Industry Statistical Handbook 1998* (Arlington, Va.: Chemical Manufacturers Association, 1998), 89.

Table 2. Toxic Release Inventory: U.S. Chemical Manufacturing (SIC Code 28) (in Millions of Pounds)

Location of Release	1988	1994	1995	1996
Total releases*	**822**	**400**	**379**	**370**
Total air	585	301	288	269
Surface water	140	27	25	32
Land	97	72	66	69
Underground injections	159	112	137	116

Source: U.S. Chemical Industry Statistical Handbook 1998 (Arlington, Va.: Chemical Manufacturers Association, 1998), 89.
* The Chemical Manufacturers' Association defines "total releases" as "total air + surface water + land." Underground injections are not included in releases to the environment.

rating twice as high as the chemical industry. The public also continues to exhibit great concern about living near a chemical plant.

The industry continues to have image difficulties, although the situation is probably getting a little better, particularly as other worries occupy the public. A Ryesta poll carried out in 2000 for CEFIC and the German Verein der Chemischen Industrie asked respondents about how much they felt the chemical industry was doing to try to reduce any harmful effects its activities might have on the environment. Respondents answering "very little" included 51 percent in France, 39 percent in the United States, and 31 percent in Germany. In that country during the mid- to late 1980s concern about environmental issues was mentioned by over 60 percent of the respondents compared with around 40 percent about employment issues. The situation completely reversed in the late 1990s, with environmental concerns at the 15 percent to 20 percent level and employment concerns heading the list at 70 percent to 80 percent.[49]

A study published in 2000 looked at Responsible Care from the standpoint of the effectiveness of an industry's self-regulation program.[50] The article acknowledged the progress the industry was making in reducing emissions, but it was ambivalent about the results of the program. It concluded first that the program, which operates without explicit sanctions for malfeasance, included a disproportionate number of poor performers and second that members do not improve faster than nonmembers. The article, which compared the chemical industry to the fishing industry, said that while a community of fishermen affect their own welfare by overfishing, "excessive polluting by chemical firms influences members' welfare only to the extent that it influences the reputational commons."[51] The article goes on to quote a reference to the effect that "Responsible Care has been disappointing and that a velvet glove may not be enough to change behavior."[52] This seems unduly harsh, given the achievements of the industry in meeting voluntary targets, such as reductions in TRI releases. *Chemical Week* magazine in July 2000 challenged the conclusions of the King study, citing the methodology used, which relates TRI data to the number of employees of the firm, and King later acknowledged this flaw in the study.[53]

Separately from the various initiatives taken in connection with Responsible Care, companies also sometimes found it useful to work directly with environmental groups, as well as with the EPA, to find beneficial solutions to environmental problems. In 1989 a group of engineers from Standard Oil Company (in Indiana) invited the EPA to work closely with Amoco engineers to put the company in compliance with benzene emissions at a refinery site. While the focus initially was on the refinery itself, where the reduction of relatively limited benzene emissions would have involved a substantial investment, the group found that much more benzene was being emitted from the process of loading gasoline barges linked to refinery operations. It was found that benzene emissions from barge loading could be substantially reduced by using a special two-nozzle hose. This involved a $6 million total expenditure as opposed to an estimated $41 million for reducing much smaller benzene emissions at the refinery itself. While initially the two groups did not trust each other, they soon found that they could work well together with notable results.[54]

As another example, in April 1999, Dow and the Natural Resources Defense Council recruited five Michigan environmental activists to take part in the Michigan Source Reduction Initiative project, begun somewhat earlier. Partly as a result of this cooperation Dow's Midland site was on track to cut generation of a specific list of wastes by 37 percent and to reduce emissions to the air and water by 43 percent. Dow's investment of $3.1 million was expected to save the company $5.4 million per year.[55]

Environmental Costs and Liabilities

Companies and their shareholders became sensitized to the fact that environmental costs and liabilities would have a pronounced effect on income statements and balance sheets. Capital charges were heavily affected by environmental outlays, sometimes amounting to half the total annual capital expenditures. Environmentally oriented operating costs also became significant as companies took all the steps mandated for regulation compliance and in many cases changed their operations to become safer and less polluting. Costs under CERCLA included a Superfund charge that came out of production costs.[56]

As an example, Dow in its 2000 annual report stated that of the $1.3 billion capital expenditures in that year, 15 percent, or $195 million, was committed to projects related to environmental protection, safety, loss prevention, and industrial hygiene.[57] DuPont in 2000 spent $140 million on environmental capital projects. It also stated that environmental expenses in 2000 were $550 million, or slightly more than 2 percent of the total operating costs.[58]

Environmental liabilities must be carried on the balance sheet. These figures generally refer to estimated future costs for environmental remediation of current or former sites. In 2000 Dow listed $325 million in such liabilities, while DuPont listed $408 million. In 1998 FMC Corporation listed $294 million in such liabilities compared with $280.6 million in reserves against these liabilities. The costs associated with cleanup of discontinued operations can be very high. In the same report FMC anticipated that cleanup work at a discontinued fiber manufacturing plant in Virginia, as agreed on with the EPA and the Justice Department, would cost $70 million.

Lawsuits brought by former workers can also be extremely expensive, as companies involved in asbestos manufacture found out. These costs have forced several companies, including Johns Manville, Babcock and Wilson, Owens Corning, Armstrong Tile, and most recently W. R. Grace, into bankruptcy. Although these companies were largely covered by insurance, it proved to be insufficient, as some lawyers exerted themselves to find more and more potential claimants. In 2001 about forty thousand new cases were filed, long after the companies being sued had stopped making or using asbestos. Included in the new claimants were children of workers exposed decades ago who brought asbestos dust into their homes. Medical opinion suggests that these children will never suffer any ill health effects. The remaining asbestos liability may be as high as $210 billion.[59] Chapter 11 bankruptcy procedures allow companies to maintain operations, with payments to claimants made from a trust established from most of the affected company's net worth. The companies can usually emerge from bankruptcy some years later under new ownership, the original equity having been wiped out.

Class-action lawsuits have become a common feature in cases where people were exposed to hazardous chemicals or involved in accidental chemical releases. Examples include asbestos, vinyl chloride, and a number of other cases involving chemical releases or dumping in landfills.

In transactions the issue of responsibility for future environmental liabilities became overriding. Potential deals were often overshadowed by environmental liability issues, to the point where such transactions were canceled because the parties could not agree. When companies decide that they should divest one of their businesses, a key decision is whether to sell the business to another firm or whether it should be spun off as a separate entity. From the standpoint of environmental liability the easiest thing to do is to spin the division off to the same shareholders who hold shares in the parent company. Thus, the shareholders of the two new companies face the same financial liability situation they had faced before the company was split into two parts. New shareholders would be informed of the respective liabilities of the two now-separate entities by reading the Securities Exchange Commission registration statements. Such spinoffs created Eastman Chemical from Eastman Kodak, Albemarle Chemical from Ethyl Corporation, Praxair from Union Carbide, and Cytec from American Cyanamid.

Environmental costs and liabilities can take another form. In several instances companies were eventually forced to withdraw a product owing to extreme adverse publicity, even when scientific facts that eventually came to light may not have supported such an action. One of the best-known cases involved a chemical ripening agent for apples, diaminozide, trade named Alar, made by Uniroyal Corporation. In 1989 the EPA announced that this chemical and its by-product, UDMH (unsymmetrical dimethyl hydrazine), could cause cancer in five of a hundred thousand persons exposed over a seventy-year period. However, the agency did not suspend use of Alar, making the announcement "in advance of suspension." While the chemical was used on only 5 percent of red apples grown in the United States, getting it off the market was not going to be easy since the chemical was almost impossible to detect. Predictably, environmental groups and the media jumped on this case. CBS aired a *60 Minutes* segment in which Alar was characterized as "the most potent cancer-causing agent in the food supply today." Shortly thereafter, the American Council on Safety and Health characterized the EPA's findings as "voodoo statistics" saying that "no evidence (has been found) of even one case of human cancer in children and adults linked to exposure to the minute pesticide residue in food."[60] A United Nations panel, which included members from the World Health Organization and the Food and Agriculture Organization, then found that Alar and its by-product "raised no special concerns regarding carcinogenicity." But for a while consumption of all apples dropped significantly: sales in the $1.1 billion-a-year apple industry were seriously affected, with an estimated $125 million of income lost. The EPA was never able to make a sustainable case against this chemical and in 1991 stated that its earlier conclusions on Alar were based on doses that were 266,000 times the normal ingestion.[61] Alar had by then been withdrawn by Uniroyal, and only in 1992 did apple sales recover to historical levels. The ironic thing is that if Alar had been a new chemical, rather than one already in use, TSCA would have put the burden of proof on the manufacturer. As it was, it rested on the EPA.

Environmental Protection in Europe and Japan

The preceding discussion concentrated mostly on environmental issues affecting the U.S. chemical industry. Because of U.S. dominance in so many aspects of chemical and petroleum processing, this emphasis seems reasonable. North American (including Canadian) environmental policies and regulatory structures often serve as models for other nations. However, while chemical industry ownership is increasingly multinational, the plants and consumer markets that compose it are local, so that the contrasts in environmental, safety, and health regulations and market preferences in each region must be understood.

While "environmental protection" is largely focused on issues of science and technology, it also relates to concerns over economics, politics, legal jurisdiction, societal values, and even ethical-religious views. Since, aside from the basic science and technology, the other aspects can differ widely in the United States, Europe, and Japan, environmental protection plays out differently among these venues. Even the role of science differs among the three. Clashes between science and these other aspects prevail, with many enigmas and confused priorities abounding in the playoff within each venue. As in any arena affected by both science and politics, these dynamics constantly change.

The notion of environmental protection (formerly called "conservation") actually encompasses three distinct objectives that often coincide or even compete: human health and welfare; ecological health and integrity; and preservation of resources, including recreational and aesthetic values. When the two objectives of mitigating risk to human health and preserving the natural ecology are in peril, it is relatively easy for the stakeholders to agree on the problem and its solutions. But sometimes the objectives are in conflict, as occurred with DDT, which was a boon to human welfare owing to larger crops and less infestation but was considered a disaster for the ecology.

To understand the differences in approaches to environmental protection, some of the differences in U.S., European, and Japanese cultures must be considered. The most important aspects of American culture and history are

- The sheer size, resources, and prosperity of the United States and Canada;
- The "endless frontier" phenomenon, in which people simply move if their surroundings become too crowded or polluted;
- Frontier and revolutionary experiences leading to a strong emphasis on individual property rights and independence (also extending to corporate entities) in culture and law;
- Layers of federal, state, and local jurisdiction, each jealous of their own priorities and prerogatives;
- A highly litigious social and business environment;
- A strong system of legislative-administrative-judicial checks and balances; and
- Standards for technology, commerce, and industry, mostly set by the private sector, or if by government through the political process and capable of being challenged in the courts.

The history of Europe in contrast has been one of

- Feudal (later oligarchical) ownership and management of land and later of major businesses and industries;

- More recently class conflict and socialistic governments;
- Careful husbanding of limited land and other natural resources (earlier, along with exploitation of distant colonies);
- Close, densely populated, but well-preserved communities and infrastructure; and
- Relatively authoritarian approaches to standard setting.

Japan historically has been

- A crowded country with very limited natural resources;
- A hierarchical, authoritarian, largely closed, but racially homogeneous and stable society, with a feudal history somewhat similar to Europe's;
- A society where communal responsibility is driven by social pressures and a sense of social responsibility rather than litigation;
- A society with highly cooperative and coordinated social, industrial, and commercial institutions; and
- A society with highly developed but compartmentalized aesthetic sensibilities.

In spite of international initiatives for nations and stakeholders to work together, these deep historical, cultural, and political differences continue to affect how environmental protection plays out in the three regions.

The different responses to genetically modified foods that have confounded biotech food developers provide a good example. Americans on the whole have been complacent about or even eagerly accepting of the introduction of biotech crops. One reason is that Americans appear to trust "the system" (government agencies) to oversee the safety of the food supply more than Europeans do. Further, with much natural wilderness left, agriculture is seen more as part of industry, whereas in Europe the open land of farms and fields is essentially the only natural world Europeans have left. Most Europeans are also far more sensitive to the authenticity and integrity of their food supply and to the artisanal quality of their food than most Americans. The Japanese are somewhere in between, willing to accept a great deal of artifice in their food and culture, yet they are steeped in a ubiquitous, aesthetic tradition that values nature in a special way (the state religion has a pantheistic quality).

In the United States broad scientific and societal value considerations, driven or limited by public pressures and special interest tradeoffs, have been embodied in laws passed by Congress. The laws set policies and objectives and sometimes promulgate lists of chemicals or other specific technical parameters for embodiment by the EPA in detailed regulations and rules. The process of developing these rules must today consider competing scientific data, economics, benchmark performances, community values, stakeholder comments, and ethnic justice. The regulations and rules may either directly apply to other government departments, industries, commercial entities, states or other jurisdictions, or to individuals, or they may require or affect rule making or permitting by other entities. Eventually, any or several of these together may sue the EPA or other entities to change or overturn the regulations or rules or force their more rigorous or even vigorous application. Some laws and regulations carry criminal penalties and in enforcing them through the courts, they may be strengthened, weakened, or reinterpreted.

The major differences between the U.S., European, and Japanese systems lie in the tendencies of the latter two to avoid much of this messy process through top-down imposition of rules (whether democratically or bureaucratically arrived at), usually involving much more direct negotiation than in the United States. In these negotiations technocrats generally have much more of a role than in the States, where lobbyists for business, environmental groups, and state and local interests compete to influence how the laws and regulations are developed and promulgated. With the European Union emerging as a united nation of sorts, integrated environmental protection is a key aspect of this new society and economy.

What the Future Holds

In March 2001 the Public Broadcasting Service ran a documentary titled *Trade Secrets: A Moyers Report*. This report presented evidence alleging that the chemical industry has a pattern of withholding evidence from plant workers and from the public about the dangers of toxic chemicals. Focusing on vinyl chloride, among other chemicals, which in the early 1960s was found to produce liver and brain cancer in plant workers, the documentary recounted the fatal battle for life of an employee who had a number of decades ago been exposed to that chemical in a petrochemical plant in Lake Charles, Louisiana.[62]

During the subsequent discussion several experts sharply disagreed about the role that the vinyl chloride industry and specifically certain producers had taken once the dangers of vinyl chloride had been established. It was stated that as soon as the dangers of vinyl chloride exposure had been understood, BFGoodrich, one of the main producers, had made the information available for an article published in the *Journal of the American Medical Association (JAMA)*. Terry Yosie, a spokesman for the American Chemistry Council, forcefully pointed out that the chemical industry has for decades been extremely responsive to the problem of identifying toxic chemicals and of making this knowledge available to the public, acknowledging, however, that this had not been the case in earlier times. But a spokesman for an environmental organization typically took the opposite side, saying that the industry continues to make toxic chemicals and uses "obsolete manufacturing processes, such as the manufacture of PVC, when (it) knows better." "For instance," this representative continued, "Dow Chemical's joint venture with Cargill will soon produce plastics from corn."[63] A medical spokesman also criticized BFGoodrich for allegedly not telling their workers soon enough about the hazards of vinyl chloride exposure, presumably expecting them to read the article in *JAMA*. It did not come out in the discussion whether and when the company did inform its workers, perhaps because nobody on the panel had this information.[64]

In a subsequent editorial in *Chemical Week*, a magazine usually friendly to the industry, editor David Hunter was also somewhat critical of the industry. He mentioned that the documentary had made the point that the industry had originally strongly opposed TSCA, Proposition 65, and other right-to-know initiatives. Then he went on to say that "the industry may have a good story to tell about its products being essential to modern life, but it does not have a good record in chemical testing and health effects research." This was a reference to a finding that Bill Moyers's blood was (perhaps typically) found to contain eighty-four synthetic chemicals, including dioxins and PCBs. Hunter's editorial urged the industry to show more progress in its testing programs and with epidemiological studies of the industry's workforce.[65]

Moyers's choice to make this documentary, which received wide publicity, is symptomatic of the fact that the industry still has a long way to go to develop a favorable image. Unquestionably, industry members and the ACC (CMA) were on record as opposing some of the strictures of the many regulations enacted in the 1970s and 1980s, as covered in this chapter. Attitudes then softened considerably, as such initiatives as CAER and Responsible Care were almost unanimously adopted. In April 2001 the Bush administration's EPA, headed by former New Jersey governor Christine Whitman, signed a Clinton-era treaty calling for the worldwide phaseout of a dozen highly toxic persistent organic pollutants, including PCBs, dioxins, furans, DDT, and other pesticides, most of which are no longer used in the industrial countries but are still used in developing nations.

But the banning of PCBs will not abate the problems associated with this chemical. Bad publicity for the industry will continue for some time, for example, as General Electric and the EPA square off regarding cleanup of PCB residues in the mud under a forty-mile stretch of the Hudson River north of Albany, New York, near two GE plants that used this chemical in its transformer business. The EPA wants GE to spend $460 million to dredge the river to remove an estimated 1.3 million pounds of historically dumped PCBs, which both sides agree have long been poisoning the fish population. However, GE contends that dredging cannot be done effectively and that doing so will cause far more harm by dispersing PCBs more widely than before. The EPA obviously disagrees.[66] Nobody can predict where this controversy will end, but the bad publicity arising from this highly public battle cannot help but hurt the chemical industry as a whole, in a sense "dredging up" an old problem (the harmful nature of PCBs, which the industry stopped manufacturing long ago).

Chemical plants are still a source of pollution—although a smaller source—and accidents will unfortunately continue to occur, again raising unfavorable publicity. Continuing concern about toxicity and the possible unknown hazards of a number of the thousands of chemicals in manufacture today will unquestionably continue. This concern may well lead to more legislation.

Recently, the European Commission issued a white paper calling for major overhaul of chemical policy and legislation, including an evaluation and registration scheme and use of the precautionary principle to ban certain previously identified chemicals "of very high concern."[67] The proposal would require registration, testing, and authorization for thirty thousand existing and new industrial chemicals at an estimated cost of $80,000 to $300,000 per substance.[68] The final version of such legislation remains to be enacted, and the chemical industry has already reacted strongly against this initiative, partly because it is itself carrying out testing efforts voluntarily.

The chemical industry will clearly continue to have to face challenges to its operations and doubts about its willingness to provide as much information as some sections of the public and perhaps government agencies want to have. While the Responsible Care program stands for a lot of the right things and the industry has undertaken a number of worthwhile initiatives to gain public trust, convincing everyone that chemicals identified as toxic substances will not cause harm will probably always be difficult. And accidents and spills will occur from time to time, reminding people about the hazards inherent in chemical manufacturing processes. Such organizations as Greenpeace will continue to hound the industry, most recently by posting on its Web site a Louisiana "worst-case accident scenario" that might develop along the lower Mississippi River, where many chemical plants are located, with distances that hazardous emissions would conceivably travel. Greenpeace uses this extremely unlikely scenario, which was actually developed by the industry itself to comply with regulatory procedures, to claim that "safer manufacturing alternatives" are required.[69] So while almost everyone will agree that synthetic chemicals and plastics are indispensable to modern life, industry-hostile groups and a section of the public that fears the toxic hazards represented by many chemicals will not change their views. It therefore seems unlikely that the chemical industry will ever gain a really warm place in the hearts of a certain part of the public.

Endnotes

1. Andrew J. Hoffman, *From Heresy to Dogma: An Institutional History of Corporate Environmentalism* (San Francisco: New Lexington Press, 1997), xvii.
2. David Buzzelli, personal communication.
3. Craig E. Colten, "Creating a Toxic Landscape: Chemical Waste Disposal Policy and Practice (1900–1960)," *Environmental History Review* 18 (Spring 1994), 85.
4. U.S. House Committee on Interstate and Foreign Commerce, Subcommittee on Oversight and Investigations, *Waste Disposal Site Survey*, 96th Cong., 1st sess, Oct. 1979, Committee Print 96-IIC33.
5. Colten, "Creating a Toxic Landscape" (cit. note 3).
6. Ibid., 95.

7. J. Clarence Davies and Barbara S. Davies, *The Politics of Pollution*, 2nd ed. (Indianapolis, Ind.: Pegasus, 1970); see also J. Clarence Davies and Jan Mazurek, *Pollution Control in the United States* (Washington, D.C.: Resources for the Future, 1998).

8. Colten, "Creating a Toxic Landscape" (cit. note 3), 100.

9. Ibid., 103.

10. "On Tap: More Spill Regulators," *Chemical Week* 107 (23 Sept. 1970), 42.

11. Peter H. Spitz, *Petrochemicals: The Rise of an Industry* (New York: John Wiley & Sons, 1988), 492.

12. Joseph L. Badaracco and George C. Lodge, "Allied Chemical Corporation," Harvard Business School case no. 379-137, 1979.

13. Sheldon Hochheiser, *Rohm and Haas: History of a Chemical Company* (Philadelphia: University of Pennsylvania Press, 1986), 169–176.

14. Lee Niedringhausen, *The Corporate Alchemists* (New York: William Morrow, 1984), 256–257.

15. Elizabeth M. Whelan, *Toxic Terror: The Truth behind the Cancer Scares* (Amherst, N.Y.: Prometheus Books, 1993), 139. See also New York State Department of Environmental Conservation, Division of Hazardous Waste Remediation, "1989 Love Canal Annual Report," Feb. 1989, and "1990 Love Canal Annual Report," May 1991.

16. Michael Fumento, *Science under Siege* (New York: William Morrow, 1993); and Mark R. Powell, *Science at the EPA: Information in the Regulatory Process* (Washington, D.C.: Resources for the Future, 1999).

17. Whelan, *Toxic Terror* (cit. note 15), 295–296.

18. Will Lepowski, "Bhopal: Indian City Begins to Heal but Conflicts Remain," *Chemical and Engineering News* 63 (2 Dec. 1985), 18–21.

19. *U.S. Chemical Industry Statistical Handbook 1995* (Arlington, Va.: Chemical Manufacturers Association, 1995), 123.

20. "Learning to Live with Ecology," *Chemical Week* 108 (16 June 1971), 11–12.

21. "It Costs More to Run a Clean Chemical Plant," *Chemical Week* 109 (8 Sept. 1971), 31.

22. Ibid.

23. Irwin Schwartz, "Environmental Control," *Chemical Week* 106 (17 June 1970), 82.

24. "Learning to Live with Ecology" (cit. note 20), 11.

25. Richard C. Fortuna and David Lennett, *Hazardous Waste Regulation: The New Era* (New York: McGraw-Hill, 1987), 16.

26. "Learning to Cope with RCRA," *Chemical Week* 127 (12 Nov. 1980), 64–71.

27. "For Disposal Firms: Big Revenues, Big Headaches," *Chemical Week* 127 (26 Nov. 1980), 31.

28. Hoffman, *From Heresy to Dogma* (cit. note 1), 64–66.

29. Ralph Nader and William Taylor, *The Big Boys: Power and Position in American Business* (New York: Pantheon, 1986), 186. See also Cathy Trost, *Elements of Risk: The Chemical Industry and Its Threat to America* (New York: Times Books, 1984), 249–251.

30. "Facing Down the Critics," *Chemical Week* 106 (15 Apr. 1970), 55–56.

31. Hochheiser, *Rohm and Haas* (cit. note 13), 177.

32. *U.S Chemical Industry Statistical Handbook 1995* (cit. note 19).

33. Daniel J. Fiorino, *Making Environmental Policy* (Berkeley/Los Angeles: University of California Press, 1995), 182.

34. "3M May Be the First to Get under the Bubble," *Chemical Week* 127 (3 Sept. 1980), 47.

35. Hoffman, *From Heresy to Dogma* (cit. note 1), 146–147.
36. Ibid., 114–115.
37. CFCs, which are not considered hazardous chemicals, were blamed for upper-level ozone depletion after results from research studies were published in 1993. CFCs attack the ozone layer that protects Earth's atmosphere from the dangerous rays of the sun, which are responsible for the development of skin cancers. Michael Brower and Warren Leon, *The Consumer's Guide to Effective Environmental Choices* (New York: Three Rivers Press, 1999), 12, 15–17; and S. Fred Singer, *My Adventures in the Ozone Layer: Rational Readings on Environmental Concerns,"* ed. J. H. Lehr (New York: Van Nostrand Reinhold, 1992).
38. Ashish Arora, Ralph Landau, and Nathan Rosenberg, *Chemicals and Long Term Economic Growth* (New York: John Wiley & Sons; Philadelphia: Chemical Heritage Foundation, 1998), 363.
39. Franklin Associates, "An Energy Study of Plastics and Their Alternatives in Packaging and Disposable Consumer Goods," Prairie Village, Kansas, Nov. 1992.
40. Niedringhausen, *Corporate Alchemists* (cit. note 14), 174–175.
41. Jeffrey F. Rappaport and George C. Lodge, "Responsible Care," Harvard Business School, case no. 9-391-135, 18 Mar. 1991.
42. Ibid., 8.
43. "Which Way Is Right?" [Editorial], *Pittsburgh Post-Gazette*, 17 Oct. 1987.
44. Emma Chynoweth and Michael Roberts, "Europe Progresses Despite Financial Hardships," *Chemical Week* 153 (8 Dec. 1993), 62–65.
45. "A Blow to OSHA's Benzene Rules," *Chemical Week* 127 (9 July 1980), 11–12.
46. *U.S. Chemical Industry Statistical Handbook 1998* (Arlington, Va.: Chemical Manufacturers Association, 1998), 87–89.
47. Keith Belton, personal communication.
48. Cheryl Hogue, "Chemical Producers' TRI Ranking Falls," *Chemical and Engineering News* 78 (29 May 2000), 46–47.
49. Bill Schmitt, "Responsible Care: Responding to Public Opinion Shifts," *Chemical Week* 162 (22/29 Nov. 2000), 24–30.
50. Andrew A. King and Michael J. Lenox, "Industry Self-Regulation without Sanctions: The Chemical Industry's Responsible Care Program," *Academy of Management Journal* 4 (Aug. 2000), 698–716.
51. Ibid.
52. Ibid.
53. Bill Schmitt, "Public Disclosure: Warts and All," *Chemical Week* 162 (5/12 July 2000), 43.
54. Gail Dutton, "Green Partnerships," *Management Review* 85 (Jan. 1996), 24–25.
55. *Dow Chemical Public Report 1999* (Midland, Mich.: Dow Chemical, 1999), 25.
56. This charge lapsed in 1995. Superfund is running out of money, and the U.S. Congress has not at this writing taken any steps to resume this type of funding.
57. "Dow Annual Report" (Midland, Mich.: Dow Chemical Company, 2000).
58. "DuPont Annual Report" (Wilmington, Del.: DuPont Company, 2000).
59. Kara Sissell, "Asbestos: The Push for Litigation Reform," *Chemical Week* 165 (5 Mar. 2003), 16–20.
60. Walter A. Rosenbaum, *Environmental Politics and Policy*, 4th ed. (Washington, D.C.: CQ Press, 1998), 122–124.
61. Whelan, *Toxic Terror* (cit. note 15), 198.

62. *Trade Secrets: A Moyers Report,* prod. by Public Affairs Television, by Sherry Jones and Bill Moyers, 1 hour and 57 min., Public Broadcasting Service, 26 March 2001.

63. Dow does not say that this material can be used as a general substitute for PVC.

64. *Trade Secrets* (cit. note 62).

65. David Hunter, "Trade Secrets" [Editorial], *Chemical Week* 163 (4 Apr. 2001), 5.

66. Kirk Johnson, "Gipper Meets 'Survivor' as GE's Image Hardens," *New York Times,* 4 Mar. 2001, Metro section.

67. David Hunter, "New Global Program on Chemicals Management," *Chemical Week* 163 (22 Aug. 2001), 3.

68. Jessica Brown, "Industry Attacks White Paper," *Chemical Week* 163 (21 Feb. 2001), 9.

69. Jeff Johnson, "Chemical Accident Debate Rolls On," *Chemical and Engineering News* 79 (9 Apr. 2001), 22.

Appendix

Important regulations discussed in this chapter are summarized and briefly discussed below.

CAA—Clean Air Act, as amended (1970): The Environmental Protection Agency (EPA), also created in 1970 by President Richard Nixon, was directed to establish national ambient air quality standards (NAAQSs) for certain identified toxic air pollutants emitted from moving (e.g., vehicles) and stationary sources. These standards were to be used by states as the basis for source emission limitations in state implementation plans. The six key pollutants initially identified under NAAQS were sulfur dioxide (SO_2), particulate matter, nitrogen dioxide (NO_2), carbon monoxide, ozone, and lead.

For chemical companies, the provisions of the CAA amendments were of particular importance with respect to emission of volatile organic compounds, such as light hydrocarbons and oxygenates, which in many cases create ozone in the lower atmosphere.

The act provided different regulations for existing stationary sources (to do what is necessary to meet NAAQSs) and for new stationary sources (to meet New Source Performance Standards using the best available technology).

CERCLA—Comprehensive Environmental Response, Compensation, and Liability Act (1980): Known as the "Superfund" law, this act was intended to cover the cleanup of existing hazardous waste sites and to address future releases of hazardous substances into the environment. A National Contingency Plan covered provisions for permanent cleanups or remediation of such sites. The term *Superfund* refers to moneys collected by the EPA to investigate sites and pay for cleanup in cases where no responsible party can be found to cover cleanup costs. "Responsible parties" could include current or future owners or operators of the sites, all facilities that provided waste for disposal at these sites, and all transporters that delivered waste to these sites. The cleanup process was required to meet all other environmental requirements during its operation.

EPCRA—Emergency Planning and Community Right to Know Act (1986): Also known as "SARA Title III" (Superfund Amendment and Reauthorization Act), EPCRA covered five items: emergency planning notification where companies were required to disclose having certain extremely hazardous substances above threshhold planning quantities; emergency release notification covering accidental plant releases of hazardous substances; community right-to-know reporting, acknowledging the fact that communities have as much right to know about releases as workers at the plant; TRI (Toxic Release Inventory) reporting, whereby companies must annually report both routine and accidental releases of certain listed chemicals from a facility plus efforts to recycle and other relevant data; and notification to customers regarding chemicals on the TRI list that are contained in products.

FEPCA—Federal Environmental Pesticide Control Act (1972): As an amendment to the 1947 Federal Insecticide, Fungicide, and Rodenticide Act (FIFRA), federal controls were extended to the application of pest control chemicals by the purchaser, including misuse, as well as registration of all pesticide products for general and restricted use. The restricted category contains those chemicals that pose a high risk to man or his environment. It also authorized the EPA administrator to monitor pesticide use and presence in the environment.

FWPCA—Federal Water Pollution Control Act (1972): Later amended by the **Clean Water Act** (1977), FWPCA was intended to restore and protect the quality of the nation's surface waters, the ultimate goal being to eliminate the discharge of pollutants from "point sources" (e.g., pipes and sewers) into navigable waters, with the interim goal of making the waters "fishable and swimmable." The surface waters covered by the act include rivers, lakes, and wetlands. Pollutants identified include "conventional" pollutants, such as biochemical oxygen demand (BOG), fecal coliform, oil, grease, and pH, and "priority" pollutants, such as various toxic pollutants.

NEPA—National Environmental Policy Act (1970): Requires that federal agencies prepare an environmental impact statement for proposed actions that may significantly affect the quality of the human environment. Effectively, private corporations would prepare the environmental impact statement in connection with the permit or licensing request they make. NEPA can also be considered as background for the establishment of the Environmental Protection Agency (EPA) by executive order as an independent agency of the executive branch of the U.S. Government with ten regional offices to provide oversight for federal environmental programs delegated to states as well as directly administering certain programs. (State and local regulations were subsequently promulgated, and, where the programs are strong enough, the EPA was set up to delegate the federal program to the states.)

OSHA—Occupational Safety and Health Act (1970): The act is intended to ensure that workers are informed about workplace chemicals and other hazards. It requires companies to develop programs that identify hazardous chemicals, training workers in plant safety, maintaining Material Safety Data Sheets that are also made available to customers, and training workers in chemical safety. It sets process management safety standards to protect workers and covers hazardous waste operations and emergency response rules intended to limit the possibility of employee exposure to safety or health hazards.

PPA—Pollution Prevention Act (1990): This act established a national policy favoring source reduction and recycling over treatment and disposal of pollutants. Facilities already required to report their releases of toxic chemicals under EPCRA were given an additional requirement to supply information about source reduction and recycling activities relating to these releases.

PPA was also intended as an important program for EPA and state environmental agencies to establish institutional relationships where the chemical industry and these regulatory authorities can work cooperatively for mutual benefit.

RCRA—Resource Conservation and Recovery Act (1976): The first substantial effort by Congress to establish a regulatory structure for the management of solid and hazardous wastes. Hazardous wastes were defined to include toxicity, corrosivity, ignitability, and reactivity. Subtitle D addresses "cradle-to-grave" requirements for hazardous waste from the point of generation to disposal. Regulations were set for hazardous waste generation, treatment, storage, and disposal. The disposal of hazardous wastes without prior treatment was proscribed (the "land ban"). The provisions of the act included requirements for comprehensive solid waste planning as well as encouragements for recycling and recovery.

Tanks and containers used to store hazardous wastes with a high volatile organic concentration were required to meet emission standards.

TSCA—Toxic Substance Control Act (1976): This act gave the EPA broad authority to regulate the manufacture, use, distribution in commerce, and disposal of chemical substances. It gave the EPA the authority to ban manufacture or place restrictions on chemicals that pose unreasonable risk of injury to health or the environment. Among the chemicals the EPA regulates under TSCA are asbestos, chlorofluorocarbons (CFCs), and polychlorinated biphenyls (PCBs). TSCA directs the EPA to balance benefits against risks in regulatory decisions.

Another provision covers premanufacturing notices for new chemicals to be made by prospective manufacturers or sold by importers, allowing the EPA to choose among a broad range of options that may limit, restrict, or prohibit manufacture, use, distribution, and disposal of the chemical substance.

Further, companies that manufacture or import more than 10,000 pounds of certain chemicals included in TSCA's list are required to report current data on production volumes and plant sites where manufacture is carried out.

Chapter 8

The Global Industry

Peter H. Spitz

I n early 2001 natural gas prices in the United States rose rapidly from typical winter levels of two to three dollars per million BTUs to eight dollars and higher, with spot prices briefly reaching ten dollars (Figure 1). A combination of an early cold winter and the growing replacement of other fuels with "environmentally friendly" natural gas had caused a sharp spike in demand for this fuel, while supplies had not recently increased much because of low drilling-rig activity. Natural gas–derived liquids, such as ethane and liquefied petroleum gas (LPG), are normally used as feedstocks for a major part of domestic ethylene production. Ethane is extracted from natural gas at eight dollars per million BTUs or higher. Ethane prices briefly rose from an early 1999 price of 18 to 20 cents per gallon to 46 to 48 cents per gallon. Corresponding to the ethane increase, ethylene prices jumped from 17 to 20 cents per pound to 30 to 32 cents per pound. For one or two quarters U.S. Gulf Coast ethylene was more expensive to produce than ethylene in essentially all other global producing regions, a complete reversal of traditional global production economics. Plants in the other regions were either using naphtha feedstock from relatively inexpensive crude oil or, in the case of Saudi Arabia, much less expensive ethane. Although in the United States naphtha is often a more expensive ethylene feedstock than ethane, U.S. ethane prices were now at levels never before seen. Key ethylene derivatives, such as polyethylene, styrene, and ethylene dichloride, were no longer competitive in export markets, while domestic customers, already facing an economic downturn, strongly resisted the substantial price increases needed to maintain production. In fact, they started to look at foreign sources.

The Gulf Coast petrochemical industry, long favored as a low-cost, export-oriented, petrochemical-producing region, was already being challenged by

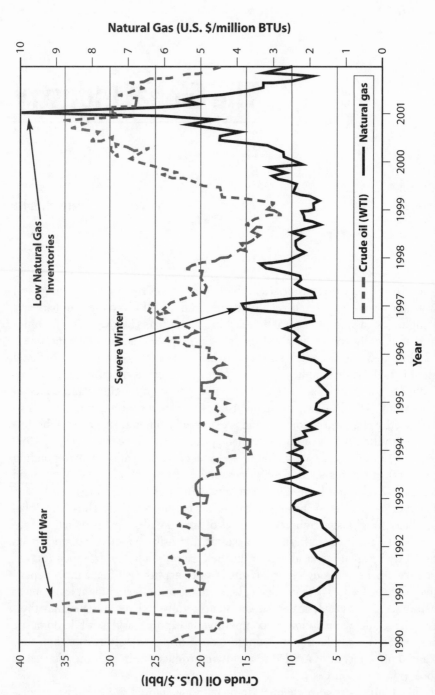

Figure 1. Natural gas and crude oil pricing trends. *Source:* Chemical Market Associates, Inc. (CMAI), by permission.

foreign plants. But for a short time it lost what was left of its comparative advantage almost overnight, and a number of U.S. plants had to be shut down.

Within a couple of months natural gas prices again declined and then during the spring dropped fairly sharply, but the crisis was far from over. The unexpected huge "blip" in natural gas pricing had merely highlighted a situation that had developed over the last several years—the accelerating decline in the once highly positive U.S. balance of trade in chemicals.

In the 1980s and 1990s this favorable balance of trade was a standout, contrasting sharply with the country's year-to-year highly negative overall balance. While part of the positive chemical trade balance was owing to large exports of fine chemicals, such as pharmaceutical or agricultural chemicals, there had also been historically a strong positive balance in petrochemicals, with exports mainly to Asia and Latin America. But these exports had been declining. Production in Southeast Asia had begun to grow fairly rapidly, and low-cost producers from the Middle East and western Canada had replaced some of the traditional exports from the Gulf Coast. The U.S. trade surplus had shrunk from $20.5 billion in 1997 to $9.8 billion in 1999 and then fell precipitously to $1.3 billion in 2001 (Figure 2). In 2001 organic chemical exports actually had a $13.1 billion deficit. Although part of the reason for the vanishing chemical trade surplus was the strong dollar, a more important reason was that a great deal of petrochemical capacity had in the 1990s come on stream in traditional export markets and particularly in Asia. Part of this capacity was created by U.S. and other multinational investment and joint ventures in traditional markets or in export-oriented locations. But a large amount of new capacity had also been built by national or privately owned companies in the Middle East, Korea, Taiwan, Singapore, Thailand, Malaysia, and India, while Mainland China was also expanding its capacity fairly rapidly (although much slower than the growth in Chinese demand). In South America, Brazil and Argentina were becoming bigger factors in petrochemicals.

During the last two decades of the twentieth century petrochemicals had joined other parts of the chemical industry in becoming truly global. The percentage of petrochemicals being produced outside the countries belonging to the Organization of Economic Cooperation and Development (OECD) had increased markedly and was slated to increase much further, as will be discussed later in this chapter.

The managements of chemical firms in the traditional exporting countries had become very aware of four key developments. First, emerging and newly industrialized countries were beginning to represent an important growth market, in contrast to traditional markets, where demand growth for most chemicals was slowing to gross national product growth rates or even lower. Second, production costs in the OECD countries were being progressively bested by exporting producers located in low feedstock cost areas. Third, plants were also being built inside these new markets by local producers, initially protected

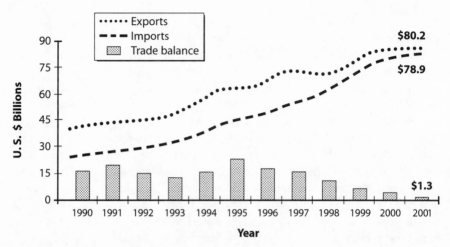

Figure 2. U.S. balance of trade in chemicals for 1990 through 2000. *Source: 2002 Guide to the Business of Chemistry* (Arlington, Va.: American Chemistry Council, 2002), 75.

by freight and tariff shields, making imports expensive. Fourth, many of the firm's customers in the OECD countries had moved some of their manufacturing activities to emerging countries and expected to be supplied with chemicals and plastics at their new locations. For example, some computers would now have their cases molded in countries in Southeast Asia, where labor was inexpensive. If these companies were using a certain type of ABS (acrylonitrile-butadiene-styrene) resin from, for example, GE Plastics, they expected to receive the same resin at all of their plants. This would mean that the resin supplier would either have to enter a joint venture with a closer supplier (for example, in Japan) or build warehouses in several regions, or perhaps build an ABS compounding plant in a new foreign location. The same held true for chemical firms supplying electronic chemicals, such as photoresists or special adhesives used in integrated circuit chip manufacture, which would have to be supplied to chip factories in locations away from the home market. Globalization of end-product manufacture thus caused a number of multinational firms to develop a global supply chain, where in many cases the large-scale manufacture was carried out in the home market with the end product made in several global locations.

All chemicals do not necessarily lend themselves to this type of globalization. Some chemicals, like commodity industrial gases (oxygen, nitrogen, hydrogen, and carbon dioxide) and chlorine, simply cannot be shipped. But most other chemicals can be sent around the globe and usually are.

Globalization of companies' manufacturing operations is most classically illustrated by the early decision of the three main Japanese car manufacturers (Toyota, Nissan, and Honda) to build assembly plants in North America and

Europe. They first sourced essentially all the components from Japan, but then began to purchase some components from suppliers in the new locations, although it was difficult for these suppliers to break into this business because of the exceptionally high quality standards of the Japanese firms.

"Globalization" would become a mantra in the 1990s for at least a portion of European and U.S. firms' product lines, which were to have a growing portion of sales shifted to Asia and other non-OECD regions and countries. While the large chemical and oil companies had established an international presence early in the century, the game and the players had now changed rather dramatically. It behooved each firm to find its own way in a global free-for-all, where many of the old concepts no longer applied.

Establishing manufacturing operations in a location other than the home country is fraught with potential problems. Companies planning to do so try to establish some sort of competitive advantage when manufacturing their traditional product lines in a new location, but they find this difficult to do. In different countries and regions certain conditions make manufacture there particularly attractive, perhaps because of low-cost raw materials or for other reasons. Companies may have enjoyed a comparative advantage of some kind in their home market but will probably not have the same situation in foreign locations. Michael Porter, in his landmark book on national competitive advantage, identified four attributes of a nation that can confer strong competitive advantage to companies operating there. They are factor conditions (skilled labor, infrastructure, and raw materials advantage for a given industry); demand conditions (relatively high home demand growth for the industry's product or service); related or supporting industries (the presence or absence in the nation of supplier industries and related industries that are internationally competitive; also called "industry clusters"); and firm strategy, structure, and rivalry (the nature of domestic rivalry, including how firms are organized and managed).[1]

Porter called the combination of these elements the "diamond," or determinant, of national competitive advantage. Ironically, he cited as an example U.S. Gulf Coast manufacture of petrochemicals, where there are (or were traditionally) inexpensive feedstocks (factor conditions); strong demand growth in the domestic market (double-digit growth in the 1950s and 1960s) (demand conditions); a network of pipeline companies, engineering contractors, and equipment manufacturers (related and supporting industries); and extremely strong competition among petrochemical producers (firm strategy, structure, and rivalry). These factors promoted innovation, continuous improvement, and skilled marketing and market development.

U.S. petrochemical companies originally investing abroad did not find such a combination elsewhere—not in Europe and certainly not in Japan. When considering investment in low raw-material cost regions like the Middle East, companies saw enough *factor* advantages to overcome a lack of Porter's other elements. This led to a number of joint ventures in Saudi Arabia and elsewhere. But investment in Asian or Latin American countries would mean facing a political as well as an industrial operating climate that would be considerably less benign and advantageous than that in the home country.

Porter's studies and viewpoint on national competitive advantage, when applied to commodity chemicals (for example, the petrochemical industry), seem to prove the point that the U.S. Gulf Coast may be the only relevant national location that is a good example. But the advantage that once made the United States such a powerhouse in petrochemicals is now overshadowed by global developments. The unique advantage of low feedstock cost, such as in the Middle East, provides all the advantage any producers there need to be globally competitive, even though all three other parts of Porter's diamond are missing. Moreover, technology diffusion through process licensing and contractor know-how has been sufficient to allow any producer in any country to build an efficient, competitive, world-scale plant; the results of continuing innovation and cost reduction have also been made available, through licensing and reengineering, to competitors located throughout the world; and local competition among firms, the final important Porter element, is not necessarily key, for example, in Asia, since competition is carried out on a global front.

Multinational companies, including chemical firms, generally recognized that in most cases they would not have true comparative advantages when investing abroad or they would have to share these with local partners. They usually opted to form joint ventures with local firms, which would provide access to local raw materials, have a strong market presence, and "know the ropes" in the region. Examples of such joint ventures include Exxon-Mobil with SABIC (Saudi Basic Industries) in Saudi Arabia, Phillips with Sinopec in China, Dow and BASF with Petronas in Malaysia, and Dow with Siam Cement in Thailand.

Companies began to invest abroad, of course, long before Porter came out with his book. The driving forces were obvious, stemming largely from the need to augment exports with local production. The decision to manufacture abroad involves many issues, including raw material availability and cost, location of key customers (an increasingly important factor as customers "globalized" their operations), planning for large future markets in the country or region, preempting competitors, and joint venture opportunities with local partners. Furthermore, companies may find it difficult to develop a new "grassroots" site or to expand significantly at an established site because of lack of land or environmental restrictions. For example, BASF built a large new complex in Antwerp, Belgium, in part because the firm found it difficult to expand at its traditional site in Ludwigshafen, Germany, a country where the "Greens" are relatively strong and very antagonistic to the chemical industry.

Robert Stobaugh, a professor at the Harvard Business School, and his collaborators studied the subject of exports and overseas investments, including their effect on international trade and balance of payments. Companies want to market outside their home country and find export opportunities in countries and regions that either have no similar production or are undersupplied by local production. Exports are good for the home country's balance of trade, but when the time comes for companies to build plants outside the home market, the balance-of-payment effect on the home country is negative unless accompanied by ever greater exports—an unlikely scenario. Companies' decisions to build abroad, however, are unaffected by balance-of-trade considerations.[2]

In one of his early books Stobaugh studied the financial results of overseas investments to determine whether the decisions underlying a number of selected investments in different industries had ultimately been financially sound.[3] The results were mixed. Nevertheless, companies have generally decided to go ahead with foreign investments, largely to protect and further build markets originally established through exports.

Building plants abroad is opposed in principle by organized labor, which also generally opposes access for imports to the home market, in both cases being concerned about loss of jobs. But companies deciding to build manufacturing plants abroad have strategic reasons to do so and will therefore invest abroad regardless of opposition by the unions. Unions are in any case less focused on chemical plants built in overseas locations since the number of workers in chemical plants is much smaller than in most other industries.

In another book Stobaugh looked specifically at international trade in petrochemicals. He looked at the time lag between the start of commercialization of a given product in the home country and the first commercial production in a foreign country, either by the same company or by a competitor. Typically, the lag was from four to seven years. Stobaugh found that variables that determined the decision for beginning production of a given petrochemical product in another country or region were market size, availability of local technology, investment climate, and shipping costs. He called shipment of exports to countries that had no production facilities "technology-gap" shipments. Then when production was begun in such a country, he found that although initially imports would drop or cease, the resulting market growth would eventually again require imports, which he called "balancing shipments," or exports from the originally exporting country or other countries (Figure 3).[4]

These studies were done at a time when plants were relatively small. Today large plants built in foreign markets may or may not eliminate imports: they certainly would not in China but would in countries with small markets, like Malaysia.

Exports do not necessarily lead to foreign investments. Their relationship to foreign investment is developed somewhat differently. A paper by J. F. Wyatt,

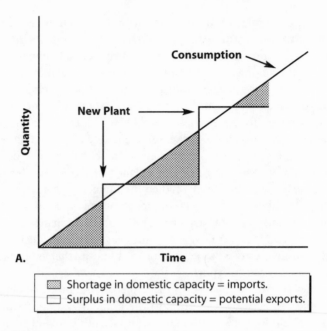

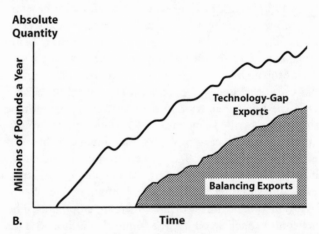

Figure 3. A: Idealized capacity (new plant) additions in a country starting to import a chemical. B: Typical pattern of U.S. exports of a petrochemical. *Source:* Robert Stobaugh, "The Neotechnology Account of International Trade: The Case of Petrochemicals," *Journal of International Business Studies* Fall (1971), 54, 57.

published in 1987, presented a cogent analysis of the motivations for exports and the construction of foreign plants and of factors underlying international trade. Wyatt divided trade flows into four categories: cyclical surpluses and opportunistic trading, cyclical deficits and market development, output from trade-driven investment, and optimization of logistics.[5]

The first group consists of companies that export their surplus product. Companies often do this, particularly when they seek to keep plants operating at higher rates and are willing to take lower plant netbacks from the export market prices as long as they can cover variable costs and, if possible, some of their cash costs. Such companies are generally not dedicated to the export market or to specific customers in foreign countries, and they will stop exporting when the domestic market improves, since domestic prices are almost always more attractive. These exports are usually handled by trading companies and usually go to the spot market.

While export market pricing is usually lower than domestic market pricing, the situation often reverses during the price "fly-up" part of the petrochemical cycle, as occurred in 1979, 1988–89, and 1994–95. When global demand meets or exceeds global supply, the export market gets short shrift because companies are generally tied into contracts with and must favor their domestic customers. This creates a shortage of material and a price spike in the export market. Every company that has any surplus material, including the domestic customers themselves, will then try to export into the spot export market. During this period export spot prices can then rise well above the domestic price.

Companies that export into the spot market under normal conditions may sell at fairly low prices and may sometimes be accused of "dumping." This is defined as selling below a company's manufacturing cost, a situation that is not always easy to prove but that may cause damage to producers in the importing country. There are always a number of anti-dumping actions being brought against exporters in various countries. In the United States these are handled by the International Trade Commission. Often by the time dumping has been established, the accused company is no longer supplying the accusing country at the low pricing that prompted the dumping action in the first place.

Wyatt's second group consists of companies that export to their established foreign customers, who depend on their foreign suppliers regardless of the conditions of the suppliers' home market or the difference between domestic and export prices. Companies in the first group and many companies in the second group often have no intention of building plants abroad.

Wyatt's third and fourth categories apply to companies that have global strategies. The third group includes exporting companies that are interested in building a market in a foreign country or region where they can eventually establish a local manufacturing plant. And the fourth group comprises exporting companies that already have several manufacturing plants and are balancing product shipments from different locations to global customers as part of their global supply chain optimization. This fourth group corresponds to the last item in Wyatt's list—optimization of logistics.

Wyatt further posited a trade-dependent rationale underlying investment decisions, which, he said, falls into one or more of the following: internationalization of markets, import substitution, the need to add value to domestic resources, and compensation agreements. Import substitution is generally the principle driving force behind investments in developing countries. Usual criteria are a reasonably sized domestic market large enough to establish economical production, acceptable logistics, and preferably availability of domestic feedstock. The presence of the local plant will often stimulate local consumption, thereby accelerating market growth. If the investment is financed through compensation payments—defined as exports from the new plant to pay back part of the investment—then the internal market will not grow as rapidly because a portion of the output is immediately exported. This occurred with plants built in Eastern Europe in the 1970s and 1980s.[6]

Trade-driven investment is best exemplified by plants constructed in the Middle East and Canada. For such investments to succeed, there must be penetrable foreign markets and a sustainable raw material cost advantage.

Companies seeking to maintain or increase their market share build new plants in their home market to preempt competitors. As plant sizes increase, companies find the export market attractive for operating the new plants at the highest possible capacity. This situation, as it applies to global market share, means that firms will build large plants in new markets and immediately seek to export part of the production, as exemplified by the new plants constructed by Dow Chemical (originally Union Carbide) and BASF in Malaysia.

International trading firms initially handled much of the exported petrochemicals, and they continue to be an important factor, although their role has changed somewhat from handling primarily spot exports. One of the earliest firms in this business was Fallek Chemical, which, according to Kurt Doldner, one of its main partners, initiated exports of benzene-toluene-xylene (BTX) aromatics to Japan in the 1960s.[7] Japanese firms were starting production of terephthalic acid for the polyester fiber market at a time when no p-xylene was being produced in Japan. Fallek imported parcels of p-xylene and o-xylene (for phthalic anhydride production) into Japan—probably the first trader to do this business. U.S. traders became very active, but they were eventually dwarfed by Japanese traders, whose role kept increasing even when global companies themselves took over a larger percentage of exports (once handled by traders) via contracts with foreign customers. By 1995 both Mitsui and Mitsubishi conducted over $10 billion worth of trades, with Itochu and Marubeni close behind.[8] These firms originally concentrated on importing petrochemical raw materials into Japan. They then expanded into international trade and later built or invested in production facilities to ensure a supply of products to trade. Today trading firms act as more than brokers and actually take positions in products (buy and sell rather than act only on commissions), including the use of toll arrangements to increase their profits. (A toll arrangement is a business arrangement common in the chemical industry where, instead of buying a product, the "buyer" supplies raw materials to the seller, which in turn produces the desired product at its manufacturing cost and charges a "tolling" fee.)

International trade gave rise to the construction of large liquid product terminals in various parts of the world. Exporters would ship such products as styrene, ethylene glycol, BTX aromatics, and acrylonitrile to their own or leased tanks located in terminals in Europe, Asia, and elsewhere. Customers would then be supplied from these terminals via barge, railroad tank cars, or road tankers.

Using container ships with many compartments allowed shippers to accept relatively small shipments of petrochemicals to foreign locations. Construction of so-called parcel tankers started during a depressed tanker market in the 1960s. Owners of such vessels would assemble several parcels of petrochemicals in lots of about 1,000 tons, which would be shipped for a cost of perhaps $30 per ton to foreign markets. The international trade in petrochemicals was thereby greatly enhanced. Parcels would include shipment of tallow, lube oils, vegetable oil, and other liquid cargoes. A 25,000-ton deadweight parcel tanker might have up to fifty independent tank and cargo handling systems, with compartments made of a number of materials and coatings, including epoxy, zinc, rubber, and polyurethane.[9]

Global Trade Started Early

Exports and the subsequent construction of plants in foreign markets have represented an important aspect of chemical industry operations for well over a hundred years. The European chemical firms were in the vanguard, and exports became an important aspect of cartel operations.

In the late 1800s two processes came into wide use in Europe and in the United States: the Solvay process for the manufacture of synthetic soda ash and Alfred Nobel's invention of dynamite as a safe detonator for nitroglycerin. In both cases international cartels were established to control these technologies and to carve up markets. The chemical industry established a truly global presence when IG Farben in Germany opened up and dominated the world synthetic dyestuffs market. Developing a worldwide network of companies, IG Farben and its affiliates controlled 88 percent of the world synthetic dyes market and a substantial percentage of other world chemical markets.[10] When IG Farben and DuPont were seen by British firms as a major threat as well as a model for global dominance, the response was to form ICI in 1926 through the merger of four British companies. From that point on the three giant firms controlled a number of international chemical markets. French, Italian, Belgian, and other European and American firms joined a number of cartels. All the firms were strongly export oriented. Some 25 percent to 30 percent of production was typically exported, including to Asia and Latin America. U.S. anti-trust legislation in the 1930s made American participation in cartels illegal, except for certain types of export cartels covered under the Webb-Pomerene Act of 1918.

While the IG Farben combine was broken up after World War II, its successor companies—notably Bayer, Hoechst, and BASF—reestablished themselves rather quickly as independent entities. By the 1950s they were already exporting about one-third of their output. ICI was slower in regaining export markets, but by the 1970s it was the most international of the large companies, marketing in nearly 150 countries and manufacturing in about 30.[11]

The American firms were slower in developing export markets, given the benefit of a large home market. DuPont, with such innovations as nylon, Teflon, and Dacron polyester, was the earliest to focus on exports and by the mid-1960s had achieved export sales of around 20 percent. Union Carbide and Monsanto followed DuPont's example. Dow Chemical, however, was arguably the first American firm to establish a strong European manufacturing presence when it built its complex at Terneuzen, Netherlands, in 1965. Chem Systems, a chemically oriented consulting firm, around that time was asked by Dow to make a detailed market survey for chlorinated solvents in Europe, including identification of market shares held by the European firms that had historically operated this business in Germany, France, Belgium, and Italy. The traditional producers were clearly unhappy to witness the entry of a new, strong producer. With the results of the Chem Systems study in hand, Dow let it be known that it aimed to get a certain market share for its chlorinated solvents in Europe and would, if necessary, slash prices to get it. In the face of these tactics opposition to Dow's entry soon softened, and Dow became an important European supplier of carbon tetrachloride, perchloroethylene, chloroform, and other solvents.

Globalization in the postwar period was, however, largely characterized by exports rather than by building manufacturing plants abroad, with foreign markets ranking well behind home markets in importance. Chemical demand in the OECD countries between 1960 and 1973 was increasing at a 9 percent to 10 percent annual growth rate, which often led to supply limitations since moderate-sized plants were then still being built. The two oil shocks and subsequent economic global slowdown then brought chemical growth rates down just when technology allowed firms to build plants of much larger scale. This created substantial overcapacity intermittently for many of the producers so that firms became much more interested in the export market. The industry crisis in the early 1980s was exacerbated for U.S. firms as foreign companies attacked the U.S. market more vigorously. Now even medium-sized and smaller U.S. companies began a more serious drive for exports. While U.S. and European firms had for a while been building some manufacturing plants in foreign markets, direct investment in foreign locations became a key strategy in the 1980s. European chemical firms invested heavily in the United States, in part

Table 1. Percentage of 1998 Sales in Home Country

Company	Percentage	Company	Percentage
BASF (Germany)	25.4	Bayer (Germany)	16.4
Hoechst (Germany)	19.8	ICI (U.K.)	18.2
DuPont (USA)	52.8	Dow (USA)	40.2
DaiNippon Ink (Japan)	47.7	Mitsubishi Chemicals (Japan)	88.4

Source: Richard D. Freeman, "The Chemical Industry: A Global Perspective," *Business Economics* 34 (Oct. 1999), 16.

through important acquisitions, such as Inmont and Wyandotte by BASF, Celanese by Hoechst, and Mobay by Bayer, or by constructing grassroots plants, as undertaken by such companies as Solvay, ICI, and Dutch State Mines (DSM).

U.S. multinationals have also invested heavily abroad. Between 1990 and 2000, U.S. direct investment abroad in what the American Chemistry Council calls "the business of chemistry" rose from $37.9 billion to $86 billion, of which 65 percent is in Europe. U.S. chemical firms' overseas investments generally gave a lower return than their domestic investments. John Roberts, author of Chapter 9 in this book, made an interesting observation. He said that U.S. firms have underestimated, or at least did not anticipate, the strength of the dollar versus foreign currencies. As a result depreciation charges for foreign plants built by U.S. companies have been higher than expected. John believes it would have been appropriate to use higher "hurdle rates" for foreign investments than were apparently employed.[12]

The strong export positions of the European multinationals was noted by Richard Freeman, who at the same time highlighted the strength of the national markets in the United States and Japan (Table 1).[13]

The major oil companies over the years continued their strategy of building primary petrochemical facilities adjacent to refineries they had already constructed in foreign countries. They thus had the infrastructure for supporting crackers or aromatics plants and would receive the benefits of the usual synergies associated with operating at the refinery-petrochemical interface. Thus, Exxon (originally Standard Oil Company of New Jersey) built crackers in France (Port Jerome), Germany (Cologne), and England (Fawley), and a large aromatics plant in Netherlands (Rotterdam). Mobil, Conoco, Amoco, and Phillips also established European petrochemical operations, while in the Pacific region Exxon and Mobil became associated in a joint venture in Japan (Toa Nenryo). Texaco and Chevron (originally Standard Oil Company of California) established Caltex, with refining and some chemical operations in the Pacific region.

Foreign firms became fairly dominant players in parts of the U.S. petrochemical industry. Japanese and Taiwanese firms invested heavily in the U.S. polyvinyl chloride (PVC) industry (Shinetsu, Westlake, and Formosa Plastics),

Table 2. Ownership of U.S. Polypropylene Capacity (1990)

Company	Capacity (in Thousand Tons)	Percentage
Foreign		
Himont	2,620	45.5
Fina	408	7.1
Aristech	309	5.4
Shell	227	3.9
Solvay	200	3.5
Domestic		
Amoco	601	10.4
Exxon	465	8.1
Phillips	227	3.9
Huntsman	213	3.7
Eastman	145	2.5
Lyondell	136	2.4
Millennium	127	2.2
Rexene	82	1.4
Total	**5,760**	**100.0**

Source: Nexant, Inc./Chem Systems.

while European and Japanese firms strongly invested in polypropylene (Shell, Himont, Fina, and Mitsubishi). Foreign ownership of U.S. PVC capacity in 1990 had Shintech with 19 percent, Formosa Plastics with 11 percent, Condea with 8 percent, and Westlake with 4 percent; domestic ownership was Oxychem with 19 percent, Goodrich with 14 percent, and all others with 25 percent. The foreign firms make up 42 percent of the total capacity. In the case of polypropylene, the high degree of ownership was in part associated with the fact that leading polypropylene technology was partly developed in Europe and Japan (Table 2).

Tax advantages were responsible in some cases when chemical operations were established outside the home market. A good example is Puerto Rico, where the local government granted outstanding tax breaks to companies locating their manufacturing activities on the island. For several years in the 1960s and 1970s Commonwealth Oil Refining Company (CORCO) was an important producer of a number of petrochemicals based on Venezuelan naphtha, with products exported largely to Europe. CORCO established several joint ventures with U.S. firms, including Union Carbide and PPG Industries. A number of pharmaceutical firms also decided to locate plants on the island.

These are still partly in operation, while CORCO went bankrupt in the early 1980s when manufacturing economics and tax treatment became less favorable and the chemical industry was in a major downturn.

Japanese firms have long been dependent on exports, relying on the major trading firms to move material around the globe. Japanese trading firms also handled a large percentage of U.S. and European petrochemical exports. Early on, the Japanese chemical and trading companies established representative offices in the United States (most frequently in New York and Houston) and in Europe (mostly in Düsseldorf, Germany). Japanese chemical firms often rotated their executives through these offices in three-year terms, then sent them home to Japan. Some Japanese firms, such as Teijin and Dai Nippon Ink, also started to invest abroad. One of the largest investments of this kind occurred in the late 1980s when a group of Mitsubishi companies headed by Mitsubishi International, acquired the chemical business of U.S. Steel, rechristened Aristech Chemical, which included plants for the manufacture of phenol, aniline, oxo alcohols, polypropylene, methacrylate sheet, and other chemicals. At that time the main driving force for this acquisition was thought to be access to products that could be exported to Japan and elsewhere. Mitsubishi sources claimed that another rationale was to create global business positions in phenol and oxo alcohol, based on technology and marketing positions of Mitsubishi Chemical, Mitsubishi Rayon, and other members of the group, but this hoped-for synergy was never realized. In 2001 Aristech was sold to Sunoco Chemicals.

Globalization was an important issue for the chemical industry in the 1980s and even more in the 1990s. The pace of U.S. investment abroad quickened as companies set goals for what percentage of their total revenues they wanted to generate in the future from non-U.S. sources, particularly from Asia. When the Asian economic crisis hit in 1998, this strategy was reexamined, but no particular change in commitment to globalization was made, according to Michael Parker, then chief executive officer (CEO) of Dow Chemical. He pointed out that Dow, in fact, became a much larger player in Asia as a result of its merger with Union Carbide, which gave Dow strong positions in Kuwait and Malaysia, adding to its already established presence in Thailand, Korea, and Japan.[14]

The United States nevertheless remains the single largest national market for chemicals and a strong magnet for continued investment by European and other foreign chemical producers (for example, the recent joint BASF-FINA cracker in Geismar, Louisiana, that started up in 2001). In 1997 a *Chemical Week* article discussed why investment in the United States is a cornerstone of many global companies' strategy. U.S. chemical firms are responsible for about a quarter of the world's chemical output. The U.S. economy is most of the time stronger than that of any other region, with the North American Free Trade Agreement (NAFTA) responsible for additional growth. U.S. firms have restructured to a greater extent than those in Europe and Japan, and U.S. employment rates are higher. America's tradition of creative thinking, its cultural diversity, and its fascination with technology all underline the key advantages of U.S. manufacturing. The U.S. market is usually the most profitable for such firms as Dow and DuPont, which accounts for the fact that U.S. firms continue to invest strongly in their home country even when investing heavily abroad in implementing their global strategies. Foreign firms now establish global headquarters for certain businesses in the United States, such as Ciba's decision to run its pigments and paper finishing businesses from the United States. Well before that, ICI, when still largely a commodity chemicals firm, headquartered its specialty businesses in Wilmington, Delaware.

While U.S. and some Japanese specialty chemicals manufacturers continue to invest in Europe, it became quite clear that the European Union had substantial disadvantages as a location for chemical investment relative to the United States. A 1998 Conseil Européen des Fédérations de l'Industrie Chimique (CEFIC) paper titled "Vicious Cycle" pointed out the reasons for the difference in profitability between the two regions, which amounted to six percentage points during the period from 1985 to 1996. The most important reasons were higher energy costs, net outflow of chemical investment capital, less innovation, higher unit labor costs, lower labor productivity, and lower energy efficiency. The relative strength of European currencies, at least during that period, also resulted in flat selling prices.[15]

Strategic Issues in International Competition

Companies developing global investments require sound strategies to support investment decisions. Even in a firm's domestic market, operations require careful strategic moves and responses. This is even more crucial when competing in international markets, since there will be more players and less history to guide the firm's strategic decisions.

Michael Porter assessed competition in global industries in one of his earlier books. He first identified firms that operate globally but in markets that do not have the essential characteristics of a global industry. Such firms as Nestle, PET Foods, and CPC International operate in many countries but have regional subsidiaries that operate essentially autonomously. Porter defined a global industry as one in which the strategic positions of competitors in regional markets are fundamentally affected by their overall global positions. He pointed out that an industry becomes a global industry when there are economic or other advantages to a firm operating in a coordinated manner in a number of national markets. He then went on to identify the sources of global competitive advantage as follows:

- *Comparative advantage:* When countries have significant advantages in factor cost or factor quality. For chemical firms this again relates to low feedstock costs in particular areas of the world. Tax advantages would also be in this category.

- *Production economies of scale:* When a firm can build a highly economic, vertically integrated production facility in one country and can then supply other countries at low cost. The large alpha-olefin production plants now operated by Shell, Chevron-Phillips, and BP Chemicals and marketed throughout the world constitute an early example of a global business.

- *Global experience:* Porter talks about proprietary experience gained by selling products in a number of regional markets, when companies benefit by sharing improvement among plants, thereby lowering production costs. Henkel's (later Cognis's) oleochemicals business would fit into this category.

- *Logistical economies of scale:* This allows spreading fixed costs by supplying many national markets and achieving economies of scale by using specialized systems. Japanese firms have achieved important cost savings by constructing and using specialized cargo ships.

- *Marketing economies of scale:* In some industries, such as engineering contracting, a common sales force is employed worldwide. Here the sales task is quite complex, with relatively few buyers so that the fixed costs of a group of skilled salespersons can be spread over a number of national markets.

- *Economies of scale in purchasing:* Achieving economies of scale as a result of bargaining power beyond what is needed to compete in national markets. For example, such oil refiners as Exxon-Mobil can negotiate very favorable supply contracts with such water-treating firms as Ondea-Nalco by giving the supplier a global contract to supply all its refineries.

- *Proprietary product technology:* Between 1960 and 1980 Oxirane (later bought by Arco Chemical) built propylene oxide plants in the United States, Europe, and Japan to capitalize on its unique coproduct process. Production economies were Arco's basis for attacking foreign markets, which then led to logistical economies and economies from purchasing feedstocks (e.g., propylene).[16]

Three examples of companies that achieved comparative advantage in Porter's terms for overseas investments follow.

Dow Chemical at Buna Sow Leuna

In East Germany a number of chemical complexes established before World War II continued operation under the Communist regime. With little attention paid to environmental issues, these plants had been responsible for a tremendous amount of pollution, going back for many decades, with chemical wastes often ankle deep within the works. Moreover, the plants were in some

cases relatively small, using outdated, polluting technology. After German uni-
fication the government wanted large international firms to take over one or
more of these plants, since they provided jobs in an economically distressed
part of the country. Enormous tax relief and grants were offered. The three
large German firms—BASF, Bayer, and Hoechst—were first asked by the
Treuhand, a government agency set up to privatize East German assets, to take
over the largest complex (Bunawerke). They all declined, saying that the country did not need this capacity and that it should be shut down. Moreover, they
had committed themselves to other sites for expansion. Dow Chemical then
stepped up and agreed to become the owner and operator of Bunawerke, along
with some plants from the vast Leunawerke and the ethylene plant in Boehlen
owned by Sächsische Olefinwerke, now collectively operated as Buna Sow Leuna
Olefinverbund GmbH (BSL). After cleaning up the site and building a number
of new units, largely with German government money, Dow by 2000 had an
essentially world-sized plant that it had built at low cost to itself. It was now
primed to supply not only central Europe but also the original countries of the
Council for Mutual Economic Assistance (Comecon; U.S.S.R., Czechoslova-
kia, Poland, Hungary, Romania, Bulgaria, and Albania), since BSL was in an
excellent logistical position.

Exxon Chemical: Refinery-Petrochemical Integration

Exxon Chemical has always stated that its liquid feedstock crackers, when built
adjacent to Exxon refineries, enjoy a cost advantage of around one to four cents
per pound over ethylene producers purchasing naphtha on the merchant market. Exxon can vary its feedstock slate to crack various liquid streams from its
adjacent refinery, always selecting those which, on a seasonal basis, have the
lowest value in the marketplace. Furthermore, Exxon Chemical is a leader in
the technology of cracking heavy liquids to olefins. Optimal refinery feedstocks
are selected through a linear program that is used off-line to optimize refinery operations as well as to select the most economical cracking feedstocks for
the adjacent olefins unit. The heavy fuel by-products produced in the cracker
are then sent back to the refinery for further processing or for blending into
fuel oil.

Exxon Chemical had been somewhat slower than BP Chemicals in establishing a petrochemical presence in Southeast Asia. Exxon had, however, a partnership in a large refinery in Singapore, where it had also constructed in
partnership with Amoco Chemicals (now part of BP) an aromatics plant. Exxon
then decided to built a large cracker in Singapore, associated with a polyolefin
and other derivatives unit. The cracker would enjoy not only the usual cost
advantages claimed by Exxon Chemical but also the tax benefits accorded to
companies building on Jurong Island. With propylene in relatively short supply, Exxon, which makes this chemical in large quantities as a by-product from

heavy liquids cracking, established a comparative advantage to other basic petro-chemicals producers in the region, which were operating smaller crackers not associated with refineries.

Methanex: Low-Cost Global Feedstock Position

Methanex, based in Vancouver, Canada, is the world's largest producer of metha-nol. It operates plants in a number of countries and regions, including Canada, the United States, Chile, and New Zealand and an offtake agreement in Trini-dad. (An offtake agreement is a contract whereby a buyer agrees to purchase a certain portion of the total production from a manufacturing plant.) It owns a number of methanol tankers and supplies customers all over the globe. The company is logistically well positioned and can ship to various locations from different plants. With sourcing from such low-cost regions as Chile and Trinidad the firm is able to optimize its overall manufacturing costs and operate more profitably than its competitors most of the time.

When natural gas prices in the United States and Canada peaked briefly in 2000–2001, Methanex reduced its output in those countries and supplied cus-tomers largely with methanol made in lower-cost facilities. At a time when petrochemical producers in North America were suffering from high natural gas prices, Methanex had exceptionally profitable financial results, given the high demand for methyl tertiary butyl ether (MTBE) in the United States, resulting from high gasoline demand and pricing.

While Porter espouses global business management, plastics polymer pro-ducers are not necessarily of one mind with respect to operating businesses on a global or regional basis. Those producers that control fundamental tech-nologies tend to operate globally. Basell, the successor to Montell (including BASF's polyolefin plants), which has plants in many regions of the world and controls fundamental production technologies, provides a good example. Montell operated on the principle that having production plants in various world markets makes the firm a reliable supplier, able to be close to its custom-ers when they move into foreign countries. Dow follows a similar course. In contrast, Union Carbide (now part of Dow) relied on exports and licensing for the most part, while Phillips and Fina selected only certain countries in which to build plants and were far from becoming global suppliers. There is little question at this writing that companies are increasingly espousing a global busi-ness model that, according to Anthony Carbone, then head of Dow's plastics business, "provided an almost immediate benefit due to more rapid decision-making, greater clarity in developing business strategies and becoming more attentive to business development."[17] At the time Carbone was interviewed, several leading European firms, including DSM and Borealis, were still pursu-ing regional business models, including some alliances.

Foreign investments are preferably made when companies believe they have some sort of sustainable advantage for protecting such investments. They then build up global positions and manage their business on a global basis.

Morgan Stanley's chemical group, headed by Leslie Ravitz and Mark Gulley, undertook in 1996 a study to identify companies that appear to have global competitive advantages and to see whether these advantages were factored into their stock price. Three of the broader conclusions they reached were

- *Power costs:* U.S. companies had a distinct advantage, with average costs at 4 cents per kilowatt hour compared with 5.5 to 6.5 cents for Europe and the Far East and 10 cents for Japan.
- *Scale:* The United States also led, with average ethylene capacity per plant at 545,000 tons compared with 450,000 tons for Japan, 390,000 tons for Europe, and 225,000 tons for Latin America.
- *Labor costs:* Europe led here with $52,000 per employee per year compared with $68,000 for the United States. However, average sales per employee in the States was $400,000 compared with $230,000 in Europe, indicating higher productivity in the United States.

A total of seventy-nine companies with global operations were studied, and financial results were compared on a number of bases. The companies were divided into eight groups: agricultural chemicals, coatings, electronic materials, fibers, gases, inorganics, petrochemicals, and specialty chemicals. Of the seventy-nine firms studied, Morgan Stanley identified twenty-one that had what it considered comparative advantage, including fourteen U.S. firms (at least one in every category), five European firms, one Japanese firm, and one Asian firm.[18]

Much of this had changed by 2001—only five years later. Some of the firms have disappeared (Courtaulds and Union Carbide); some others would probably no longer be considered to have comparative advantage (Novartis, Air Products, Praxair, Sealed Air, and PPG Industries). The conclusions of the Morgan Stanley study were reasonable for the time the study was conducted, but the industry moves on.

To what extent does the *size* of multinational companies that are globalizing their business confer an important advantage? McKinsey, a leading strategy consulting firm, recently concluded that high market capitalizations are an essential characteristic of successful global companies.[19] Large market caps give companies a real edge in acquiring other companies and capturing global growth opportunities, while also protecting them from acquisition. The firm cited General Electric, Microsoft, Coca-Cola, Procter and Gamble, American International Group (AIG), and Citibank as characteristic companies with enormous market caps.

In the chemical industry Dow Chemical, DuPont, BASF, and Bayer have among the highest market caps. The major oil companies with chemical operations—Exxon-Mobil, Shell, and BP—are even much larger. It remains to be

Table 3. Asia's Per Capita Consumptions

Product	Per Capita Consumption (kg)	
	Asia	World Average
Petroleum	50–110	560
Ethylene	2–10	14
Plastic products	1–5	17
Fiber	3–5	7

Source: Annual Asian Petrochemical Summit 1997.

seen, however, whether the large chemical and oil companies enjoy greater benefits from globalization than some of their smaller competitors. Certainly they are much less likely to be acquired.

The McKinsey study also found that even with the tremendous advantage conferred by their high market caps, these giant firms have problems in developing a winning global strategy. Among the problems they are dealing with are more opportunities than they can handle and a perceived step change in competition. All of them consider globalization an "urgent issue."[20]

The Focus Shifts to Asia

The most important change in the landscape of the global petrochemical industry is the emergence of Asia as a rapidly growing manufacturing and consuming center. Before 1980 little petrochemical capacity was found in Asia outside Japan, although India, Taiwan, and to some extent Korea had already established relatively small petrochemical facilities. But in the 1980s Thailand, Malaysia, and Singapore began to build large petrochemical projects, soon followed by Korea and Indonesia. In Singapore the existing well-established refining center provided the feedstock since, similar to Japan, there is no natural gas available for feedstock purposes there. In Thailand and Malaysia newly discovered offshore hydrocarbons spurred these countries' petrochemical development, while Indonesia had long-established crude oil and natural gas resources.

The emergence of the four Asian "tigers"—Taiwan, Korea, Singapore, and Hong Kong—has been well documented elsewhere. These countries, which at the start of their rapid growth period enjoyed a combination of favorable government policies, low-cost labor, and available investment capital, planned (with the exception of Hong Kong) on becoming important petrochemical producers. Hong Kong, with no good sources of feedstocks and limited land, selected a number of other areas for emphasis, with banking as a focus for growth.

Taiwan was earliest in establishing a petrochemical industry, with participation by both government and private industry. Environmental problems and lack of cheap feedstocks have kept Taiwan from becoming an even larger player

Table 4. Buildup of Ethylene Capacity in Asia

Location	Production Capacity (in Thousand Tons)			
	1980	1990	1995	2000
South Korea	805	1,205	3,880	5,480
Taiwan	514	889	1,015	2,365
ASEAN*	0	755	2,330	4,275
China	588	1,918	2,840	4,630
Japan	6,200	6,320	7,070	8,000
Asia total†	7,230	11,670	17,700	27,270
World total	47,300	63,600	80,800	102,700
Asia as percentage of world total	15	18	22	27

Source: Nexant, Inc./Chem Systems.
* Association of Southeast Asian Nations includes Malaysia, Thailand, Indonesia, Singapore, the Philippines, and Vietnam.
† Includes other Asian capacity not shown above.

in the industry, though Taiwanese firms, such as Formosa Plastics and Westlake, invested heavily in the U.S. petrochemical industry.

The Asian market for such petrochemicals as plastics, synthetic fibers and rubber, solvents, and adhesives is obviously huge, although it is still far from developed and will take a long time to come even close to maturity. Only a few years ago, in 1997, Asian consumption of petrochemicals and petroleum products was still well below the world average, in some cases far below. China and India represent by far the largest market opportunities (Table 3).

An indication of the explosive growth of petrochemical production in Asia can be gleaned from the statistics in Tables 4 and 5, which also highlight how the polyester business and its precursors have strongly shifted to Asia, a result of the region's dominant position in the manufacture of apparel and other textiles.

The emergence of a large Asian petrochemical industry, including plants built in the Middle East, has dramatically affected the global industry in at least two important ways. First, exports of many commodity petrochemicals to this region from the United States, Western Europe, and Japan have been substantially reduced. Second, petrochemical pricing in the region has at times been even more depressed than in other regions, as new producers such as Korea periodically flooded the market with surplus material and as some of the other new, local producers periodically engaged in damaging competition.

It is not easy to build a petrochemical industry essentially "from scratch." How this was done in three Asian countries is discussed in the next section.

Table 5. Buildup of Pure Terephthalic Acid Production Capacity in Asia

Location	Production Capacity (in Thousand Tons)			
	1980	1990	1995	2000
South Korea	175	1,150	2,140	4,500
Taiwan	490	1,200	2,440	3,390
ASEAN*	0	150	525	2,535
China	0	711	1,111	2,075
Japan	600	1,430	1,580	1,520
India	0	150	250	1,750
Asia total†	**1,265**	**4,790**	**8,050**	**16,170**
World total	**3,560**	**8,420**	**13,120**	**24,450**
Asia as percentage of world total	**36**	**57**	**61**	**66**

Source: Nexant, Inc./Chem Systems.
* Association of Southeast Asian Nations includes Malaysia, Thailand, Indonesia, Singapore, the Philippines, and Vietnam.
† Includes other Asian capacity not shown above.

Initiating a Petrochemical Industry

Singapore

In the 1960s the Singapore government of Lee Kwan Yew embarked on a strategy of encouraging multinationals to set up export-oriented businesses on the island and becoming a "first world" base in a "third world" area, with standards of administration, education, and infrastructure consistent with those in OECD countries.

To facilitate investments, the government created the Economic Development Board (EDB), which would participate financially in projects. In the late 1970s the EDB and Japan's Ministry of Trade and Industry (MITI) jointly helped to finance the first petrochemical complex, known as the Petrochemical Corporation of Singapore (PCS). The owners were Shell Eastern Chemical and a consortium of Japanese firms led by Sumitomo Chemical. A 300,000-ton-per-year world-scale cracker was constructed on Ayer Merbau Island and started up in 1984. It was based on naphthas from several refineries in Singapore. (Shell, Esso, Mobil, and Singapore Refining Company had a total refining capacity in Singapore of close to a million barrels per day, making Singapore the world's third largest refining center after Houston and Rotterdam.) Ethylene, propylene, and other cracker products were converted to polyolefins, ethylene glycol, acetylene black, MTBE, and other derivatives. A number of other firms, including Phillips Chemical, Denka, Kureha, and Hoechst, participated in the PCS complex.

Given the relatively small population of Singapore itself, these products were primarily destined for the Southeast Asian export market. While the economics of this naphtha cracker were not outstanding relative to plants in other countries that were cracking inexpensive gas liquids, the profitability of the PCS complex was superior to that of comparable plants in Japan, which were also cracking naphtha, a more expensive material in that country. Moreover, PCS and its derivatives plants had freight advantages over shipment of comparable products from Japan, which had traditionally been an important supplier to Southeast Asian markets.

Within a few years the EDB and MITI had sold their shares to the operating companies, so that the complex became largely privatized. Other plants followed, including two aromatics complexes (Mobil and Exxon Chemical–Amoco) and isopropanol. A second cracker was built by PCS and started up in 1997, while a third, much larger cracker was built and started up by Exxon-Mobil in 2000. By the end of the decade Seraya Chemical (Shell-Mitsubishi) had started a propylene oxide–styrene plant, Sumitomo Chemical an acrylate plant, Eastman Chemical an oxo alcohol plant, and Celanese a large acetic acid plant. A number of other derivatives plants were also under construction or in operation. These plants were generally built with project financing, involving Japanese or U.S. banks.

Singapore is considered the safest and in many respects the most attractive place in Southeast Asia to invest in chemicals. This is because of government tax incentives, stability, the existence of a large infrastructure of plants and supporting facilities, and Singapore's access to attractive regional markets. However, the economics of PCS, largely based on naphtha, are less favorable than, for example, newer plants based on Malaysian or Thai natural gas liquids. Further, derivatives coming into the region from the Middle East can be sold for considerably less, or carry a higher profit margin, than comparable products made in Singapore.

To the extent that other countries in the region (Thailand, Malaysia, and Indonesia) are also now substantial petrochemical producers, the plants based in Singapore have less and less of the market to themselves. However, Mainland China will be a major importer for a long time.

Regardless, Singapore is firmly established as the largest petrochemical center in Southeast Asia. Unlike petrochemical initiatives in some of the other countries in the region, the establishment of production in Singapore proceeded similarly to that in the established economies of the world. It represents an excellent example of how multinational companies from the United States, Europe, and Japan can work closely with local governments and financing banks to establish a solid industrial base where nothing had existed before.

South Korea

Korea's petrochemical industry was created by chaebols (a Korean term for industry groups) as the country drove to industrialize with strong initial emphasis on exports and a low-cost textile industry. Small derivatives plants producing PVC and polystyrene were already in operation in the early 1970s, as well as a small naphtha cracker operated by Korean Oil (now Yukong) in Ulsan. In 1979 a 350,000-ton naphtha cracker was built by Daelim (Honam Ethylene) and several Japanese joint-venture derivatives producers at Yeo-Chan. In the following years Lucky (Goldstar), Samsung, and Hyundai also became major players in the petrochemical industry. A large number of derivatives plants were built by these firms and their subsidiaries, mostly without participation by foreign firms. By that time Korea was the second largest petrochemical producer in Asia.[21] The chaebols discussed here are very large, with a slate of highly diversified products, including steel, electronics, automobiles, aerospace, and shipping. Samsung and Hyundai had sales exceeding $40 billion in the early 1990s. By then the large chaebols, including Yukong, Daelim, Lucky, and Han Yang, were building an additional two million metric tons per year of new ethylene capacity.

By the early 1990s Korea had lost two original advantages—low wages and favorable exchange rates—leaving it with fewer national advantages for a petrochemical industry apart from a fairly large internal market. But the government continued to encourage strongly the construction of the country's industries, including petrochemicals, by providing assistance with bank financing and various forms of credits, particularly for income earned from exports.

Unlike Singapore, Korea's petrochemical industry is essentially owned wholly by private local firms. While much of the technology is licensed from multinational firms, the Korean government and the chaebols have invested heavily in research and development. Samsung, Lucky, and Hyundai spend between 1.3 percent and 2.6 percent of sales on R&D. Education in Korea is at a high level, and there is a highly skilled workforce. Strong rivalry exists among the chaebols, thus providing some of the advantages identified by Porter (except for factor advantages).

While the Korean government took a strong role in the development of its petrochemical industry, it eventually backed off. Overcapacity resulted in excessive competition between the chaebols. By 1995 ethylene production capacity in Korea amounted to 3.8 million tons, equal to that in France and close to that in Germany (5.5 million tons). Petrochemical production rose from 2.2 million tons in 1987 to 9 million tons in 1995, while the "self-sufficiency rate" increased from 68 percent in 1987 to 125 percent eight years later.[22] With so much overcapacity Korea desperately sought larger export markets, with northern Mainland China the most attractive. With economically highly leveraged plants, companies had to do everything possible to gain hard export currency, and so they flooded the Southeast Asian market between 1991 and 1994 with products surplus to domestic needs and at very low price netbacks.

Korea's example shows that a country can build a petrochemical industry without direct investment by foreign multinationals if feedstocks are available (for example, from local refineries), the government provides incentives and assists with financing, technology can be obtained through licensing, and strong local entrepreneurial firms are willing to invest in a new industry. The last point particularly fits the Korean culture. The country and its chaebols would have built a more stable, less vulnerable industry if there had not been such fierce competition among the large local firms, which created the overcapacity problem, and if the government had been more careful with its assistance to the eager chaebols.

The Korean petrochemical industry is and will remain at a cost disadvantage, being based on imported technology, imported crude oil, and an end market now with relatively high-cost labor. While the country did export $9.4 billion worth of petrochemicals in 2000, there was actually an unfavorable balance of trade in chemicals since Korea has very little fine and specialty chemical or pharmaceutical manufacture.[23] Chastened somewhat by the country's and industry's financial problems in the late 1990s, the Korean petrochemical industry will now grow more slowly. However, it is firmly established as a major producer in the region.

Thailand

The Thai petrochemical industry owes its existence primarily to the discovery of natural gas and condensate offshore in the Gulf of Thailand in the 1980s. With feedstocks becoming locally available, both private and government sources began to plan petrochemical assets. Thai Petrochemical Industry (TPI) had begun producing low-density polyethylene (LDPE) at Rayong in 1982, initially based on imported ethylene. Meanwhile the Thai government began to implement Thailand's first petrochemical complex by building an ethane-based cracker nearby at Map Ta Phut, with the Petroleum Authority of Thailand (PTT) having a controlling interest. Private firms built derivatives plants downstream of the cracker, which began operating in 1989.

To ensure the success of these ventures, the Thai government decided to price feedstocks and ethylene on a cost-of-service basis. It also established relatively high import duties to help the new industry. Chem Systems worked closely with the government in evaluating transfer mechanisms and in providing information on global competitor economics.

Start-up of Thailand's first complex coincided with a rapid increase in demand because of the buoyant Thai economy, while the improved political climate increased investor confidence and attracted some multinational investors. The government's Board of Investment granted various incentives, including tax holidays, accelerated depreciation, and waiver of taxes on imported equipment.

A second cracker was built by Thai Olefins Company and was referred to as NPC-II (National Petrochemical Public Company Ltd.). It started up in 1995 with a capacity of 385,000 tons per year of ethylene. This was followed by an aromatics plant (Aromatics [Thailand] Public Company Ltd., or ATC) based on naphtha and condensate from the Gulf of Thailand. In the private sector Siam Cement in 1999 started up a 600,000-ton cracker. The firm is in a joint venture with Dow Chemical involving the production of polyethylene. A number of other derivatives units have been constructed.

With petrochemicals production firmly established in large plants, Thailand was able to conform to the Association of Southeast Asian Nations' (ASEAN, which includes Malaysia, Thailand, Indonesia, Singapore, the Philippines, and Vietnam) Common Effective Preferential Tariff Plan, under which import duties began to be gradually reduced. Thailand is somewhat limited in feedstocks for future plants, but it is sourcing some from neighboring countries, such as Malaysia and Indonesia.

The Asian recession in the late 1990s hit Thailand hard, slowing down the growth of the country's petrochemical industry. The country had an oversupply of petrochemicals in 2001, but nevertheless wants to turn PTT into an industry leader in Southeast Asia. To this effect, it plans to invest 740 million bahts in petrochemical production, in part to make PTT and its three affiliates—the Thai Olefins Company (TOC), NPC, and ATC—more attractive to investors when the planned privatization occurs. It is doing this with a wary eye on expansions for low-cost petrochemical plants in the Middle East, which historically target Southeast Asia with their exports.

With high projected gross domestic product growth, feedstocks, infrastructure, and a population of 70 million people in the domestic market, Thailand was and remains a logical country to produce petrochemicals. Moreover, it abuts Mainland China, which is the world's largest export market. While Singapore remains a favorite location for foreign investment, building inside a large market has substantial advantages. Such firms as Union Carbide, BASF, and Dow in the late 1990s therefore considered Thailand and Malaysia the best potential locations for petrochemical investment. With its acquisition of Union Carbide, Dow is now producing in both countries, as well as importing to the region from Kuwait.

This chapter would be too long if it were to include descriptions of such other Asian countries as Indonesia and Malaysia jump-starting their petrochemical industry. Southeast Asia has now indisputably become a major petrochemical hub, an exporter rather than an importer, and an important actor on the world petrochemical stage.

Mainland China—Growth with Continuing Imports

China started impressive economic growth in the 1990s, after reforms made by Deng Xiaopeng during the previous two decades. With a strongly rising gross domestic product and a rapidly increasing level of consumer income, China's demand for chemicals and plastics soared. The country quickly became the largest global importer of these materials, with its demand so large that it strongly affected the entire global market. For example, Chinese demand for agricultural polyethylene film dwarfed the demand for this material in most other countries, and most of it continues to be imported. But Chinese buying of petrochemicals has historically had a tendency to fluctuate, partly for political reasons. When China largely stopped importing polyolefins in the summer of 1989, the 1988–89 global petrochemical price fly-up came to a sudden end as material intended for export to China was backed into domestic markets.

China will continue to be a large importer of polymers for at least the next five years. By 2005 polypropylene imports are expected to reach 3 billion tons, and LDPE imports should reach 2.5 billion tons. Polystyrene, high-density polyethylene, and PVC imports should be around one billion tons.[24] These figures will hold despite a vigorous program to build domestic capacity to augment existing production now carried out in many, mostly small, units. In 2001 there were sixteen ethylene producers with capacities ranging from 140 to 550 kilotons per year; there were twenty-six polyethylene producers, over eighty polypropylene producers, twenty-two polystyrene producers, and eighty-five PVC producers. At its March 2000 Annual Petrochemical Conference, CMAI (Chemical Market Associates, Inc.) projected that China's demand for ethylene (including net equivalent imports destined for exports in derivatives and products) would rise from slightly over 6 million tons in 1996 to 18 million tons in 2008, amounting to a 10 percent compounded annual growth rate.[25]

China's petrochemical industry is mostly controlled by Chinese National Petroleum Company in the north and by Sinopec in the south, both of these being petroleum companies with petrochemical participation. These firms are planning or proceeding with large joint-venture complexes with multinationals, including Shell, BP, BASF, and Dow. Other multinationals have for some years engaged in smaller joint-venture projects, including acetate filter tow (for cigarettes) plants with Celanese and high-density polyethylene plants with Phillips Chemical (now part of Chevron-Phillips).

The Chinese chemical industry has long been protected by tariffs, but these are now falling as China gets ready to join the World Trade Organization (WTO). For example, styrene tariffs will fall from 20 percent in 1995 to 3 percent in 2005, and polypropylene tariffs will drop from 25 percent to 9 percent over the same period. Impending membership in the WTO will bring much additional foreign investment to China, with Chinese provinces and municipalities outdoing each other in concessions.

China is anxious to attract foreign firms and their technology and marketing skills. But these foreign firms are meeting increasingly fierce competition from domestic Chinese companies. While there are still many small chemical and fertilizer plants that will not be able to withstand the competition caused by falling tariffs, domestic firms are now building world-scale plants and have the advantage of local knowledge: low local input costs and an established position in a large and fast-developing local market.[26] Nevertheless, the multinationals are determined to be present in this huge market and seem to be satisfied with the progress they are making. BP Chemicals is operating a 200,000-ton-per-year acetic acid plant at Chongqing, started up in 1998, to be doubled in 2003. An 80,000-ton acetate derivatives plant is now under construction.

The Middle East—Exporting Powerhouse

In 1978 the secretariat of the Organization of Petroleum Exporting Countries (OPEC) held a meeting in Vienna. One of the topics was the desire of oil-rich countries to build large downstream businesses in petrochemicals. Present at the meeting were executives from a number of multinational firms who, on the one hand, were vitally interested in obtaining crude oil from the OPEC members but, on the other hand, were concerned that construction of a number of large petrochemical plants in the Middle East and elsewhere would exacerbate the problems of an already oversupplied business. Chem Systems was invited to speak as a "neutral" consultant. In a speech to the secretariat, Peter Spitz, managing director of Chem Systems, acknowledged that oil-producing nations had every right to industrialize their countries by building high-technology downstream plants, but he added a note of caution about supply-and-demand issues.[27] In other words, he suggested that OPEC members building petrochemical plants should be careful not to flood the market at times of overcapacity.

Saudi Arabia soon became a major exporter of petrochemicals, with SABIC set up by the government to be the Saudis' main constructor and marketer of petrochemicals, as well as of fertilizers and metals. SABIC's joint venture partners included Exxon, Shell, Mobil, Celanese, Mitsubishi, ENI, and other multinationals. Other Middle Eastern countries, including Abu Dhabi, Qatar, and Bahrain, also built plants, using inexpensive domestic natural gas. These were largely joint-venture plants as well, with such multinational partners as Atochem, Phillips, and Borealis.

The Middle Eastern countries, when contrasted with Thailand and Korea where initial petrochemical plants were small and needed substantial tariff protection, built large export-oriented plants right away. Because of the multinational firms' interest in partnering in manufacturing facilities based on very low-cost feedstocks, these firms vigorously assisted their local partners in building large, efficient export-oriented plants.

Instead of flaring the natural gas and gas liquids associated with millions of daily barrels of crude oil produced in the kingdom, the Saudis built collection systems and made ethane and propane available as feedstocks for several joint-venture crackers that came on stream in the 1980s. As mentioned above, SABIC's multinational partners participated in a number of petrochemical projects, including production of methanol and MTBE, as well as such ethylene derivatives as polyethylene and ethylene glycol. These plants were built at two huge new sites—Al Jubail on the east coast and Yanbu on the west. Aramco, one of the world's largest refineries, which had been taken over by the Saudi Arabian kingdom, was not allowed to use its feedstocks to make petrochemicals. SABIC later built some plants without international partners, and as time went on, some private Saudi entrepreneurs also built chemical plants.

While Saudi natural gas is no longer priced at the original 50 cents per million BTUs, its price is still around 75 cents, and along with the extracted ethane (which is priced at the same level as the gas), it continues to be a very attractive petrochemical feedstock. While domestic demand for Saudi petrochemicals remains low, exports have no difficulty finding markets in Southeast Asia, Europe, and even the Americas. Shipping costs are not particularly high, and tariffs have been coming down all over the world. Thus, even on a landed basis, Saudi polyethylene and ethylene glycol are at the low end of the cost curve.

SABIC has become one of the two or three largest global producers of ethylene glycol. Even during difficult times of oversupply or economic slowdown Saudi products based on natural gas continue to be bought on the basis of price as well as quality.

When the first group of Middle Eastern petrochemical plants came on line, multinationals were very concerned about the marketing methods and savvy that these new producers would show. Initially, much of the material was marketed by the partners, but SABIC and the others soon showed considerable skill in marketing so that the initial concern abated rapidly.

Two points need to be made to put the Middle East situation in perspective. First, the region has no inherent advantage with respect to crude oil–based feedstocks and therefore with respect to production of aromatics. This statement may surprise some readers, given the fact that most Middle East crude is inexpensive to produce and could therefore arguably be processed into inexpensive aromatics. But the Saudis clearly have no incentive to discount their crude oil into aromatics processing units when they can sell all they want into the crude oil market at full prices. They do discount condensate for petrochemical use, which has supported aromatics plants built by Chevron and SABIC. Second, customers for Middle East petrochemicals are remote, far away from the plants where export material is made. Therefore a substantial part of Saudi and other Middle Eastern exports are "commodities," that is, such specification products as ethylene glycol, styrene, purified terephthalic acid (PTA), and linear LDPE (LLDPE). For more differentiated petrochemicals like special forms of polyethylene, customers tend to want to be closer to their source of supply, allowing greater interaction between buyer and seller. There is now a move in Abu Dhabi and Qatar and in new Saudi projects to differentiate some grades of polyethylene.

Hydrocarbon-rich countries around the Arabian Gulf have already built a large number of petrochemical plants and have many projects and plans in various stages of development. The most dramatic new slate of projects is in

Iran, which despite significant existing capacities has a large domestic market and has therefore not yet become an important exporter. But its enormous low-cost gas reserves are now being used to support a series of mega-projects for olefins and derivatives that could soon rival Saudi capacities. The impact of the Middle East on global petrochemicals seems set to continue to grow.

The Alberta Advantage Fades

Before the oil shocks in the 1970s Canada was not a major producer of chemicals, although it had a relatively diverse chemical industry, often with plants that were smaller than world scale. It also imported a large amount of its chemicals, primarily from the United States. Several multinational companies, including Exxon (Imperial Oil), Celanese, BASF, ICI, DuPont, Union Carbide, Dow Chemical, and Sun, had invested in the Canadian industry over the years, either in Montreal, Quebec, or Sarnia, Ontario. Polysar was a large Canadian firm that grew out of the World War II synthetic rubber program.

While Canada does not have large crude reserves, it has a great deal of natural gas in Alberta and British Columbia. Alberta, after the oil shocks, initiated a program to build an export-oriented petrochemical industry based on inexpensive ethane extracted from this gas. Ethylene production was based on cost-of-service pricing mechanisms to achieve attractive cash costs. The first cracker started up in 1979. The ethylene was then converted into polyethylene (Canadian Industries Limited [CIL], an ICI affiliate), ethylene glycol (Dow), vinyl acetate (Celanese Canada), and vinyl chloride (Dow). The second cracker came on stream in 1984, in conjunction with a styrene plant (Shell Canada). Methanol also became an important export product. These products were shipped either to Asia or across the border into the U.S. Midwest. With Canadian ethane cheaper than that in the United States in most cases, and because much of the U.S. petrochemical industry is located on the Gulf Coast, Canadian material was generally competitive in its U.S. markets.

The Canadian petrochemical industry gradually shifted to Alberta because feedstock costs in Montreal and Sarnia were more expensive and so much of the new investment capital went to Alberta. Ethylene was often in excess supply, and some was shipped to Sarnia through the Cochin pipeline, which had been built primarily to send excess ethane to Canada's eastern petrochemical centers as well as to the United States.[28]

Alberta continued to receive most of the petrochemical investments in the 1990s, including a large new joint-venture cracker built by Nova Chemicals and Union Carbide, which had previously built large ethylene glycol plants jointly with Far Eastern partners. Consolidation of the industry eventually reduced the number of participants. By 2001 the main players in Alberta were Nova Chemicals and Union Carbide (now part of Dow), with Shell, which produced styrene, and BP Chemicals, which produced alpha-olefins, also owning large plants. Methanex had earlier taken over the Alberta Gas Chemical methanol plants in Alberta and British Columbia but had shifted its new investments to areas with lower feedstocks costs, principally Chile and Trinidad.

Canada has for decades been investing heavily in converting Alberta's tar sands to synthetic crude oil, with a new 155,000-barrel-per-day plant expected to come on stream in 2002. This plant, as well as earlier tar sands–based production, creates chemical feedstocks but not in very large amounts.

Unfortunately for Canada the rich sources of natural gas in the western provinces are beginning to dry up, so that most likely only one more new cracker can be built based on ethane extracted from this gas. More important, Canadian natural gas is now sold in large quantities to the United States, where utilities are switching to this clean source of fuel and where much of the new generation capacity will use natural gas. In fact, 94 percent of new electrical-generating capacity in the United States will be based on natural gas, which has a much higher percentage of its total capacity based on gas than is the case for Canada. So the Canadian gas price is now close to that of domestic U.S. gas by simple supply-and-demand considerations. When in early 2001 gas prices in the United States peaked (see beginning of the chapter), Canadian gas prices also rose sharply, reaching peaks not far below those in the lower forty-eight states. With Albertan plants world scale or larger and gas still slightly cheaper, firms like Nova still have some advantage in cash costs, although shipping costs from Alberta to its historical destinations eats up some of this advantage.

Canada now sees the projected gas pipeline from Alaska as its best hope for additional, relatively inexpensive natural gas and gas liquids. The federal and provincial governments as well as private industry hope and expect that Alaskan gas will follow a NAFTA-based exploitation, with Alberta a partial beneficiary. If not, they say that Alberta may have seen its last big petrochemical investment.

There is some hope that natural gas found offshore of Sable Island in Nova Scotia can "breathe some life" into the Montreal and Sarnia chemical industry. If not, some basic chemicals will sooner or later probably be phased out in both these locations, which will undoubtedly continue to manufacture higher-value chemicals.

South America—Considerable Progress

Chemical industry developments in South America over the last two decades have been greatly overshadowed by the explosive growth of chemicals production in Asia. This is in some respects surprising given some of the similarities between the two regions: unfulfilled demand (for example, low per-capita consumption of chemicals and plastics); raw materials availability from Venezuela, Mexico, and Brazil; and the presence of entrepreneurial groups in Brazil, Mexico, Argentina, and Chile, among others.

An important difference is that in all the South American countries cited, there was already an established, although in many cases not world-scale, chemical industry, in contrast to such countries as Korea, Malaysia, and Thailand, where the governments were bent on establishing a world-scale industry as quickly as possible with heavy government support. Moreover, periodic serious problems with currency, inflation, and changes in governments have affected the pace of industrialization in Latin American countries.

Since Brazil is the largest chemicals producer in the region, developments in Brazil are of greatest importance. Perhaps the most significant development was the Brazilian government's decision in 1991 to lower import duties drastically and to eliminate price controls. This step forced local companies to become more competitive and has also led to a complete restructuring of ownership, as well as to consolidations. The number of companies in Brazil has decreased to about one-third of what it was in the early 1980s. Most of the original local private groups are still in positions of ownership, the strongest being Odebrecht, but also including Suzano, Unipar, Ultra, Mariani, and Ipiranga Petroquimica.

The much lower import duties have increased imports from various sources. However, Brazilian producers alert importing companies that exporters are not necessarily reliable suppliers over the entire cycle, since many firms export on an opportunistic basis. Thus, Oxiteno, which is Brazil's only local producer of ethylene oxide and derivatives (including ethylene glycol used for PET bottle resin production), ethanolamines, and ethylene oxide adducts, can hold its own against imports and also has a substantial export business.

Creation of the Mercosur trading bloc was another significant development; the bloc includes Brazil, Argentina, Paraguay, and Uruguay. This coincided with privatization initiatives in Brazil and Argentina, where the government holdings in petrochemicals were largely sold to local and multinational entities. Dow Chemical, which already had some chemical operations in the region, bought the Bahia Blanca complex in Argentina and thus became one of the largest petrochemical producers in Mercosur. BASF and Bayer have also invested strongly in the region. And Petrobras, after selling most of its petrochemical participation (in Petroquisa) several years earlier, is reentering the scene with the new Rio Polimeros cracker in Duque de Caixas, Brazil, based on offshore gas liquids. Several of the private groups, including Unipar and Suzano, have joined to build derivatives plants downstream of this cracker. Petroquisa's once very substantial holdings in forty or more companies were by 1998 in the hands of private groups, except for 15 percent to 17 percent participation in the Petroquimica Uniao, Copene, and Copesul crackers; 28 percent in linear alkyl benzene (LAB) producer Deten; and minority participations in several other firms, including Metanor, Ciquine, and Nitrochlor.[29]

After the recent auction of Copene in Bahia, Odebrecht, the winner, consolidated its interests with the Mariani Group in a company called Braskem. This company is now the thirteenth largest global polyolefins producer and the eleventh largest PVC producer in the world. The Copesul cracker, partly owned by the Ipiranga group and Odebrecht, may be consolidated with the Rio Polimeros complex.[30]

Brazil now has a substantial capacity for olefins and polyolefins at its three existing petrochemical poles and must export its excess capacity. The recent devaluation of the *real* has made these exports much more competitive. However, Brazil still has an unfavorable trade balance in chemicals, given the relatively modest production of fine and specialty chemicals.

Argentina has been ramping up its petrochemical capability since the late 1990s when Dow Chemical acquired the Bahia Blanca cracker. Now a mega-cracker with a capacity of 450,000 tons per year will be built by a joint venture that includes Petrobras (34 percent), Repsol (38 percent), and Dow (28 percent). YPF, the Argentine oil monopoly, has recently built a 400,000-ton-per-year methanol plant, while Repsol-YPF and Agrium constructed a $600 million urea plant.

Venezuela, where Pequiven was contemplating large projects in José and elsewhere up to several years ago, based on its virtually limitless hydrocarbon resources, delayed these projects as a result of the change in government. Given the low price of gas in Venezuela (as low as 50 cents per million BTUs compared with three dollars or more in the United States), the country should eventually become a major exporter of ethylene derivatives and methanol. Exxon-Mobil is likely to proceed eventually with a million-ton-per-year ethane cracker, feeding large polyethylene and ethylene glycol plants, but probably not before 2007. Chevron-Phillips is teaming with Pequiven for a styrene plant to be built in the mid-2000s. One nagging issue is the proposed increase in royalties on hydrocarbons that the government is likely to charge.

Mexico Poised for Growth

Mexico's petrochemical industry made little progress over the last two decades owing to continuing uncertainties and inaction regarding Pemex, the government oil monopoly that controls all feedstocks and operates the country's crackers. Now, with a change in government, the country is on the verge of a real breakthrough. Pemex will now be run by Raul Munoz Leos, the former DuPont Mexico CEO, and Pemex Petroquimica (PPQ) will most likely be run by another executive from the industry. While Pemex for historical reasons will not soon cede control of the country's hydrocarbon resources, it is now almost certainly going to have PPQ enter into joint ventures with private groups and multinationals operating or building derivatives plants. This will allow Mexican petrochemicals to be effectively vertically integrated, similar to most such complexes in other countries. However, large new investments in feedstock production will be required, as Pemex has underinvested in this area for many years. The forthcoming 49 percent privatization of Pemex with Mexican partners is a positive step.

Mexico's relatively buoyant economy and the new government policies have caused multinationals to look at the country as a good area for investment. Local private groups are also encouraged, and there are plans for a very large (975,000-ton) naphtha-fed cracker to be built in Altamira by Corporacion Serbo.[31]

At this writing the strong peso, high energy costs, and low petrochemical prices have created problems for the country's chemical industry. Imports from the United States have increased substantially because of NAFTA, the strong Mexican currency, and the loss of Asian markets. Mexico's chemical deficit grew from less than $1 billion in 1995 to over $5 billion in 2000.

With a large market (over 100 million people), good raw materials availability, and a new government that strongly favors private enterprise, Mexico is likely at last to become an important global player in petrochemicals.

* * *

This chapter could be much longer, but in combination with Chapters 1 and 10, it should still provide a reasonable overview of a very large and diverse global industry. Its emphasis has largely been on petrochemicals, and it is acknowledged that there are important omissions, such as the large presence of India and China in fine chemicals, dyestuffs, and other specialties. Further, little has been written about the chemical industry in the former Comecon countries, which are becoming more important factors in the global industry.

Perhaps the most important point is the one made at the beginning of the chapter—the fact that what was once largely a regional industry with some global participation by multinationals has now become a truly global industry.

Endnotes

1. Michael E. Porter, *The Competitive Advantage of Nations* (New York: Free Press, 1990).
2. Robert Stobaugh, personal communication. (Stobaugh, a professor at Harvard Business School, has studied and taught extensively in the area of foreign trade in chemicals.)
3. Robert B. Stobaugh, *Nine Investments Abroad and Their Impact at Home* (Cambridge, Mass.: Harvard University Press, 1976).
4. Robert Stobaugh, *Innovation and Competition: The Global Management of Petrochemical Products* (Cambridge, Mass.: Harvard Business School Press, 1988).
5. J. F. Wyatt, "Changing Trade Flows and Investment Needs," European Chemical Marketing Research Association conference, Barcelona, Oct. 1987.
6. Ibid.
7. Kurt Doldner, personal communications.
8. Gregory D. L. Morris and Suzanne McElligott, "Traders Cozy to Suppliers: Producer Ties Eclipse Spot Trading," *Chemical Week* 158 (18/25 Dec. 1996), 25–29.
9. Roy Neresian, *Ships and Shipping: A Comprehensive Guide* (Tulsa, Okla.: Pennwell Publishing, 1981), 18ff.
10. Richard D. Freeman, "The Chemical Industry: A Global Perspective," *Business Economics* 34 (Oct. 1999), 16–22.

11. Ibid., 18–19.
12. John Roberts, personal communication.
13. Freeman, "The Chemical Industry" (cit note 10), 21.
14. Mike Parker, personal communication.
15. *1998 Barometer of Competitiveness* (Brussels: Conseil Européen des Fédérations de l'Industrie Chimique [CEFIC], 1998).
16. Michael E. Porter, *Competitive Strategy: Techniques for Analyzing Industries and Competitors* (New York: Free Press, 1980).
17. Gregory Morris, Robert Westerwelt, and Sylvia Pfeifer, "Plastics' Clear Objectives: Regional Players Challenge Lenders," *Chemical Week* 159 (18 June 1997), 27–30.
18. Morgan Stanley and Company, "Chemical Industry Equity Investment Research Report," 7 Nov. 1996.
19. Lowell Bryan et al., "Corporate Strategy in a Globalizing World," *McKinsey Quarterly* 3 (1998), 7–19.
20. Ibid.
21. "Production Trends in South Korean Petrochemicals Industry," *Chemical Economy and Engineering Review* 18 (Oct. 1986), 24–27.
22. Bank of America, *Guide to Petrochemicals in Asia* (Hong Kong: EFP International, 1997), 175–186.
23. Jean-François Tremblay, "Change Arrives in South Korea," *Chemical and Engineering News* 79 (6 Aug. 2001), 15–19.
24. Andrew Swanson, "China Petrochemicals after WTO," Chem Systems Annual Conference, Houston, Jan. 2001.
25. Mark A. Berggren, "Lessons Learned by Post-Crisis Asia," Chemical Market Associates, Inc., Fifteenth Annual World Petrochemical Conference, Houston, 29–30 March 2000.
26. Jean-François Tremblay, "Rapid Changes Come to China," *Chemical and Engineering News* 78 (21 Aug. 2000), 25–34. See also Balaji Singh and Irma Tan, "Global Watch: The Rise of Petrochemicals in China," *Chemical Market Reporter* 262 (15 July 2002), 25–27.
27. Peter Spitz, keynote speech, Organization of Petroleum Exporting Countries (OPEC) Secretariat, Vienna, Oct. 1978.
28. C. L. Dmytrk, "The Petrochemical Industry in Alberta," European Chemical Marketing Research Associaton conference, Barcelona, Nov. 1987.
29. Pedro Wongtschowski, *Industria quimica: Riscos e opportunidades* (The chemical industry: Risks and opportunities) (Sao Paulo, Brazil: Edgard Bluecher, 1998).
30. Kara Sissell, "Streamlining Brazil's Chemicals Ownership," *Chemical Week* 163 (31 Oct. 2001), 21.
31. Kara Sissell, "Crafting a New Pemex," *Chemical Week* 163 (20 June 2001), 23–24; and Sissell, "Private Sector Gets a Voice in Government," ibid., 26–29.

Chapter 9

The Financial Community Takes Charge

John Roberts

Over the last two decades the influence of the financial community on the chemical industry and its management became broad and in many respects pervasive. Investment and commercial banks, which form a part of this community, have traditionally been very important to the industry, rendering various financial services. What has changed dramatically, however, is the importance of stock prices.

The price of a company's stock is now much more important to executive compensation. And stock prices are a key driver of the significant acquisition and demerger activity that is reshaping the global chemical industry. Companies split up their businesses to "unlock" value, and many mergers and acquisitions (M&As) are done, or not done, because of the potential implications for the companies' stock prices.

This chapter attempts to show how the financial community views the chemical industry and highlights areas where the financial perspective may be different from the perspective of the managers who operate within the industry. Also discussed are some of the key drivers of chemical stock prices and potential "potholes" in analyzing company financial results that are particularly relevant to chemical firms.

The evaluation of chemical stock prices is done primarily by institutional investors, who manage 401K accounts, pension funds, and direct investment in mutual funds that collectively own the majority of public companies, and now exert much more influence on the management of chemical companies. As more and more people participate in the market through these vehicles, the vehicles themselves now dominate the market and can be said to hold the fate of companies largely in their hands. As will be seen later in this chapter, institutional investors own 60 percent to 70 percent of some chemical companies. If the managers of these funds become disenchanted with the performance of specific companies whose shares they hold in their portfolios, they may decide to sell these shares. If a number of large funds decide to sell shares as a result of their own analysis or owing to security analysts' recommendations, the value of these companies' shares may be seriously affected. Moreover, large funds find it increasingly difficult to invest in companies with a low market capitalization, since these funds make large specific investments, but they prefer not to hold more than a small percentage of stock in a given company. Accordingly, small companies have a more difficult time attracting investment from large funds. How these investors gauge the value of companies' shares is a key part of this chapter.

Given this situation, chief executive officers (CEOs) and other top management executives spend a great deal of time explaining to security analysts and institutional investors why their firms represent good investment vehicles. Conversely, the large investors have made it clear to company management that they expect firms to deliver positive shareholder value. They now insist that firms use accepted financial tools, such as economic value-added (EVA) analysis models to substantiate regarding their financial performance and to identify parts of the business where shareholder value is actually being destroyed.

Other segments of the financial community have also assumed a more critical role as the chemical industry has undergone massive consolidation and restructuring. While investment bankers have traditionally helped firms with M&As, the pace of such transactions has greatly accelerated as firms seek more rapid growth and a chance to reduce fixed costs by combining a number of functions duplicated in both of the original two firms. In earlier days companies themselves usually decided on transactions and then used investment banks to carry them out. More recently, however, many transactions were actually conceived by the bankers, who showed firms how these could be realized, often with highly creative financing schemes. The fact that such schemes often loaded firms with a great deal of additional debt made these firms even more dependent on their financial advisers.

Another important development over this period was the rapid growth of private investment partnerships (so-called buyout funds), which were prepared to acquire companies or more often divisions or businesses via leveraged buyout (LBO) transactions. This created a large new market for businesses that no longer fit the long-term strategies of firms and could be spun off and made private. Such businesses could then be restructured—often achieving major cost reductions—and expanded under the management of these funds. Because chemical businesses are relatively capital intensive, they tend to have large cash flow characteristics, often allowing the LBO firms to pay back the acquisition debt fairly rapidly. The fund's equity can then grow steadily, providing a good return to the fund when several years later the business is either sold to the public in an initial public offering or is bought by a public company that sees the business as having a good fit with its existing operations.

The underlying theme is that the fate of companies is now much more dependent on stock prices, which are in the hands of money managers who must decide every day whether to buy, sell, or hold the shares of these firms. How they make these decisions is described in some detail in the rest of this chapter, which represents a security analyst's view of the industry.

Union Carbide: High Earnings Target Leads to Acquisition

"Four dollars in EPS (earnings per share) at the next trough." That financial target, introduced to the investment community in October 1997 by William Joyce, newly promoted chairman and CEO of Union Carbide, may be one of the most historic in the industry. But at that multi-hour presentation in midtown Manhattan, Joyce did not announce something perhaps as important; rather, it came out in response to a question. If the company was acquired or merged before the four-dollar mark was achieved, then management would be compensated as if they had achieved that target as long as the deal valued Union Carbide stock at a minimum of $64 per share ($4 times a "normal" price earnings [P/E] ratio of 16 times). In August 1999, with Wall Street expecting EPS of only $2 for 2000 (which later proved far too optimistic), Union Carbide agreed to be merged into Dow Chemical. Union Carbide traded above $64 that month, and Joyce received $24 million in compensation when the deal closed—essentially what he would have earned had the $4 peak EPS target been achieved.

There undoubtedly were a number of factors that led to the disappearance of Union Carbide—one of the earliest chemical companies listed on the New York Stock Exchange (NYSE) (1 March 1926)—as an independent company. But the importance of EPS and stock prices should not be minimized, and Wall Street has played a major role in the reshaping of the chemical industry. Other recent high-profile examples of the price of a company's stock changing its fate include the liquidation of Dexter (the oldest company of any industry on the NYSE at the time) and the pending sale of Hercules, and the sale of Monsanto to Pharmacia. Company earnings and stock price performance have also played a significant role in the large number of spin-offs in the chemical industry.

First Call, a service that tracks the Wall Street analysts covering the stock market, lists fifteen analysts (including this author) covering DuPont. These analysts cover most of the other publicly traded U.S. chemical companies. The extent to which these analysts influence a company's stock is a source of constant academic debate, but few would disagree that this is a constituency of high importance to the CEOs of essentially all public U.S. chemical companies. Almost every company hosts a conference call (which are now open to the public as a result of the Securities Exchange Commission's Regulation FD [full disclosure]) each quarter with this community. Most companies host all-day or multi-day meetings at least every few years with the analysts (along with institutional investors), as well as periodic field trips to visit company operations. And senior managers from most chemical firms participate in annual conferences held by the largest brokerage firms.

The topics discussed in this chapter are relevant to a number of industries, but the examples are specific to the chemical industry. In particular, the focus is on the pitfalls of many financial analyses when applied to chemical companies. It appears that little academic attention has been paid to how financial observers should view highly cyclical companies. Before the chemical industry is analyzed, however, the next section discusses how Wall Street's view of the chemical industry may differ from the industry's view of itself.

Wall Street's View of the Chemical Sector

Celanese stock trades at the lowest valuation of any company in the U.S. major chemicals group. It trades at two times cash flow compared with ten times for Dow, and at 20 percent of the estimated replacement value compared with 100 percent for Dow. While there are many reasons for valuation differences, such an extreme gap probably comes from the difficulty Wall Street has in categorizing Celanese.

The largest percentage of Celanese's sales, assets, and employees are in the United States, but the company is almost exclusively covered by European analysts. Celanese is a hybrid of a growth business (for example, specialty plastics and chemicals), a cyclical commodity business (for example, petrochemicals), and a leveraged buyout, or LBO, type of business (for example, cigarette filter fiber). This situation results from the acquisition of U.S.-based Celanese in the 1980s by Hoechst (now part of Aventis), a highly diversified company, and the subsequent spin-off at the start of 2000. To make the spin-off tax-free to Hoechst's German investors, Celanese was domiciled in Germany, even though its heart is in the United States.

Wall Street views the chemical sector along several dimensions that may differ from the views of other observers, and it tries to organize securities neatly along these dimensions. The financial markets are quite interested in the owners of a company, and so private companies are often treated differently from public companies, for example. And even for public companies there is often a different view of a company's debt from its equity or stock. Some companies have public debt but no public equity, like Huntsman Chemical and Dow Corning (which has equity owned by Dow Chemical and Corning Inc.). And other companies have public equity and no debt, such as Cabot Microelectronics.

There are also different flavors of equity (for example, voting versus non-voting stock), different flavors of debt (for example, public bonds and bank credit lines), and combinations of both (for example, convertible preferred).

Table 1. U.S. Chemical Sales by Ownership

Type of Company	Estimated Sales in 2000 (U.S.$ Billions)
U.S.-based publicly traded chemical firms	100
U.S.-based private chemical firms	50
Divisions of U.S.-based diversified firms	50
U.S. subsidiaries of foreign companies	75
Total U.S. chemical industry	275

Source: John Roberts, "And Then There Was One? Consolidation to Remain Major Industry Theme," Merrill Lynch Equity Research report, 6 Mar. 2001.

And the stocks of chemical companies are often characterized along different dimensions, such as growth versus value stocks or large capitalization versus small capitalization.

Furthermore, foreign companies are often treated differently than U.S. companies, although the differences are small enough that this chapter deals almost exclusively with U.S. companies. At the major U.S. brokerage firms, none of the U.S. analysts cover foreign chemical stocks, and vice versa. And U.S. investor holdings in foreign chemical stocks are still very small overall.

The Many Sides of Wall Street

Table 1 provides a breakdown of the sales of the U.S. chemical industry along a dimension of interest to Wall Street. The sales are divided into publicly traded U.S.-based companies whose primary business is chemicals, chemical divisions of diversified companies (both public and private), private U.S.-based chemical companies, and U.S. subsidiaries of foreign-based chemical companies, which may be public outside the United States. Only the first category, publicly traded U.S.-based companies whose primary business is chemicals, are of significant interest to most U.S. investors.

There are several foreign companies with stock listings in the United States, but the value of the U.S.-based holdings is very modest. And U.S. investors have holdings overseas in foreign chemical companies, but these are also very modest. While the chemical markets themselves may be global, most investors in chemical companies are still largely domestic, or at best multiregional.

Although this chapter draws heavily on U.S. data, many of the conclusions are also applicable to international markets. The British chemical industry, for example, is consolidating even more than the U.S. industry (frequently by U.S.-based acquirers), but for the similar rationale to create larger companies, which is discussed later. Public markets for chemical companies in emerging economies (for example, China) have also developed in the last decade or so. The first Chinese company ever to go public on the NYSE was Shanghai Petrochemical, in 1993. Focusing on specific international markets, however, is beyond the scope of this chapter.

Another difference between financial and nonfinancial market observers is that the financial markets do not consider the pharmaceutical industry part of the chemical industry, even though pharmaceuticals are part of the government's SIC (standard industry classification) group 28. Government data and information from the American Chemistry Council (formerly the Chemical Manufacturers' Association) include the results of the pharmaceutical industry in chemicals. This is also true for such consumer products as toiletries and cleaning products. These areas are generally not covered by the chemical investment community.

Another interesting characteristic of the U.S. chemical industry is that, even excluding pharmaceutical and consumer product sales, the first group of companies shown in Table 1 represents only 36 percent of industry sales. So most of Wall Street is focused on a relatively narrow part of the chemical market. Among the largest companies in the industry are privately held Huntsman Chemical and the chemicals divisions of GE and ExxonMobil, for which the chemical divisions of the latter are not a significant point of interest for most investors. And the largest U.S. chemical by volume is sulfuric acid, which is not a key product for any company in the first category. The largest product among the first group of companies is ethylene, and even here the U.S. public chemical companies represent only 40 percent of the North American market and only 15 percent of the global market.

The Equity Market View

Table 2 shows the global sales and market capitalization for the first group of companies shown in Table 1. Investors care more about the market capitalization of a company than about its sales. Although the equity market is only one facet of the financial markets, it is the one most commonly referred to when discussing company performance. Debt is a fixed obligation with no ownership right to company's earnings beyond the payment of interest and return of principal. So except for the least creditworthy companies, changes in a company's financial performance are evident first and to a greater degree in the performance of a company's stock. And it is stock that has become a prime driver of executive compensation.

Our discussion of debt issues will therefore be limited to its role in valuation. But for the least creditworthy companies, their debt takes on more equity-like characteristics. And in the extreme of bankruptcy the debt holders become equity holders. There are various hybrid securities with debt and equity characteristics, but for the sake of this discussion debt and equity will be treated as distinct and the focus will be on the equity markets.

Table 3 compares the equity market value of the U.S. chemical sector with a number of other sectors and with the market overall. Although chemicals are ubiquitous in the economy, there are many mutual fund managers and other institutional investors who simply ignore the sector. At its current size the chemi-

Table 2. Equity Market Values of U.S. Publicly Traded Chemical Companies
(as of 20 June 2001)

Company	U.S.$ Billions	Company	U.S.$ Billions
DuPont	49.16	Solutia Inc.	1.39
3M	45.36	Hercules	1.26
Dow Chemical	29.79	Crompton Corporation	1.22
Air Products	9.98	Celanese AG	1.14
Monsanto	9.03	Nova Chemicals	1.12
Praxair	7.48	Millennium Chemical	0.97
Rohm and Haas	6.98	RPM	0.93
Ecolab	5.23	PolyOne	0.83
Avery Dennison	5.06	Minerals Technologies	0.81
Eastman Chemical	3.63	Ferro Corporation	0.77
Engelhard Corporation	3.19	Olin Corporation	0.73
Sigma-Aldrich	3.16	H. B. Fuller	0.68
Cabot Corporation	2.77	Wellman Inc.	0.57
FMC Corporation	2.22	MacDermid	0.55
Valspar	1.78	Arch Chemical	0.50
Lyondell Chemical	1.67	Park Electrochemicals	0.48
Cytec Industries	1.56	Georgia Gulf	0.48
Great Lakes	1.51	ChemFirst	0.38
Cabot Microelectronics	1.42	Polymer Group	0.08

Source: John Roberts, "And Then There Was One? Consolidation to Remain Major Industry Theme," Merrill Lynch Equity Research report, 6 Mar. 2001.

cal sector is relatively unimportant to an investor trying to beat the overall market.

The investor view of the maturing of the U.S. basic materials industry, of which the chemical industry is the largest part, can be seen in Table 4. Chemical stocks today make up about 47 percent of the basic materials stock index, not very different from its leadership position at the beginning of the table. But the basic materials sector as a whole has declined to about 2 percent of the S&P 500, down from 15 percent in 1968. The shift toward so-called new economy sectors has been surprising to many since most new economy products are dependent on "old economy" materials and chemicals.

Table 4 may also reflect slower growth in U.S.-based chemical operations compared with that of international companies. The areas of growth globally have been concentrated in the emerging markets as they target self-sufficiency and in the Organization of Petroleum Exporting Countries (OPEC) that have low-cost raw material positions in petrochemicals.

Table 5 lists the percentage of the largest chemical stocks owned by institu-

Table 3. Standard & Poors Market Capitalization by
Sector (as of 31 March 2001)

S&P 500 Sector	Percentage of S&P 500
Chemicals	1
Other basic materials	1
Capital goods	10
Communications services	5
Consumer cyclicals	9
Consumer staples	13
Energy	7
Financials	18
Health care	13
Technology	18
Transportation	1
Utilities	4

Source: John Roberts, "And Then There Was One? Consolidation to Remain Major Industry Theme," Merrill Lynch Equity
Research report, 6 Mar. 2001.

tional investors. These owners are mutual funds and institutional investors who
manage company retirement plans, foundations, and so forth. Collectively, they
own about 70 percent of the stock of the companies listed, probably a similar
percentage of the total publicly traded U.S. chemical industry.

The high level of institutional ownership has placed a greater emphasis on
balancing short-term performance against long-term objectives—and the long
term becomes a series of short terms. A number of chemical industry executives have described the management pressures they face as "running a series
of fifty-yard sprints." The need to deliver short-term earnings performance
makes it increasingly difficult to sustain spending on such long-term programs
as capital spending and especially R&D (which is expensed immediately rather
than capitalized).

Results for institutional investors are published quarterly in a number of
journals, and investors can and do reallocate their funds on the basis of these
results. Funds facing net redemptions are forced to sell stocks to generate the
cash outflows. The underperformance of a sector, such as chemicals, can feed
on itself since redemptions in the underperforming funds cause more selling of
the stocks that have already underperformed. Stock valuations often move to
extremes that differ significantly from long-term trends. This volatility is evident in that the fifty-two-week high for the average NYSE stock is about 40
percent higher than the fifty-two-week low, year in and year out, and chemical
stocks are no different.

Table 4. S&P 500 Composition over Time

S&P 500 Sector	Percentage of Total Market Cap					
	1968	1980	1986	1990	1994	1998
Basic materials (including chemicals)	15	8	7	7	7	3
Capital goods	14	11	11	10	10	8
Communications services	10	6	8	9	9	8
Consumer cyclicals	13	10	14	11	12	9
Consumer staples	8	8	12	17	16	15
Energy	14	27	12	13	10	6
Finance	7	6	10	8	11	14
Health care	3	6	7	10	9	12
Technology	7	10	9	7	10	19
Transportation	3	2	2	1	1	1
Utilities	6	6	8	7	5	3

Source: John Roberts, "And Then There Was One? Consolidation to Remain Major Industry Theme," Merrill Lynch Equity Research report, 6 Mar. 2001.

Table 5. Institutional Ownership of the Top Chemical Companies (as of 31 March 2001)

Company	Percentage Institutionally Owned
3M	70
Air Products	80
Cabot Corporation*	60
Dow Chemical	65
DuPont*	52
Eastman Chemical	75
Engelhard Corporation	85
FMC Corporation	73
Lyondell Chemical	87
Nova Chemicals	53
Praxair	81
Rohm and Haas*	75
Average	**71**

Source: Spectrum Data Services.
* Excludes significant founder ownership.

Table 6. Executive Compensation for Top 200 Industrial and Services Companies

Type of Compensation	Percentage	
	1989–1990	1999–2000
Stock options	25	58
Long-term incentive	24	14
Annual incentive	21	19
Salary	30	9

Source: Association of Executive Search Consultants, http://www.aesc.org/presentations/5, 15 October 2003.

Executive Compensation

Table 6 shows the increase in stock ownership for CEOs over the past ten years. A similar increase has occurred in the chemical companies as well. Management is increasingly taking the equity market's view of their companies as their compensation is increasingly aligned with that of shareholders. The growth in the importance of stock compensation has outpaced almost all other measures of growth in the chemical industry, as it has in most other industries. In some extreme cases CEOs have been paid totally in stock or options, or both, with no base cash salary for a year or more, as with both Thomas Gossage of Hercules and James Malcolm Edward (Ted) Newall of NOVA Corporation, before the merger of its pipeline business with TransCanada and simultaneous spin-off of NOVA Chemicals.

Minimum stock ownership guidelines are becoming commonplace, and stock ownership is being driven down to front-line employees. Many permutations and combinations of stock compensation exist. Dow Chemical executives, for example, were granted stock options in 1988 that expired worthless in February 2003 since Dow's stock did not trade above $50 per share for at least thirty days prior to its expiration.

While some stock-based compensation is granted on the basis of achieving a certain stock price itself, most stock-based compensation is granted on the basis of achieving operational goals. These can include growth targets for earnings or sales, or both; return-on-investment targets; or such event-driven milestones as completing an acquisition or achieving targeted cost savings. Improving economic value added, or EVA, has also become an increasingly popular compensation target.

Specialties versus Basics

Sell-side investment analysts categorize chemical stocks in broad terms into specialty and "other" chemical companies. But this is only a U.S. phenomenon of the past fifteen years and has not existed outside of the United States. Very few firms on the buy side of Wall Street make this distinction. The sell side consists primarily of brokerage firms who sell stock and recommendations, while the buy side consists of money management firms that do the actual investing in the stocks of chemical companies. These "other" nonspecialty companies include commodity chemical producers as well as diversified companies and specialty companies that have been demerged from diversified companies but have not been reclassified by Wall Street as specialties. They are sometimes referred to as "major chemical" companies, which is a vague term reflecting the diversity of the chemical companies.

Increasingly, the buy side of Wall Street is organizing less by industry and more by investment style, such as growth versus value or large cap versus small cap. These investors care less which industry a company is in (and much less about such subcategories as specialty versus basic chemicals). So a company like DuPont is increasingly compared more with other large Dow Jones Industrial Stocks, like GE and Alcoa, and less with other chemical companies.

The coverage of stocks by sell-side analysts has not kept up with the changes in the industry, and so the delineation between *specialty* and *basic* is losing its meaning. For example, Monsanto, which once had a sizable basic chemical business (now known as Sterling Chemicals and parts of Solutia), is still primarily covered by basic chemical analysts, even though Monsanto today is a crop sciences company (agricultural biotechnology plus Roundup herbicide).

Most specialty companies are also smaller, and the focus on larger capitalization stocks has driven specialty analysts to focus on the few larger companies that straddle the specialty-basics world. Industrial gas companies (what is more generic than air separation?), which are highly capital intensive, were once mostly covered by basics analysts but are now mostly covered by specialty analysts, although the majority has not been large in either direction. Rohm and Haas is another company that has been historically covered by basic analysts but has begun to draw coverage from specialty analysts since its acquisition of Morton International. Rohm and Haas is no longer any more of a specialty company than DuPont, with the latter exclusively covered by basic analysts.

The delineation between *specialty* and *basic* has also become blurred as specialty markets have become larger and matured. Former specialty companies have been posting cyclical earnings, including losses, with more frequency. So former specialty companies are acting more like basics, and former basics companies are becoming more like specialties as they shed their basic businesses and invest in specialty businesses. Even Dow Chemical today claims about 50 percent of its sales from specialty products.

Specialty products are generally thought to mean higher growth; more value added; more of a sales, service, and technology orientation; and less capital intensity. Later in this chapter the pitfalls of analyzing the financial results of chemical companies will be discussed, but there is no operational-financial definition that would filter out the companies identified as "specialty" by Wall Street. If a software program were written to sort companies by ratios such as

SG&A (selling, general, and administrative costs) as a percentage of sales, sales-to-assets ratio, and so forth, it would show significant overlap between what are generally considered specialty companies and the rest of the chemical industry.

The Loss of the Basic Chemical Stock: Dow Stands Alone?

Pharmaceuticals, the most specialty-oriented end of SIC group 28, is no longer considered part of the chemical industry by investors. The low end, basic chemicals, is also being lost from this sector as far as investors are concerned.

Table 2 showed the market capitalizations for U.S. chemical stocks. Among the commodity stocks Dow Chemical is the only large capitalization name, the next largest stock being only about $2 billion. And even Dow Chemical has claimed for many years that about 50 percent of its sales can be categorized as "specialty."

Two phenomena are contributing to the disappearance of U.S. commodity chemical stocks. The first is that these stocks are simply too volatile and cyclical for most institutional investors to own long term. The reduced liquidity in cyclical downturns should require outperformance over time, but there is no evidence that chemical stocks outperform the market over long periods. The second reason is that basic chemical businesses can often be more highly valued as parts of oil and gas companies (what is known as the "halo" effect on valuation) or as private companies. The chemical divisions of ExxonMobil or its competitors do not appear to affect significantly the valuation of those stocks (which means most U.S. basic chemical stocks would be more highly valued if they were owned by oil companies, since oil companies trade at premiums to chemical companies).

Measuring Financial Performance: Return on What?

Lyondell Petrochemical promotes its ranking as one of the most profitable chemical firms as measured by sales per employee. Eastman Chemical prefers to promote its leading return-on-equity statistics. Dow Chemical, Millennium, Olin, and many others have adopted programs in recent years based on economic profit or value added, and they now promote that as their leading measure of profitability. Most specialty chemical firms continue to focus on return on sales (ROS). All this leads to an unending debate about which is the most profitable company in the industry.

The two most common measures of financial performance are ROS (typically gross margin or operating margin) and return on investment (for example, return on assets and return on equity). This section highlights a number of problems with most ROS (that is, margin) analyses and strongly favors return on investment as the key measure to watch. The preferred return-on-investment measure to use is cash flow return-on-replacement value.

Table 7. Higher Return on Sales and Lower Return of Investment

Period	Revenue	Cost	Return on Sales	Total Profit
A	$100	$50	50%	$50
B	$167	$100	40%	$67

Source: John Roberts, "When Margins May Not Matter," Merrill Lynch, 28 Jan. 1997.

Return on Sales

ROS analyses can be difficult in a process industry like chemicals where issues of vertical integration and raw material cost volatility can distort comparisons. The raw material volatility also often leads to agreements with customers to pass through certain cost changes, which further distorts margins. Vertical integration is also sometimes achieved through alliances that create equity income without corresponding revenue, which inflates net margins. Royalty revenue is another common factor that inflates margins, since it generally has little offsetting current cost and therefore nearly a 100 percent margin.

Table 7 presents the financial results for a hypothetical company during two periods. The example is exaggerated for illustration, but investors are often sensitive to even a few basis point changes in margins. The only difference during the two periods is a doubling in costs in the second period and a 67 percent increase in prices, which more than passes through the increase in costs (because costs are initially only 50 percent of revenues, prices only have to increase half as much as costs to maintain earnings). While the company's absolute profits increase 33 percent, the company's ROS declined from 50 percent to 40 percent. Investors focused on ROS as a measure of profitability would view this as a negative turn of events when in fact it is quite positive. A converse scenario could also be developed where declining costs and profits result in rising ROS.

The above scenario is common in two situations in the chemical industry. The first is when petroleum prices surge, which happens from time to time. The second is when operating rates are very high and the marginal source of supply shifts from increasing the output of existing plants to building new plants. Marginal costs then leap from variable costs (mostly raw materials) to replacement costs. The latter includes a return of capital (depreciation) and return on capital, which can be very large relative to variable costs since this is a capital-intensive industry.

When ROS Simply Measures Vertical Integration

Table 8 presents financial results for three companies in the same industry. Company U makes an upstream intermediate, which is sold to company D to

Table 8. Effect of Vertical Integration of Return on Sales

Company	Investment	Revenue	Cost	Return on Sales	Return on Investment
U (upstream)	$1,000	$1,000	$900	10%	10%
D (downstream)	$500	$1,100	$1,000	9%	20%
I (integrated)	$1,500	$1,100	$900	18%	13%

Source: John Roberts, "To Be or Not to Be-Integrated," Merrill Lynch, 13 Oct. 1997.

process into a downstream end product. This example assumes the return on investment for the downstream process (20 percent) is higher than that for the upstream process (10 percent). Company I makes the same downstream end product as company D, but is vertically integrated backward (that is, upstream) to make its own intermediate. While companies D and I sell the same product at the same price, the ROS for the integrated company I (18 percent) is significantly higher than that for the unintegrated companies U (10 percent) and D (9 percent). Investors focused on ROS would be attracted to company I, but in fact company D is the most profitable company as measured by the more meaningful return on investment.

The above situation occurs most commonly when dealing with the chemical divisions of the oil companies, which are among the largest chemical operations in the world (ExxonMobil's chemical division is the third largest chemical company in the United States after DuPont and Dow Chemical). The transfer prices on intermediates and the classification of shared refinery assets can significantly affect the ROS from publicly reported data.

One clue to when ROS analysis can be distorted by vertical integration is a high degree of intersegment and intrasegment sales. Intersegment sales are reported in the segment information section of a company's annual report, but intrasegment sales can be harder to identify for an outsider. For example, Dow's high degree of vertical integration results in a higher ROS than if all the company's intermediates were sold to outside customers.

No comment has been made here on whether it is better to be more or less vertically integrated or when vertical integration is better because that is outside the scope of this chapter. With the increased sophistication of contractual arrangements between companies, the concept of "virtual integration" has emerged. It has even become commonplace for companies to divest process units within a large integrated complex, so that almost all assets are separable to some extent. The issue of being vertically versus nonvertically integrated continues to decline in importance (except that investors need to account for this when looking at ROS ratios).

Specialties and Commodities

This discussion so far may appear to deal primarily with commodities since they tend to be more cost based and more often deal with issues of vertical integration. Most specialty businesses price on value rather than cost and are people rather than asset intensive so that ROS is relatively stable. However, the ROS for many specialty businesses can also be distorted by the similar effects of volatile raw material costs or vertical integration.

A variation of the vertical integration effect is a high degree of raw material cost pass-through to customers in the form of higher prices. Two notable examples in specialty chemicals are metals-based (including precious metals) chemicals and plastic compounders. These specialty businesses tend to operate on a value-added margin with customers absorbing most of the volatility in metals and plastics prices. Further, the specialty chemical divisions of many large, diversified chemical companies may have a significantly different ROS than an independent specialty chemical company because of vertical integration issues.

Product Leverage: The Largest versus the Most Focused

Investors are often looking for how to get "the biggest bang for the buck" in a specific product or raw material. But a common mistake investors make is to look to the largest producer or consumer of that material or to look for the company with the largest exposure per share. The correct approach is to look for the company with the largest capacity per dollar of stock value.

Table 9 shows three hypothetical companies producing the same product. Company A has a hundred units of capacity and a hundred shares outstanding trading at $2 per share. The capacity per share is 1.0, and the capacity per dollar of stock value is 0.5.

While company B is the market leader, with twice the capacity of company A, it is less leveraged to the product of interest. Company B has four hundred shares outstanding at $2 per share, and so it is four times the size of company A. The product of interest is obviously only a small part of company B's business, and so company A is obviously a much more focused investment.

Company C is the same as company A except that it has split its stock in two, so that there are twice as many shares trading at half the price. Since the market value is unchanged, the capacity per market value is the same as that of company A. But the capacity per share is half the level of company A, and so

Table 9. Product Leverage

Variable	Company A	Company B	Company C
Capacity (units)	100	200	100
Shares	100	400	200
Price (U.S. dollars)	2	2	1
Capacity/share	1	0.5	0.5
Capacity/market value*	0.5	0.25	0.5

Source: John Roberts, "To Be or Not to Be-Integrated," Merrill Lynch, 13 Oct. 1997.
* Market value is shares times price.

Table 10. Recent Fifty-Two-Week Trading Range for the Ten Largest U.S. Chemical Companies

Company	52-Week High	52-Week Low	Percentage of Change
	(U.S. Dollars per Share)		
DuPont	50.69	38.19	33
3M	127.00	80.50	58
Dow Chemical	39.67	23.00	72
Air Products	48.70	29.25	66
Monsanto	38.47	19.75	95
Praxair	54.00	30.31	78
Rohm and Haas	38.70	24.38	59
Ecolab	45.69	33.25	37
Avery Dennison	70.63	41.13	72
Eastman Chemical	55.65	35.06	59

Source: Wall Street Journal, 30 June 2001, p. 18.

focusing on capacity per share would lead to the incorrect conclusion that company C was a less focused investment than company A.

Looking at capacity per share is useful for understanding what the change in earnings per share might be from a change in price or cost of a specific product. But it is misleading to compare capacities per share across companies to determine which is a more leveraged investment to a specific product.

Valuation: An Imprecise Science

This chapter began with an example of the valuation of Union Carbide and a discussion of the importance of stock values to CEO compensations. Union Carbide believed it was worth $64 per share, or sixteen times what turned out to be an unrealistic trough EPS target of $4. When the company announced this target in October 1997, investors at the end of the month valued the company at $46 per share. To paraphrase an old saying, "value is in the eye of the beholder."

One thing to remember about valuation is that it is an imprecise science. Table 10 shows the recent fifty-two-week highs and lows for the ten largest U.S. chemical stocks. On average the highs are 50 percent above the lows, which is a fairly typical trading range. So predicting stock prices within the approximately 20 percent bandwidth that normally delineates buy or sell opinion is a fairly challenging exercise.

Table 11 lists the most popular investment methodologies from a recent survey done by Merrill Lynch. This list is an aggregate and would be different for different industries and different for stocks within an industry. Which valuation measure sets the price of a stock depends on the situation: there is no one-size-fits-all. If a stock is a potential takeover candidate, then its value may be set more by the replacement value (a build versus buy decision), the recent or all-time high (the price to provide a gain or reason to sell for most investors), or the enterprise value-to-EBITDA (earnings before interest, taxes, depreciation, and amortization) (what the company could be recapitalized for). Without the potential to be taken over or taken private, trading values are more typically set by P/E multiples, or in the absence of an "E" (earnings) during the downturn, the price-to-cash flow or yield (dividend or distributable cash flow) may set the valuation.

The Importance of Size in Valuation: And Then There Was One?

A practical issue in valuation is the scarcity of large capitalization chemical stocks. For example, if all public U.S. chemical companies were consolidated into one large conglomerate, it would still rank only fifth in the Fortune 500 in terms of market capitalization (Table 12). Total sales for the combined public U.S. chemical companies are similar in size to GM, Ford, WalMart, or ExxonMobil. The global market shares of this hypothetical company would be about 20 percent or less in ethylene, all major ethylene derivatives, industrial gases, chloralkali, and most other major chemical products.

Table 11. Top Reasons Investors Buy or Sell Stock

Rank	Reasons	Percentage of Respondents
1	Earnings surprise	54
2	Analyst's earnings revisions	48
3	Price–to–cash flow ratio	48
4	Projected five-year earnings-per-share growth	46
5	Earnings momentum	41
6	Relative price strength	39
7	Price-to-earnings ratio	34
8	Price-to-book ratio	34
9	Analyst's opinion change	34
10	Earnings variability	31
11	Dividend discount model	30
12	Neglected stocks	25
13	Beta (relative volatility)	14
14	Earnings estimate dispersion	14
15	Dividend yield	12

Source: Richard Bernstein, "Quantitative Strategy Review," Merrill Lynch, 30 June 2001.

Figure 1 shows the level of M&A activity in recent years. So the industry remains relatively fragmented despite a high level of consolidation. But not all M&A activity reflects consolidation. In the terms of the game Monopoly much of this activity reflects buying and selling houses without building hotels. We believe the industry will continue to undergo consolidation for years to come.

Table 13 shows the enormity of the largest U.S. mutual funds, which contributes to the pressure for small public companies to get larger. There are some practical constraints that affect the size of stocks a fund will hold. One is that a fund manager can only keep track of so many stocks, and so most funds try to limit themselves to no more than a hundred positions (that is, no more than 5 percent of the fund), and often many fewer. Another issue is that many firms want to limit the positions of all their funds combined to no more than 5 percent of a company's stock, above which a fund has to report publicly to the Securities Exchange Commission any changes in its position. Positions above 5 percent make it difficult for a fund to change its position without revealing to other investors what that fund is doing. Investors also look for stocks with average trading volume that would allow the fund to accumulate or liquidate its holdings over a relatively short period.

There are few U.S. chemical companies in which the largest fund holders could own enough stock to make up at least 5 percent of the fund and still not be over 5 percent of the outstanding stock for the company. In fact, for many institutional investors DuPont and Dow Chemical are the only companies with

Table 12. U.S. Chemical Industry Market Cap Relative to the Fortune 500

Rank	Company	Market Capitalization
		(U.S.$ Billions)
1	General Electric	480
2	Wal-Mart	235
3	Cisco	230
4	Microsoft	225
5	Chemical Industries	215
6	Intel	205
7	HSBC	140
8	Procter & Gamble	100
9	Wells Fargo	98
10	Tyco International	95

Source: John Roberts, "And Then There Was One? Consolidation to Remain Major Industry Theme," Merrill Lynch Equity Research report, 6 Mar. 2001.

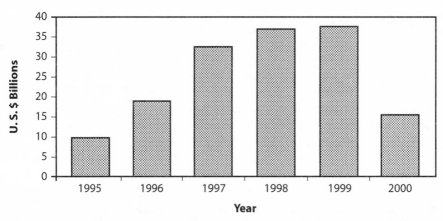

Figure 1. Value of recent merger and acquisitions activity for 1995 to 2000. *Source:* Young and Partners data as reported in Karol Nielson, "The Last of the Big Deals? M&A Values Begin to Slide," *Chemical Week* 162 (13 Sept. 2000), 31.

stocks large enough for them to hold. Because of this scarcity DuPont and Dow trade at substantial premium valuations to their peers.

So the definition of critical mass is very different from an investor's viewpoint than from a company's viewpoint. Companies typically think of critical mass at the plant-level size (for example, a world-scale plant) or at a size large enough to support a critical mass R&D organization (as in crop protection chemicals). The higher valuations afforded larger companies will probably drive companies away from smaller, focused business models toward larger, more diversified business models.

Table 13. Top Ten Largest Mutual Funds

Rank	Largest Mutual Funds	U.S.$ Billions
1	Vanguard 500 Index	113.09
2	Fidelity Magellan	111.55
3	Janus Worldwide	49.22
4	Fidelity Contrafund	44.36
5	Fidelity Growth & Income	39.66
6	American Funds Investment Company	30.17
7	American Funds Washington Mutual	29.58
8	American Century Ultra	29.46
9	Janus	29.22
10	American Funds Growth Fund	24.90

Source: Wall Street Journal, 30 June 2001, p. 18.

P/Es Are King

While "cash is king" and therefore multiples to cash flow are perhaps more fundamental than earnings multiples, P/E multiples are still the most commonly used measurement to assess valuations. Comparability provides a key reason for the focus on EPS rather than cash flow, as most of the financial reporting systems that collect estimates from analysts (for example, First Call, IBES, and Zachs) collect only EPS estimates. So there are systems that automatically report "consensus" EPS estimates for each company and easily allow investors to compare P/Es across companies within an industry and even across industries.

EPS may be the correct way to look at earnings for less asset-intensive companies or where the physical lives of the assets match their accounting lives (that is, capital spending is stable around the level of depreciation). But since most chemical companies have overly depreciated assets, cash flow would be a better measure of earnings. Unfortunately the cash flow estimates of analysts are not collected in the same routine manner as EPS estimates, and so comparing cash flows across companies or across industries is a lot more work for investors.

Even something as basic as EPS is a challenge for most industry observers to grasp. There are debates about whether to use the P/E multiple on trailing earnings, current earnings, estimates of next year's earnings, or peak earnings. The large amount of restructuring in the chemical industry leads to frequently recurring "one-time" items, and there are debates among investment professionals on whether earnings before or after these items are the right measure of profitability.

Some industry observers introduce another level of complexity by using "normalized earnings," which then allow cyclical stocks to be "shoehorned" into growth stock analyses. There has never been a good operational definition of normalized earnings, but most methods typically involve a least-squares regression of earnings over at least a full cycle and then using this relationship to determine where current earnings should be.

Normalized earnings achieve little additional predictability from such a time-consuming exercise, and normalized estimates are difficult to track, which allows forecasters considerable latitude in making revisions. Investment recommendations on stocks are typically made with at least 20 percent upside expectations, and the variance of most earnings regressions for cyclical chemical companies are well outside this level of accuracy.

The cyclicality of earnings has important implications not just for investors but also for how companies manage their portfolio of businesses. Public companies come under significant shareholder pressure to buy or sell cyclical businesses at exactly the wrong points in the cycle (acquire at the top of the cycle and divest at the bottom). It can be near-term accretive (that is, a positive effect on share price) to EPS to divest a business at the bottom of the cycle since the benefit of any proceeds can offset the weak earnings (or even losses) of the divested business. Similarly, it can be near-term accretive to EPS to buy businesses near the top of a cycle since the benefit of the peak earnings can offset the debt or equity cost of making the acquisition.

Each cycle tends to have somewhat different levels and durations for the peak and trough EPS. Typically the spikier the peak, the shorter the duration of near-peak earnings, and the more muted the peak, the longer the duration of peak earnings. So it is the area under the curve (that is, the cumulative earnings across the peak) that matters more than the absolute level of the peak. And a similar observation applies for trough earnings as well.

Cash Flow

Discounted cash flow and its relative, the dividend discount model, are the most fundamental ways to estimate the value of a company. This exercise is too time consuming for most investors, which is why it did not even rank in the earlier list of valuation tools investors use. Discounted cash flow also suffers from the fact that most of the value comes from a terminal value assumption many years into the future, where there is little certainty (just look at the difficulty investors have in forecasting near-term financial results).

Discounted cash flow can also provide a wide range of results depending on the assumed shape of the cycle. In Table 14 the first cash flow series starts at a peak of $150, declines by $50 per year for two years, then recovers by $25 per year for four years back to the peak, and then repeats. The second line shows the present value (PV) at 12 percent for the first six-year cycle, then the second six-year cycle, and so forth. The final value at the right-hand side is the approximate total present value if the cycle repeated indefinitely. The third line repeats the exercise at a discount rate of 8 percent. A 4 percent drop in interest

Table 14. Discounted Cash Flow at Different Points in the Cycle (in U.S. Dollars)

Year	1	2	3	4	5	6	7	8	9	10	Total
Peak-to-trough											
Cash flow	150	100	50	75	100	125	150	100	50	75	
PV @ 12%	417	189	85	39	17	8	4	2	1	0	761
PV @ 8%	466	272	159	93	54	32	18	11	6	4	1,114
Trough-to-peak											
Cash flow	50	75	100	125	150	100	50	75	100	125	
PV @ 12%	391	177	80	36	16	7	3	2	1	0	713
PV @ 8%	447	261	152	89	52	30	18	10	6	4	1,068

Source: John Roberts, "What a CROC!" Merrill Lynch, 1 May 1998.
PV = present value.

rates equates to over a 40 percent increase in the present value. This slightly exaggerates the interest rate decline of recent years, but the powerful effect it has on valuations is obvious.

The second series is the same as the first, except it starts at the trough instead of the peak. Most investors would probably say that the primary reason for the volatility in many chemical stocks has been the cyclicality in the earnings more than changes in interest rates. But the difference between the two scenarios at the lower interest rate is only about 4 percent; for the higher interest rate it is about 6 percent because of the greater importance of the first few years, or where the company is in the cycle. For a six-year earnings cycle that triples from the trough to the peak, this change in valuation is probably less than most investors would guess.

Many investors use semilogarithmic charts of price-to-book (P/B) ratios versus returns on equity (ROE) to compare the stock prices of companies, but they are unfamiliar with the simplified discounted cash flow relationship behind the analysis and what it can say about cyclical companies like chemical stocks.

For example, the initial investment of B, which generates a rate of return of ROE for T years, at maturity has a rate of return that drops down to the cost of equity (r). The present value (P) of this is equal to $P = B \times \exp(T \times (ROE - r))$, or alternatively, $\text{Log}(P/B) = T \times ROE - r \times T$. This is a linear relationship between log (P/B) and ROE and is the basis for the familiar semilog plots of price-to-book versus return on equity. One interesting use of this relationship is to show why P/Es, which are more commonly used for stock price comparisons, tend to cycle inversely with earnings for cyclical companies like chemicals, which will be discussed later.

By definition $ROE = (P/B) / (P/E)$, so that P/B can be eliminated in the above equation to leave a relationship between P/E and ROE. Readers can prove to themselves that for low values of T, there is an inverse relationship between P/E and ROE, while for higher values of T there is a flat to upward-sloping relationship (seen with growth stocks). T is the period a company can sustain returns above its cost of capital, which for a drug company with patent protection could be over a decade, but for a commodity company could be two to three years out of a cycle.

The Yin and Yang of Value: Earnings and Valuation Are Countercyclical

Over this most recent cycle (1992 trough–to–current trough expectations), estimates for Dow Chemical's earnings have changed from $3.00 to $1.50. The P/E on these earnings expectations has changed from eight to twenty-four times. But the stock has never traded as low as eight times $1.50, or as high as twenty-four times $3.00. Earnings and valuation cyclicality are closely intertwined (inversely related), and in this last section thoughts on these types of cyclicality will be integrated from earlier sections.

Chemical stock cyclicality, when viewed broadly, has two dimensions: demand-side cyclicality and supply-side cyclicality. Demand cyclicality affects almost all chemical stocks, while for basic chemicals the supply cyclicality can be much more important. But the biggest misunderstandings occur with supply-side cyclicality, while demand cyclicality is relatively universal and well recognized.

Demand: Cycles within Cycles

Chemical companies often have minicycles within overall macroeconomic cycles. These minicycles are typically driven by inventory adjustments and by the lag between product prices and raw material costs. The inventory lag also occurs in many other industries, while the raw material–cost lag is more specific to chemicals because of its high dependence on petroleum feedstocks.

Oil and gas prices can change minute to minute, while most chemical product is sold on a contract basis, with prices fixed for various periods. First-order derivatives of petrochemicals—olefins, aromatics, and polyolefins—have the shortest cycles, or periods, for repricing contracts, and even then prices are generally fixed for at least a month. Many specialty chemical products have annual pricing cycles, with the potential to adjust prices midstream only under very unusual circumstances.

When energy prices spike, investors know to sell stocks in the chemical, transportation, and other energy-intensive industries quickly. What many investors often forget, however, is the chemical industries' historical success in recovering these costs as pricing cycles allow.

This phenonemon was addressed in the discussion of the pitfalls of ROS analyses. Because sales margins decline in a rising feedstock environment, investors often assume that feedstock costs have not been recovered, which is not true.

Normally the lags on the upside and downside balance out; so chemical companies are not hurt by changes in energy prices in the long term. There are exceptions when energy prices reach extremes, since product prices tend to be sticky at their previous all-time highs and lows. And sharp rises in energy prices can require several chemical pricing periods to allow companies to recover fully. If raw materials change direction and turn down before margins have been fully recovered, then it can be very difficult to get further price increases, resulting in some permanent loss of profitability.

Supply Cycle: Long Duration

While the supply cycle may be more volatile than the demand cycle for the most basic products, the good news is that at least it is more predictable. The lead time for building a new ethylene plant is four to five years, and almost all

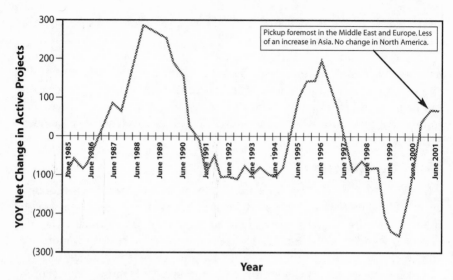

Figure 2. Net change in active petroleum projects worldwide. *Source:* John Roberts, "Quarterly Segment Review," reprinted by permission, © 2001, Merrill Lynch, Pierce, Fenner & Smith Incorporated.

major chemical projects have lead times of at least a year. So for investors trying to forecast this year and the next, the amount of new supply coming is typically well known.

Figure 2 presents a leading indicator of the overall supply cycle. This is the rate of change (year over prior year) of the number of chemical projects in the pipeline, or backlog. The source of data is a survey done three times a year of the major engineering and construction firms serving the chemical sector. It is pretty clear to us that the profit cycle drives the capital spending cycle. The last peak in earnings in 1995 and the 1998 Asian crisis, which caused a large number of project delays or cancellations, are particularly noticeable in the subsequent level of project activity.

The Prisoner's Dilemma: Somebody Stop Me

So if it is so easy to see any coming overcapacity, then what keeps the companies from helping themselves by adding capacity in a more orderly fashion? The reason is a decision-making situation known as the "prisoner's dilemma,"

Table 15. The Chemical Industry's "Prisoner's Dilemma"

	Competition Waits	Competion Builds
You wait	Maintain share; improve margins	Lose share; lose/sustain margin
You build	Gain share; lose/sustain margin	Maintain share; lose margin

Source: John Roberts, "The Prisoner's Dilemma," Merrill Lynch, 23 Sept. 1998.

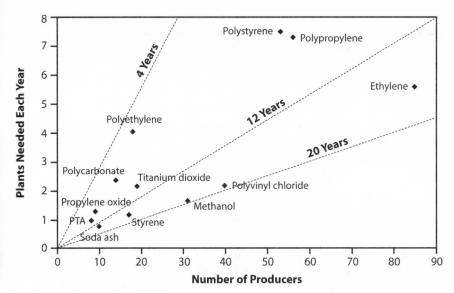

Figure 3. New plants needed per year versus the number of producers worldwide. PTA = purified terephthalic acid. *Source:* John Roberts, "The Prisoner's Dilemma," reprinted by permission, © 1998, Merrill Lynch, Pierce, Fenner & Smith Incorporated.

which has application in a number of business situations, particularly the decision chemical companies face when adding new capacity.

For example, Table 15 shows the decision matrix facing two competitors when the industry needs just one new plant. The best outcome from a profit perspective is for both companies to wait to add capacity. Margins should rise, and each competitor maintains share. However, the most likely outcome is for both companies to build a plant, driving down margins because of overcapacity. The worst outcome occurs because neither company wants to take the risk of potentially losing share while margins are only sustained at best. The dilemma is even worse when there are more than two competitors, which is often the case in chemicals.

On the vertical axis of Figure 3 is the number of new world-scale production units that would be required each year to keep up with worldwide demand growth. On the horizontal axis is the number of key competitors worldwide. The upward sloping lines represent how many years of demand growth are needed to absorb one new plant by each competitor.

For example, the world needs just over one new styrene plant each year. This assumes a base market of 40 billion pounds per year, demand growth of 5 percent, and the scale of a new world-class plant at 1.75 billion pounds per year. There are seventeen producers who already operate capacity at least equal to half of a world-scale plant. If each of these producers built a new world-scale plant, this would total 30 billion pounds per year of new capacity. It would take twelve years of average demand growth to absorb this much capacity. Or in

Table 16. Reasons Companies Increase Capacity during a Peak

Opportunity Cost Reasons for Building Now

Poor alternatives for cash flow

Make acquisitions in an auction environment

Raise dividend to a level unsustainable at the next trough

One-time dividend benefits past shareholders at the expense of future shareholders

Buy back stock at a cyclical high

Cannot inventory cash

Low investment returns

Acquisition vulnerability

Source: John Roberts, "The Prisoner's Dilemma," Merrill Lynch, 23 Sept. 1998.

other words, to maintain a stable supply-and-demand balance, each producer should wait an average of twelve years to add a new world-scale plant. Since this is longer than a typical business cycle, it is unlikely most competitors could resist building a new plant for that long.

A key reason why some chemicals are less cyclical is that competitors do not face the dilemma of building now or building later. Most specialty chemicals and some large-volume products like polycarbonate are this way. Either demand growth is high, capacity increments are small, or there are few competitors. These producers can add capacity fairly regularly. And if the industry does get a little out of balance, it does not take much restraint to bring it back into balance quickly.

Polyethylene and purified terephthalic acid (PTA), both produced in greater quantities, used to be this way until the number of competitors began to rise. The normal maturation is for products to migrate to the northeast in Figure 3.

This is obviously not news to most producers, so why can rational competitors not stop themselves from being "put in jail." Compounding the "prisoner's" penalty for building is an opportunity cost penalty for not building, as outlined in Table 16. The prisoner is between a rock and a hard place. When a company does not build, it has to deal with excess cash. And while most investors quickly say simply give the cash back to the shareholders, it is not that easy. The "make-or-buy" alternative to building is acquisitions, but deals can become very expensive when most competitors are potential bidders.

There are three ways to give back cash, each of which has its problems. The first two involve the dividend, which is tax inefficient for many shareholders (since the company pays tax when the dividend is earnings and shareholders pay tax again when it is distributed) and lowers the earnings growth rate relative to reinvesting the cash. If the regular dividend is raised too high, it may not be sustained at the bottom of the cycle. And in a period of very low interest rates the dividend may not provide much support to the stock price. Instead a one-time dividend could be paid, but this has the undesirable effect of churning the shareholder base since the types of investors who want the stock before and after the special dividend tend to be different. The last and increasingly popular alternative is to buy back stock. But stock prices also tend to be high when companies have a lot of cash. The stocks of many chemical companies are now well below where their management repurchased shares.

The best alternative might be just to keep the cash and wait to invest off-cycle, when opportunities are priced more reasonably. Private companies are more apt to do this since a lot of cash may make a public company, particularly a smaller one, vulnerable to being acquired. The other argument against sitting on the cash is that the returns are very low, and so you have to endure sub-par returns up front in the hope of superior returns a few years down the road. Few institutional investors have that type of long-term patience.

Two ways out of this prisoner's dilemma are to reduce the number of competitors and to reduce the scale of the investments.

Wall Street is doing its part to help the producers on the competitive intensity dimension. Figure 1 showed the increasing M&A activity in the chemical industry. This has been a key factor in stabilizing the titanium dioxide (TiO_2) pigment business. The businesses of ICI, Rhône-Poulenc, and Bayer have each been acquired by competitors. In the face of declining prices for almost all commodities, TiO_2 pricing has been increasing for two years now and is near its historical peak, despite Asian demand being down 30 percent. This shows just how powerful industry concentration is to profitability. This is not about monopolies, but about structurally balanced markets where each new player adding a plant is not a major problem.

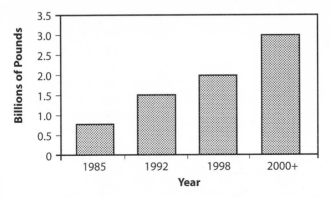

Figure 4. Planned, or eventual, scale of ethylene plants. *Source:* John Roberts, "The Prisoner's Dilemma," reprinted by permission, © 1998, Merrill Lynch, Pierce, Fenner & Smith Incorporated.

Reduced scale was the second way out of the prisoner's dilemma, and process developments appear to be working against the companies here. Figure 4 shows the progression of world-scale plants for new ethylene projects. The new joint project in Alberta, Canada, between NOVA Chemicals and Union Carbide is targeted to ultimately reach 3 billion pounds per year. There are a few examples of scale being rolled back, such as Rohm and Haas's modular emulsion plant technology, but scale continues to increase for the vast majority of processes. This is because volume goes up with a cubic while surface area goes up with a square. So it takes a lot less steel (surface area) per unit of volume as scale increases, which reduces capital cost per unit volume. It is hard to argue with geometry.

The Chemical Industry at the Millennium

Peter H. Spitz

This concluding chapter is not meant to be a prognosis. It is primarily intended to summarize and relate to the future important themes that have been discussed in earlier chapters and that will influence how the industry evolves over the next period. Important themes include restructuring, consolidation, globalization, regulation, and maturity, all of which have had a strong hand in shaping industry developments over the last twenty years. Two hopeful new themes are sustainable development and the possible rebirth of innovation. Together, these themes will most likely dominate the next decade or two.

At this writing, shortly after the turn of the millennium, the managements of most chemical firms in the Organization for Economic Cooperation and Development (OECD) countries have a lot to worry about. A global recession cut sharply into the demand for chemicals, which resulted in low operating rates. Overcapacity in many parts of the industry has strongly contributed to the problem and has made it likely that the next petrochemical price fly-up, once projected for 2002–2003 will occur a couple of years later. Chemical stocks continue to be out of favor with investors, as industry growth prospects are still in serious question. With low stock prices, companies cannot sensibly use their shares as acquisition capital, as has often been the case in the past, yet firms recognize that practically the only way to grow meaningfully is through mergers and acquisitions. The leveraged buyout (LBO) community also finds it difficult to make acquisitions, as companies do not want to sell at times of low valuation, while banks have become more leery of highly leveraged financing models and demand more equity input by the investors, making returns for these acquisitions that much less attractive. Most worrisome for petrochemical firms that have most of their plants in their home countries is the fact that the new producers in the Middle East and Asia have been building large, highly economical plants, often using indigenous low-priced feedstocks.

During the last two decades of the twentieth century the global chemical industry reached a somewhat ambiguous maturity. In the industrialized world, even when countries were not in an economic recession, demand for most chemicals slowed significantly, and development and commercialization of new "breakthrough" process technology almost ceased. In contrast, in the developing world manufacture of chemicals, as well as of synthetic fibers, plastics, and other polymers, was growing rapidly, as per-capita consumption of these materials was on the rise in these large markets and as local firms as well as multinationals seized opportunities to convert locally available hydrocarbons into these more valuable products. Asia became a petrochemical powerhouse because of the rapidly growing number of high-volume global chemicals being produced there. This trend will continue unabated and will keep diminishing the amount of traditional commodity chemical exports by multinationals from plants in the OECD countries. The multinationals are reacting to this by building many of their new commodity plants in the Middle East and East Asia: they are not constrained by balance-of-payment considerations in their home country because their shareholders expect management to build new plants in the most logical locations, wherever these might be.

As noted in other chapters, the U.S. balance of trade in chemicals is now negative after many years of strongly positive balances. The problem is not only loss of exports but also steadily increasing imports, of both chemicals and chemical-derived products. In 2001 more than a billion pounds of polyethylene bags were imported into the United States, mostly from Asia. This occurred because between 1996 and 2001, the price differential for high-density polyethylene resin rose from 0 to the range of 10 to 15 cents per pound in favor of Asia. The story is similar for low-density polyethylene. The dramatic developments in the production cost of ethylene, which now strongly favors Middle Eastern and some other Asian producers, are shown in the projected cost curves for 2001 and 2006 (Figure 1). Economics for U.S. and Western European ethylene plants (with the United States using liquefied petroleum gas using feedstocks based on an assumed $3.50 per million BTUs for natural gas and with Europe using $21.50 per barrel of crude oil) are well to the right (the unattractive high side) of the two curves, with most of the Japanese plants near the top right. Many of the U.S. and European plants that are still operating have capacities below 200,000 tons per year as compared with new world-scale ethylene plants now being built in Asia, which are often in the range of a billion tons per year.[1]

Recently, a number of important issues served to define or reinforce the characteristics and likely future of the industry, as discussed in greater detail in the preceding chapters:

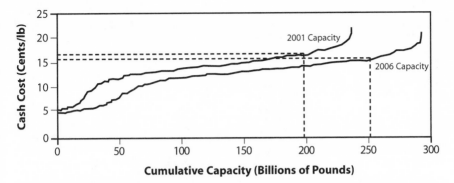

Figure 1. World ethylene cost curve: normalized forward prices (U.S. gas at $3.50 per million BTUs; West Texas Intermediate crude oil at $21.50 per barrel). These curves show cumulatively the production costs of several hundred global ethylene plants. *Source:* Graham L. Copley and Hassan I. Ahmed, "U.S. Chemicals: The End of the 'Good Old Days,'" International Petrochemical Conference of the National Petroleum Refiners Association, San Antonio, Texas, 25 Mar. 2002, by permission.

- Demand growth for most major chemicals in the developed world slowed to the level of gross national product (GNP) growth, in the range of 2 percent to 3 percent annually. This essentially defines an industry becoming mature.

- Industry cyclicality was endemic, in terms of operating rates and pricing for most commodity chemicals, occasioned by economic factors, periodic overbuilding, extreme competition in such a highly fragmented industry, and the consequent inability of major producers to provide pricing leadership. Companies continued to reinvest during and after short periods of high profits, and there is no evidence that this trend, which typically gives the industry one or two great years and several very poor years per decade, will not continue. Figure 2 gives an excellent picture of this situation, showing the wide range of operating cost margins between 1981 and projected 2005, and indicating periods when profitability was or is likely to be high enough for reinvestment.[2] Chemical Market Associates, Inc. (CMAI) recently forecast that the global recession would push the next cyclical peak out a bit further (Figure 3).

- Industry restructuring continued as companies sought to define their proper roles, building on core strengths and achieving top-line growth through acquisitions that often did little to improve shareholder value. Many old-line companies disappeared as mergers and acquisitions took their toll. Restructuring within individual industries became significant. For example, the number of companies accounting for half of the global ethylene capacity shrank from twenty in 1995 to eleven in 2000, while total capacity rose from 80 million to 100 million tons.[3]

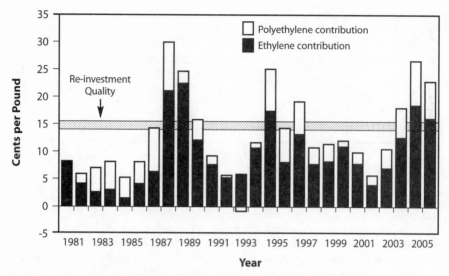

Figure 2. U.S. integrated ethylene and polyethylene margins for 1981 to 2005.
Source: Chemical Market Associates, Inc. (CMAI), by permission.

- Evidence was mounting that the future of commodity petrochemicals would largely be in the hands of the chemical divisions of major oil companies and the chemical arms of national oil companies in hydrocarbon-rich nations. Lack of back-integration into raw materials could eventually become a fatal problem for all but a few nonintegrated chemical firms.
- Dramatic changes were taking place in the role, performance, and size of specialty chemicals producers. With such end-market industries as steel and refining maturing, the demand growth for specialty chemicals serving these industries slowed from double-digit to much lower levels. This made specialty chemical companies serving these industries less attractive to investors and forced consolidation.
- Although leveraged financing slowed down somewhat, the role of so-called financial buyers kept increasing, as chemical firms spun off divisions and various other businesses they did not consider strategic to their future. Private firms using leveraged financing, such as Huntsman and Ineos, in most cases bought commodity or pseudo-commodity businesses to build up large companies that might eventually go public. The more typical financial buyers were investment funds operated by such partnerships as Kohlberg, Kravis and Roberts (KKR), CVC Capital Partners, Blackstone, AEA Investors, Morgan Grenfell, Apollo, and Ripplewood, as well as the merchant banking groups of such firms as J. P. Morgan Chase, Deutsche Bank, and Goldman Sachs, who bought businesses with the intention of improving and expanding these businesses and then re-

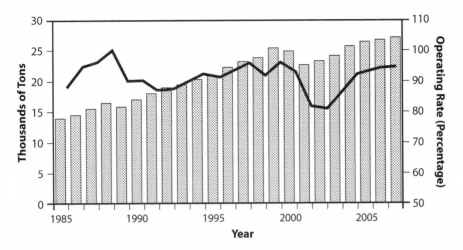

Figure 3. U.S. ethylene demand forecast for 1985 to 2007. *Source:* Chemical Market Associates, Inc. (CMAI), by permission.

selling them to strategic buyers (public operating companies) or to the public.

• During the last thirty years the industry became highly regulated, as federal, state, and local authorities passed a number of laws and set up agencies to protect the public and workers from the hazards of manufacture, to reduce progressively the emission of toxic and other objectionable materials from plants and to monitor and control the use of chemicals. Industry leaders, under the banner of such initiatives as Responsible Care (and corresponding programs in many other countries), reached out to the public to show sensitivity regarding these concerns, while taking a multitude of steps to alleviate or eliminate problems associated with the manufacture, transportation, use, and disposal of chemicals.

This chapter not only summarizes important industry issues at the turn of the millennium but also takes a cautious look ahead. Clearly, the chemical industry is no longer resting on its laurels. The tremendous strides it made in the hundred years just completed have nevertheless left it in a parlous state, mature and not as confident of its future as its successes in the past might otherwise suggest. Still, it is useful to look back at the just-completed century to see how far the industry came during the roughly seventy-year span before the period covered in this book, which saw the arrival of a host of problems. The period from 1900 to 1970 was characterized by the start of continuous versus batch processing, the creation of a petrochemical industry, and a wave of innovation unlike that seen in almost any other industry to that time. Here are some highlights from each decade.

1900–1910

- BASF develops a continuous sulfuric acid process.
- Courtaulds starts production of viscose fibers.
- Fritz Henkel puts a household detergent (Persil) on the market.

1910–1920

- General Bakelite begins large-scale production of a synthetic plastic, Bakelite.
- Fritz Haber and Carl Bosch produce synthetic ammonia at BASF.
- IG Farben and B. F. Goodrich synthesize polyvinyl chloride (PVC) from acetylene-based vinyl chloride.

1920–1930

- Standard Oil of New Jersey produces the first true petrochemical, isopropyl alcohol, from propylene.
- Union Carbide begins production of ethylene and acetylene from natural gas liquids in Tonawanda, New York, and later in Charleston, West Virginia.
- Four British chemical firms combine to form Imperial Chemical Industries (ICI).

1930–1940

- IG Farben develops synthetic elastomers.
- Dow Chemical begins to produce polystyrene.
- DuPont begins production of polyamide fiber (nylon).
- ICI chemists discover polyethylene.

1940–1950

- Geigy introduces DDT (dichlorodiphenyltrichloroethane), the first synthetic chemical pesticide.
- DuPont produces nylon and Teflon.
- Dow Corning begins production of silicones.

1950–1960

- DuPont begins production of polyester fiber (Dacron).
- Hoechst initiates the first commercial production of high-density polyethylene and Montecatini of polypropylene.
- Universal Oil Products (UOP) commercializes the platforming process for making BTX (benzene-toluene-xylene) aromatics.

1960–1970

- Amoco Chemicals begins producing pure terephthalic acid from p-xylene.
- Arco Chemical and Halcon commercialize a new propylene oxide process, with styrene or tertiary butyl alcohol (later MTBE) as major co-products.

Shortly after 1970

- Sohio develops a process to make acrylonitrile from propylene.
- Monsanto makes acetic acid via methanol carbonylation.
- Mitsubishi Chemical produces acrylic acid from propylene.[4]

This snapshot of a vibrant industry captures the better part of a century of tremendous progress. But the early 1970s awakened management to the fact that the industry had to do more than invent, produce, and market its chemicals. It also needed to develop a conscience, learn to live comfortably with its neighbors, and deal with the many agencies set up to monitor and supervise its activities.

The Lineup in 2000

The consolidation and restructuring moves that took place over the twenty years covered in this book led to the disappearance or complete transformation of a number of old-line chemical companies, notably such firms as Union Carbide; Allied Chemical; Stauffer; Hoechst; ICI; Rhône Poulenc; Monsanto; National Distillers and Chemicals; the chemical divisions of Texaco, Arco, and Amoco; Ciba Geigy; and Sandoz, among others.

One of the most dramatic areas of restructuring has involved companies focusing for a time on the so-called life sciences. Beginning in the mid-1990s, several of the largest chemical firms concluded that their destiny was in the area of pharmaceutical and crop sciences (agricultural chemicals), where profitability was usually much higher than in other areas of the chemical industry, and which were similar to each other in terms of developing complex, potentially very profitable molecules. Concentration in this area would finally liberate these firms from the cyclicality of their other chemical businesses and presumably increase their value in the stock market. To some extent this trend had begun a decade earlier when Sir Denys Henderson at ICI had "demerged" the firm into a commodity part (the old ICI) and a specialties and life sciences part called Zeneca. This was to some extent occasioned by a new way of thinking about corporate management: it was determined that certain parts of a firm's business needed quite different management talents than the rest of the business. Specialties and pharmaceutical businesses with higher growth prospects, fewer capital-intensive assets, and the need for more entrepreneurial

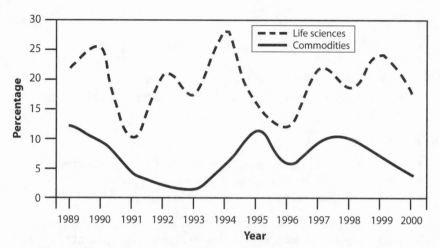

Figure 4. The attraction of life sciences over commodity chemicals (annual return on capital—percentage). *Source:* Jim Langabeer and Amol Joshi, with Demantra Inc., "Margin and Performance Improvement: Using Supply Chain Excellence to Improve Competitiveness," *Chemical Market Reporter* 260 (3 Dec. 2001), 29, by permission.

management would be better off in a separate company that was managed quite differently from commodity businesses. Such a firm's shares would enjoy higher price-earnings (P/E) ratios and therefore attract more investor interest. Taking such a step would thus "unlock" shareholder value.

In the late 1990s a few firms, such as Monsanto and later Hoechst and Rhône Poulenc, spun off their entire chemical operations to concentrate totally on their life sciences businesses, the latter two companies joining to form Aventis. Previously, Sandoz and Ciba Geigy had likewise combined their life sciences businesses into Novartis, after spinning off their chemical operations into Clariant and Ciba Specialties, respectively. Hoechst then divested its specialties into Clariant, making that company a much larger firm.

It is easy to see why life sciences were viewed as more attractive than commodities (and many specialties). During 1992 pharmaceutical and crop sciences–major pharmaceuticals far outperformed other sectors of the industry (Figure 4).

Up to the late 1990s several of the other large firms, notably Bayer, BASF, and DuPont, believed that they could be players in their traditional chemical businesses but could still pursue their life science activities, including either or both pharmaceutical and crop sciences. Two other large companies, Astra Zeneca and American Home Products (which had purchased American Cyanamid), also believed they could be life sciences players. Dow, however, had earlier spun off its pharmaceutical business (Marion Merrill Dow) to Hoechst. Dow, strongly pursuing its crop sciences business, bought out its partner Eli Lilly's participation in this business.

Very soon it became clear that a life sciences strategy involving both pharmaceutical and agricultural (life sciences) activities was unlikely to be successful—that there are insufficient synergies between the pharmaceutical and agricultural chemicals businesses in spite of the fact that both usually involved synthesis of complex molecules and both were beginning to explore biotechnology. These two types of life sciences were now seen as relatively incompatible because they had entirely different market requirements, which should have been obvious from the start. Moreover, profits in the pharmaceutical sector, aimed at health care, tend to be much higher than those generated in selling crop chemicals to farmers. Pesticide and herbicide sales are also cyclical, depending on the weather and commodity prices.[5] Novartis now decided to spin off its crop sciences operations into a new firm, Syngenta, which included Zeneca's business in this area. BASF had purchased American Home's crop sciences business but divested its pharmaceutical operations. DuPont had meanwhile sold its pharmaceutical business to Bristol-Myers Squibb, keeping crop sciences. The new lineup in crop sciences has Bayer as the leading firm, with revenues of almost $8 billion, assuming completion of the Aventis deal. If Bayer decides to sell its pharmaceutical operations, as has been rumored, the industry will have abandoned the life sciences concept almost entirely. Syngenta, Monsanto, DuPont, BASF, and Dow Agro-Sciences remain as large crop sciences players.

The Lineup at the Turn of the Century

Table 1 lists the twenty-five largest global chemical companies in 1999, while Table 2 shows the twenty-five largest chemical firms in the United States in 2000. Some observations should be made. First, only eight out of twenty-five global companies are headquartered in the United States. Only Dow and BASF may eventually be large enough to survive and prosper as non-back-integrated, still largely petrochemical companies. Their size tends to approach the advantages of back-integrated firms with petrochemical arms, such as ExxonMobil, BP, TotalFinaElf, and Shell. Shell's chemical operations have shrunk somewhat because a couple of years ago the firm divested a number of its downstream derivatives operations, such as Kraton polymers and epoxies. Considerably more restructuring is ahead for the industry, as firms strive to become large enough to compete with other growing firms and to attract or keep investors. SABIC became a petrochemical powerhouse and will soon join this group: it continues to build new crackers and derivatives plants and in 2002 purchased the petrochemical operations of DSM. It negotiated for, but failed to acquire, the remaining petrochemical arm of ENI.

In Japan the proposed Sumitomo Chemical–Mitsui Chemical merger would put the combined firm into the top ten globally. Other restructuring moves will have to occur in that country to shut down uneconomic capacity and to combine smaller firms into more viable entities.

Table 1. The Twenty-Five Largest Chemical Companies in 2000

Company	2000 Chemical Sales (U.S. $ Millions)	2000 Total Sales (U.S. $ Millions)
BASF (Germany)	30,790	33,117
DuPont (U.S.A.)	28,406	31,831
Dow Chemical (U.S.A.)	23,008	23,008
ExxonMobil (U.S.A.)	21,503	232,748
Bayer (Germany)	19,295	28,534
TotalFinaElf (France)	19,203	105,543
Degussa (Germany)	15,584	18,672
Shell (U.K., Netherlands)	15,205	149,146
ICI (U.K.)	11,746	11,746
BP (U.K.)	11,247	161,826
Akzo Nobel (Netherlands)	9,364	12,901
Sumitomo Chemical (Japan)	9,354	9,659
Mitsubishi Chemical (Japan)	8,976	16,212
Mitsui Chemicals (Japan)	8,720	8,720
Huntsman Corporation (U.S.A.)	8,000	8,000
General Electric (U.S.A.)	7,776	129,853
Chevron Phillips (U.S.A.)	7,633	7,633
DaiNippon Ink & Chemicals (Japan)	7,512	9,110
Equistar (U.S.A.)	7,495	7,495
DSM (Netherlands)	7,295	7,453
Henkel (Germany)	7,215	11,773
SABIC (Saudi Arabia)	7,119	7,119
Syngenta (Switzerland)	6,846	6,846
Rhodia (France)	6,835	6,835
Sinopec (China)	6,791	39,008

Source: "Global Top 50," *Chemical and Engineering News* 79 (23 July 2001), 24.

In Europe, ICI keeps shrinking as it divests the balance of its commodities. Henkel sold Cognis, part of its chemical operations, to a financial buyer. Rhodia is said to be seeking a merger partner. Europe has seen several combinations to form specialty firms with revenues of over $5 billion (Clariant, Ciba Specialty Chemicals, and Degussa), a trend that began in the United States with the Rohm and Haas–Morton International merger. DSM sold its petrochemical business to concentrate entirely on specialties, having acquired Gist-Brocades several years ago.

The commodity petrochemicals business in Europe will be under particular

Table 2. The Twenty-Five Largest U.S. Chemical Firms in 2000

Company	2000 Chemical Sales (U.S. $ Millions)
DuPont	28,406
Dow Chemical	23,008
ExxonMobil	21,543
Huntsman Corporation	8,000
General Electric	7,776
BASF (U.S.A.)	7,756
Chevron Phillips	7,633
Equistar	7,495
Union Carbide	6,526
PPG Industries	6,279
Shell Oil	6,265
Rohm and Haas	6,004
Eastman Chemical	5,292
Air Products	5,238
BP Chemical	5,100
Praxair	5,043
TotalFinaElf	4,800
Honeywell (Allied Chemical)	4,055
Lyondell Chemical	4,036
Nova Chemicals	3,916
Monsanto	3,885
Occidental Petroleum	3,795
ICI Americas	3,314
Akzo Nobel	3,313
Solutia	3,185

Source: "Movers and Shakers," *Chemical and Engineering News* 79 (7 May 2001), 28.

threat from the large, continuing buildup in the Middle East, given the fact that after Southeast Asia, Western Europe is the next largest market for Middle East producers. According to one projection, Middle Eastern ethylene capacity will rise from around five million tons per year in 2000 to over fifteen million tons by 2010. By that year Saudi capacity will be around 7 million tons, while Iran will have over 4 million tons and Qatar and Kuwait 2 million tons or more.[6]

In the United States more restructuring will mean further consolidation, with more acquisitions or mergers of specialty chemicals producers and more shakeouts of smaller petrochemical companies. The acquisition of Aristech Chemicals by Sunoco Chemicals is indicative of the fact that this oil company is a higher-value owner of petrochemical intermediates than the several Mitsubishi companies that bought Aristech (the old U.S. Steel chemical division) a decade or so ago.

It is difficult to project the destiny of the various commodity and specialty businesses and companies now owned by private operators and investment groups. The larger firms, such as Huntsman and Ineos, will eventually seek public ownership or may be bought as a whole or in pieces by strategic buyers. The smaller firms, including many specialty companies and businesses, will to some extent be bought by strategic or financial investors or will combine in some manner.

It is now clear that LBO firms have played a very important role in the industry and will continue to do so, with the emphasis now increasingly in Europe and perhaps soon in Japan. McKinsey recently pointed out that LBO firms in chemicals over the last ten years have substantially outperformed publicly traded companies and the Standard and Poor 500 (24.2 percent compounded total shareholder return for a chemicals LBO versus 18.5 percent for the S&P 500 and 15.8 percent for publicly traded chemical companies).[7] There appear to be two main reasons for this. First, LBO firms can use much greater leverage than publicly traded firms and, provided their acquisitions do not go bankrupt, which has also occurred, they are in a much better position to create value for the equity. Second, there were and still are a number of poorly managed companies for which an LBO firm can substantially reduce costs and sell poorly performing businesses to more qualified buyers, thereby achieving higher EBITDA (earnings before interest, taxes, and depreciation-amortization), which can lead to successful sale or initial public offering.

Future restructuring of the chemical industry, at least in the United States and Europe, will most likely see more rather than fewer LBOs. Chemical companies tend to be asset-rich with relatively high depreciation cash flows, which facilitates debt repayment by a leveraged purchaser. That, along with the fact that mutual funds, pension funds, and other institutional buyers appear less and less interested in chemical companies, leaves buyout firms as logical owners in many cases, particularly for firms with market capitalizations in the range of $500 million to $2 billion.

McKinsey believes that "chemical corporations could unlock significant value by applying the lessons of LBO firms to many businesses in their portfolio." The firm suggests that this could be done via "internal LBOs," joint venturing with an LBO firm or executing a leveraged recapitalization for the entire firm, as was actually done, for example, by FMC in 1986.[8]

The chemical industry in the United States, the European Union, and Japan is still quite fragmented compared with a number of other industries, such as the automotive, detergent and soap, and pulp and paper industries. Given increasing global competition, the need for size, and the difficulty for smaller firms to attract and hold stockholders, the case can be made for a much greater degree of chemical industry consolidation. There is an important overall industry trend toward consolidation into just a few companies, certainly in specific countries and now even globally. For highly capital-intensive businesses, such as automobiles, steel, and computer chips, this trend has been evident for some time. Consider, for example, that in memory chips there are only a few global players left: Micron Technologies, Toshiba, Hynix (Korea), and Infineon (Germany). These firms use economics of scale to bring prices down further and further, making it more and more difficult for competitors to stay in business. Once plant investments are committed, there is a temptation to run these large plants flat out, even in a recession, charging only rock-bottom marginal prices. Furthermore, chip yield is the most important variable, and here the experience curve comes in again: the more experience, the higher the yield. The losers in a downturn often cannot afford to build the next generation of plants and drop out. One or two of the firms mentioned above may not make the next round.[9]

How does the structure of the chemical industry compare with that in other mature industries? In March 2001 John Roberts and Karen Gilsenan, at that time lead equity analysts at Merrill Lynch, published a tongue-in-cheek report that looked at a scenario in which all U.S. major and specialty chemical firms then under coverage by Merrill Lynch were combined into one firm named "Chemzilla." (This included twenty-four majors and fourteen specialties, comprising most of the important U.S. chemical manufacturers, excluding fertilizer and pharmaceuticals producers.) The fictional firm would have about 59 percent of total U.S. chemical sales, the balance being largely served by the chemical divisions of oil companies, private companies, U.S. divisions of foreign companies, and imports. It would have a 10 percent share of global ethylene capacity, 18 percent of polyethylene capacity, and 23 percent of industrial gas capacity. Surprisingly, the market cap of Chemzilla was still less than that of Walmart, Cisco, and Microsoft and only about double that of Procter and Gamble! Chemzilla, although very large, would still only be tied with Walmart and ExxonMobil in total revenues.[10] Such a consolidation will not come to pass, but there could eventually be several smaller Chemzillas.

Business Issues and New Solutions

In many respects the managements of chemical companies have faced the same issues for a number of years, a true case of *plus ça change, plus c'est la même chose.* The challenges to grow, to improve profits, to raise the share price, and to deal with regulations and environmental issues and with the demands of their many stakeholders will remain the same over the next decade or two. There is room for both optimism and pessimism: optimism owing to the many opportunities to develop new chemical and other useful technologies, to serve emerging markets, and to find more and more uses for the Internet and e-business; and pessimism when considering the likely imposition of new regulations, the historic decline in innovation, the uncertain economic climate, and the circumstances that have brought about too many years of low profitability for the industry.

One of the biggest challenges facing companies is the need to resume top-line growth in an industry that is on the whole growing no faster than the GNP. This slow growth has led to an intensive search for the best, most appropriate business model to fit the firm's capabilities and position within the industry. Given the industry's relative maturity, general commoditization of products, and high degree of competitiveness, companies are vigorously seeking to find a way to differentiate themselves to get back somehow to a believable growth and profits model. The days of just building plants and selling the products were over some time ago. Growth via acquisitions, with no specific game plan other than to get bigger, has been shown to be a questionable approach that has not yielded the desired results, mostly because the hoped-for synergies that would achieve a "one plus one equals three" result have not happened. This is true both in the cases where revenues have grown substantially through large acquisitions (for example, DuPont-Conoco, Rohm and Haas–Morton International, and Lyondell-Equistar) and in cases of so-called mergers of equals (for example, Crompton-Witco and Geon-M. A. Hanna), although the situation will improve in the future for some of these.

Two consulting firms have given a great deal of thought to the study of past and possible future business models. Accenture studied a group of close to a hundred global chemical firms of various sizes and looked at their financial performance against the hypothesis that companies should either be one of the lowest-cost producers in their industry or be a firm that achieves higher margins by differentiating its offerings through technology or service to customers. Accenture called the first group "operators" and the second group "solution providers." The financial results of these firms over a five-year period in the late 1990s, in terms of total return to shareholders, was then plotted as a function of gross profit over total productive asset investment. The best operators would conceptually have high shareholder return despite high asset intensiveness, while the successful solution providers would have high shareholder return owing to higher margins and a much lower asset base. There was also a large group in between, which Accenture called "hybrids." Successful operators (for example, Exxon Chemical, Dow, Air Liquide, and Union Carbide) achieved good results owing to intensive efforts to reduce costs through operating scale to become efficient, low-cost producers rather than trying to differentiate themselves in the market through services and knowledge-based solutions. (This relates to most of their operations: even operators have some specialties.) Successful solution providers are specialty firms that have found a

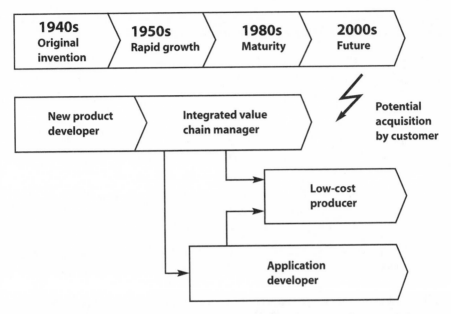

Figure 5. Evolving or expanding business models as market dynamics change. *Source:* Florian F. Budde et al., eds., *Value Creation: Strategies for the Chemical Industry* (Weinheim, Germany: Wiley-VCH, 2001), 59.

successful model based on downstream value plays (Avery Dennison and Sigma Aldrich), technology (3M), service (Ecolab), or "changing the rules of the game" (Valspar, which bypassed the traditional paint seller channel used by paint resin manufacturers). Hybrids are trying to be in both camps, which is strategically quite difficult, although they can also be successful, as BASF and the UCB Group (formerly Union Chimique Belge) have shown. Thus, companies should choose which of these paths to try to follow and to set their strategies accordingly.[11]

McKinsey has also given much thought to business models and came up with six winning models in the specialty chemicals industry. The firm terms these "new product developer," "application developer," "systems-solution provider," "process technologist," "value-chain integrator," and "low-cost producer." These models have to be matched with the type of chemicals produced, the needs of the firm's customers, and the segment's state of maturity. Thus, a "new product development model" could not thrive in a mature, highly competitive market where customers mainly value low-cost producers. McKinsey says that the "key success factor is the ability to anticipate changing market dynamics in order to be a first mover in developing a new and more successful model. How a firm's business model may evolve under changing market dynamics is shown in Figure 5.[12]

McKinsey also identified the following potentially useful growth strategies particularly applicable to specialty chemical firms:

- Increasingly focus on service-based business models and develop entirely new service-based businesses (for example, systems-solutions provider, process technologist, and value-chain integrator);
- Aggressively seek out and develop new technologies, both inside and outside the company (for example, establish important aggressive technology search-and-licensing functions);
- Become a leader in e-business in chosen segments (increased market reach, lower costs, and improved customer value propositions);
- Enter new parts of the value chain, typically downstream elements closer to the end consumer (although this can add significant value, it tends to be the most difficult to implement).[13]

Two companies can serve as examples of how to increase shareholder value greatly, even in mature businesses, using growth strategies in line with these concepts. Valspar adopted an acquisition strategy in its own area of coating resins and formulations, and also found a way to get around the entry barriers erected by traditional paint sellers like Sherwin Williams and Benjamin Moore. Valspar executed a channel shift strategy by offering an important value proposition to large discount retailers like Home Depot and Walmart, who could put their own brand names on the paints they sold to consumers rather than offering established brands. This end run around traditional paint sellers was coupled with close linking of Valspar's supply chain system to those of customers like Walmart who only deal with suppliers able to cooperate in this manner.

Ecolab offered a new value proposition to its institutional cleaning customers by bundling a range of products and services and then offering to keep their customers' kitchens and housekeeping services supplied and cleaned—a one-stop shopping model that worked. Valspar and Ecolab have been cited as demonstrating the ability to achieve excellent top-line growth in the maturing chemical industry, where the top-tier companies were able to increase revenues an average of 38 percent from 1995 to 1999 while achieving average shareholder returns of 210 percent.[14]

While most industry characteristics have stayed the same, the ways of doing business have changed drastically over the last years because of maturing markets, greater global competition, consolidations of suppliers and customers, various technology enablers, and the Internet. To succeed in the new business order, companies must do whatever they can to change their historic business practices and to benefit from the use of the various new available tools. Companies that fail to do this will most certainly fall by the wayside. Those that seize the opportunities may nevertheless be acquired, but their shareholders will be better rewarded because of their firm's better historical financial performance.

Specialty chemicals firms, once the darling of investors, are now under the

Table 3. Forces of Change in the Specialty Chemical Industry

Business is growing more complex.

Supply chain is taxed by breadth of markets, products, and geography.

Growth from new products and business development is risky owing to high costs and high failure rates.

Markets are maturing, population growth is slowing, and population is aging.

Health, safety, and environmental regulations are tightening.

Commoditization of specialties is contributing to price and margin erosion.

Source: Nexant, Inc./Chem Systems.

gun. In early 1999 Chem Systems launched a multiple subscriber study titled "Characteristics of a Successful Specialty Chemicals Company in the Next Decade."[15] The study was intended to provide management with an overview of some of the important forces driving changes in the industry (Table 3) and to identify what Chem Systems believed would be the characteristics of companies that understood how to change their traditional ways of doing business. Almost four years later the conclusions still look pretty good and could be applied in many ways to commodity companies as well. Some of the more important conclusions are summarized below.

- *Customer relationship management:* Improve and use customer relationship management to differentiate competitive positions by satisfying customer needs, acquiring and maintaining profitable customers, and building loyalty in the most effective manner.

- *Innovation and product development:* Focus on innovation as a key growth option. The innovative process will be closely linked through knowledge management systems to incorporate research and information from markets, customers, and partners that collaborate in the process.

- *Core competencies:* Use these as a basis for deploying resources in areas of sustainable competitive advantage, using outsourcing for activities best left to others.

- *Business transformation:* Share practices throughout the firm to facilitate changes in work processes and organizations.

- *Supply chain management:* Integrate and optimize supply chains extensively, enabling the company to redefine partnerships and to streamline its raw materials, finished products, and vendor network.

- *e-Business:* Integrate information technology infrastructure to link it to suppliers and customers and to the company's organization to facilitate commerce, information exchange, and collaboration.

- *Asset management and economic value added:* Achieve enhanced financial performance through improved asset turnover and more efficient resource deployment.

- *Portfolio management:* Systematically restructure the portfolio periodically to renew business growth and improve financial performance.

Companies are now using many new techniques to improve performance in various areas of their business. For example, a great deal of effort has been expended on customer segmentation and channel strategies, using information technology such as data mining as enablers. Typically, customers can be divided into a number of groups: for example, those wanting service along with the product, those buying primarily on price, and those buying on a seasonal basis, in addition to segmenting by size of orders. Studying customer orders and buying patterns with data mining can also identify other likely customer needs, which can lead to additional business. (This is now being done extensively by online retail merchants, such as amazon.com.) Companies also now have the tools to study their entire product line to see which products (often expressed as stock-keeping units, or SKUs) can be dropped because of insufficient business levels or uneconomic operations. Procter and Gamble in the early 1990s sliced its product mix by a third because 85 percent of sales came from just 8 percent of the company's products. For some commodity chemical companies 80 percent to 90 percent of sales come from just 10 percent to 20 percent of their products.[16]

Another area that has changed drastically is procurement. Companies have gone in the direction of reducing—sometimes greatly—the number of vendors they deal with in order to facilitate procurement and to exert greater pricing leverage. Several companies have decided periodically to team up to pool their MRO (maintenance, repair, and operating articles) purchases in order to develop a much larger order and again to apply greater pricing leverage. An increasing amount of procurement is carried out over the Internet, part of it in auctions managed by business-to-business exchanges. There has also been a strong trend toward vendor-managed inventory, where suppliers monitor the level of product in the customer's tank and, under long-term contract, refill the tank and send an automatic electronic invoice to the customer.

An enterprise resource planning (ERP) system that works well allows companies to link their supply chain to those of their suppliers and customers, making it possible to develop effective partnerships. This is particularly useful, for example, when an important change in the supply chain is anticipated or suddenly occurs (for example, an unexpected plant shutdown or unexpectedly large new orders). Increasingly effective supply-chain software, from firms like Manugistics, i2, Aspentech, and SAP, have made demand forecasting considerably more effective when these programs are linked to a well-functioning ERP system.

In 2000, Accenture identified a number of key traits for companies hoping for success in the new environment. "Strategic options" included such items as new sources of growth and asset effectiveness; "operational tactics" included process effectiveness and people effectiveness (for example, change-receptive culture); and "impact of e-commerce" included new business models.

Relative to the last point, an interesting difference of opinion exists regarding the effect of the Internet and the networked world on companies' competitive strategies and on how companies might shape themselves for the future. Some strategists, such as Don Tapscott, believe that partnerships and the Internet are delivering a new breakthrough in strategic thinking. This is in sharp distinction to Michael Porter's views, believing as he does that companies should favor vertical integration over partnering and that business leaders "should abandon thoughts of new business models" or e-business strategies that he says encourage managers "to view their Internet operations in isolation from the rest of their business."[17] Tapscott, who has written broadly on outsourcing, reshaped his ideas on how companies should plan their business, and he takes sharp issue with Porter.

> In the future, strategists will no longer look at the integrated corporation as the starting point for creating value, assigning functions, and deciding what to manage inside or outside a firm's boundaries. Rather, strategists will start with a customer value proposition and a blank slate for the production and delivery system. There will be nothing to "outsource" because, from the point of view of strategy, there's nothing "inside" to begin with. Instead, managers, using new tools of strategic analysis, can identify discrete activities that create value and parcel them out to the appropriate web partners.[18]

It is perhaps not too far a stretch to conceive of new companies, or more likely, new subsidiaries being created that take new technologies—for example, the results of combinatorial research in specific areas carried out by such firms as Symmix—and outsource all production and many administrative functions to other parts of the organization and outside partners, while maintaining the marketing and R&D function. This approach would seem to combine both Porter's and Tapscott's thinking. New chemical entities can be formed to create maximum value while minimizing capital investment and ownership of assets.

However, it is unlikely that many firms in the chemical industry will be created in this manner. More likely, the successful firms will take advantage of the various tools now at their disposal in an attempt to grow and thrive in the new, increasingly competitive global environment, and they will outsource many or all the functions that others can do better and cheaper, provided that nothing is compromised in the process.

Dow has been a leader in the use of information technology. Under William Stavropoulos, the company embarked on a long-range plan to transform itself into an even stronger global competitor. The important elements of Dow's strategy are shown in Figure 6.[19] Dow's transformation began with a major reengineering effort in early 1992, when it also embraced the concept of value-based management. Dow became a chemical industry leader in information technology implementation, including substantial use of the Internet for sales and for customer relationship management. A substantial part of Dow's information technology needs is outsourced to Accenture. Dow is now engaged in

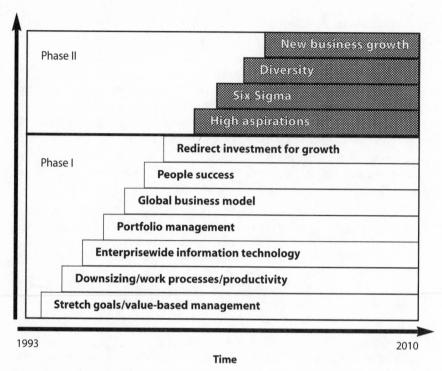

Figure 6. Staged evolution of Dow Chemical, 1993–2010. *Source:* Dow Chemical Company, by permission.

phase II, using Six Sigma for continuous improvement and further reengineering and cost reduction, and it is again looking for new business growth.

DuPont in the late 1990s in some respects led the industry by outsourcing much of its information technology work to Computer Sciences Corporation (CSC) and Accenture. There is now also a growing trend toward outsourcing a number of administrative functions, such as finance and administration and human resources. IBM, Accenture, CSC, and Electronic Data Systems (EDS) are considered leaders in providing outsourcing services for industrial clients.

A New Technology Vision

The chemical industry has always been supported and transformed by technological innovation. What remains to be seen is which companies will seize the opportunities offered by the new technologies, thereby becoming more competitive and more profitable and to some extent reinventing themselves.

In 1996 a study group comprising the American Chemical Society (ACS), the American Institute of Chemical Engineers, the Chemical Manufacturers Association (now the American Chemistry Council [ACC]), the Council for Chemical Research, and the Synthetic Organic Chemical Manufacturers Association (SOCMA) published a report that examined the factors affecting the competitiveness of the U.S. chemical industry in a rapidly changing business environment and attempted to develop a vision for the industry's future.[20] Although this study is several years old and only covered the U.S. industry, its conclusions are relevant to the scope of this chapter.

The study group saw six forces changing the topography of the business landscape: increasing globalization, sustainability, financial performance, customer expectations, changing workforce requirements, and increased collaboration.

Somewhat paralleling this assessment, made six years ago, are a number of megatrends recently identified by Jim Mack, CEO of Cambrex Corporation. He believes these trends will influence the course of the chemical industry over the next decade or two. They included, among others, consolidation, sustainable development, an aging population, the human genome, and nanotechnology.[21]

The technical recommendations of the ACS study, aimed directly at the chemical industy, are summarized below.

Technology Area 1: New Chemical Science and Engineering Technology

- Chemical synthesis: Develop new synthesis techniques incorporating the disciplines of biology, physics, and computational methods. Other recommendations relate to surface and catalytic science, structure control of complex molecular architecture, and development of chemistry in alternative reaction media.
- Bioprocesses and biotechnology: Develop more powerful and efficient biocatalysts, broaden the knowledge base for bioprocesses, and seek government support for biotechnological R&D.
- Materials technology: Industry to work with academe to promote interdisciplinary approaches to material science, including the integration of computational technologies, work with federal laboratories and academe to develop practical materials synthesis and processing technologies, and define common approaches for disassembly and reuse of materials.
- Enabling technologies: This covered such areas as process science and engineering technology, chemical measurement, and computational technologies.

Technology Area 2: Supply Chain Management

Given the high cost associated with supply-chain issues (as much as 40 percent of sales value internationally), the industry should increase its competitiveness by

- Implementing the ACC's Responsible Care initiatives throughout all logistics operations;
- Benchmarking studies identifying best practices and performance targets;
- Including operations optimization tools worldwide;

- Harmonizing packaging, labeling, documentation, handling, storage and shipping, and working with government agencies; and
- Developing superior information systems related to this area.

Technology Area 3: Information Systems

- Encourage technology providers to develop open systems.
- Facilitate the use of modeling and simulation technologies, working with the government to enhance the transfer of process simulation and modeling techniques.
- Encourage the development of expert systems and intelligent decision-support tools flexible enough to model multinational, multiproduct enterprises for use in business decision making.

Technology Area 4: Manufacturing and Operations

The study pointed out the need for advances in six areas of manufacturing operations: customer focus, production capability, information and process control, engineering design and construction, improved supply chain management, and global expansion.[22]

Some of the items covered in these lists are discussed in earlier chapters dealing with technology (Chapter 2) and information technology (Chapter 6).

It is difficult to extrapolate from these recommendations how the industry might actually change over the next decade or two. Many of the recommendations in the report relate to R&D issues. This is significant because, as has been pointed out in earlier chapters, most chemical firms have cut back on R&D to reduce overhead costs and increase earnings. Moreover, much of the remaining R&D is application research supporting existing product lines. Yet the above recommendations relate much more to process and engineering research, including fundamental R&D, and implementing this would presage a rather strong shift in companies' R&D spending toward what they were doing up to the 1980s. Favoring a move toward spending more on R&D are statistics showing that innovative companies fare much better than average, information that was recently presented by one of the top equity analysts for the chemical industry, Mark Gulley. Using as a sample the specialty chemicals firms he follows, he found that the "best in class" companies spent 7 percent of sales revenue on R&D (versus 4 percent for the average) and that the leaders had 35 percent new products over the last five years (versus 25 percent for the average). Gulley goes on to show that such firms as Sigma Aldrich, Pall Corporation, and 3M trade at significantly higher valuations (enterprise value/EBITDA) than their specialty chemical peers.[23] This brings into question the wisdom of managements that cut back on R&D to increase earnings: valuations may be hurt rather than helped.

Sustainable Development

Over the last several years the global chemical industry has embraced a relatively new concept called "sustainable development," which may be defined as forms of progress that meet the needs of the present without compromising the ability of future generations to meet their needs. In contrast to environmental policy, sustainable development is broader, emphasizing the links between ecological, economic, social, and cultural development. The World Business Council for Sustainable Development (WBCSD) includes over 150 international companies, whose leaders believe that the pursuit of sustainable development is good for their enterprises, the planet, and its people. Significantly, the executive committee of WBCSD includes such chemical industry leaders as Rodney Chase of BP, Jurgen Dormann of Aventis, Charles Holliday of DuPont, Egil Myklebust of Norsk Hydro, Bill Stavropoulos of Dow Chemical, and Phil Watts of Shell. The organization states that while many different cases can be made for sustainable development (moral, ethical, religious, and environmental) and its leaders deeply believe in those cases, it is the business case that is emphasized. The organization states that during the five years before 2001, the so-called Dow Jones Sustainability Index (DJSI),[24] made up of companies in twenty-one countries seen as leaders in sustainable development, outperformed the Dow Jones Global Index (DJGI), with 15.8 percent compounded annual return for the DJSI versus 12.5 percent for the DJGI. The business case is also seen as being more entrepreneurially driven, by companies becoming more competitive by staking out realistic value opportunities related to sustainable development.

Another form of business case can be made from a defensive standpoint by an industry constantly under attack by environmentalists and other sections of the public. In a speech to the ACC Thomas Burke, a well-known environmentalist as well as a consultant to BP, laid out this aspect of the case. He cited a pan-European survey carried out in 1998 that found "a slow but broad and continuous decline in the chemical industry's reputation as regards the benefits of the industry, its risk management aspects . . . a deteriorating climate in the public's view regarding the industry's license to operate."[25] While these conclusions are not necessarily borne out in other surveys, which show, for example, that Responsible Care has started to turn around public opinion about the industry, Burke's views are worth noting. Although Burke acknowledges the industry's considerable success in reducing emissions and decreasing energy consumption per unit of output, he also finds a so-called trust deficit between the industry's accomplishment and real-world opinion. This in his view affects issues that can be potentially damaging to business, such as access to resources and markets, license to operate, and cost of market entry. He lists a series of actual and potential policy impacts: compulsory environmental reporting, extensive use of environmental taxation; fully financed "closure bonds," tougher enforcement of existing policies, substitution of hazardous products, product-driven legislation, and a mandatory environmental liability regime.

To emphasize his position, he cited a vote, unfavorable to the chemical industry, taken in connection with a chemicals policy review by the Environment Committee of the European Union that took place in late October 2001. Emphasizing this vote, he ended his speech by saying that while the "case for the chemical industry" in the twentieth century could be termed "legal and profitable," the case for the twenty-first century was still to be made. To this effect, he said: "The industry must deliver continuous improvement in environmental and social performance. . . . It must ensure that it captures the reputational benefits of that improved performance in order to enhance its ability to influence public policy outcomes that might adversely or favorably impact on the business."[26]

Sustainable development should be a key part of industry vision and strategy. Whether industry's reasons for embracing sustainable development are "being on the side of the angels" or have a solid business foundation, there is no question that companies must do more than pay lip service to the concept. The following descriptions of what several leading firms are doing are largely taken from a recent article.[27]

- Dow Chemical established in 1996 a broad series of ecoefficiency goals for 2005. The company developed a sustainability index where overall performance is measured against economic, environmental, and social metrics. It conducts sustainable development workshops where twenty-one measurements are used to evaluate business performance and the progress of each global business against the three objectives. Dow is also integrating sustainable development principles into the strategic planning process.

  The DJSI identified Dow Chemical as the leading sustainability-driven company in the chemical industry.
- DuPont tracks annually the value of its environmental initiatives, with a corporate recognition program to reward individuals. Sustainable development goals for 2010 include holding energy use to 1990 levels, reducing greenhouse emissions by 65 percent from 1990 levels, sourcing 10 percent of its energy from renewable sources, and generating 25 percent of total revenues from nondepletable sources.
- BASF began to develop eco-efficiency analysis in 1996, which has been used to analyze a hundred products and production processes. It established a sustainability council in June 2001 headed by an executive board member and comprising the presidents of operating and functional divisions.

The company's *Verbund* structure, which links units at production sites, facilitates the design of products and services for sustainable development. The company also uses life-cycle analyses from extraction of raw materials to disposal of products, including behavior of end users. An ecological fingerprint provides a picture of the environmental effect of a product.

- Shell Chemical's strong tradition in scenario planning (see Chapter 4) has helped the company stay attuned to gradual as well as sudden changes in the business environment. Shell divides a sustainable development metric into economic, environmental, social, and governance values and has a number of cross-business teams developing metrics across these elements of their business. Going beyond introducing a "cost of carbon" (dioxide emissions) into project investment economics, Shell Chemicals requires all dimensions of sustainable development to be addressed in its capital investment proposals.

Shell sees the need to contribute to sustainable development as a new, enduring element of securing society's license to operate and to grow.

- BP, in an effort to drive progress in meeting sustainability goals, has focused on climate change, lower carbon energy, clean fuels, biodiversity, water, and social investment. BP has devised a global social investment strategy with emphasis on job and wealth creation, backed by commitments to human rights; partnerships with governments, international bodies, and nongovernmental organizations; and participation in policy debates at every level.

Sustainable development is also strongly promoted by CHEMRAWN (Chemistry Research Applied to World Needs), a group that has been active for several decades and is a program of the International Union of Pure and Applied Chemistry. In June 2001 two hundred scientists from over thirty countries convened at "CHEMRAWN XIV: Toward Environmentally Benign Processes and Products." The conference discussed the current state of the art in green chemistry and the role of chemical research and science policy in advancing global environmental protection and sustainable development. The conference identified "Twelve Principles of Green Chemistry," including less hazardous chemical synthesis, safer solvents, energy efficiency, renewable feedstocks, design for degradation, and real-time analysis for pollution prevention. Some interesting areas of research were identified, including p-xylene oxidation to terephthalic acid in supercritical water, which would be less energy intensive and would preclude the use and combustion or disposal of acetic acid solvent used in current technologies; production of polytetrafluoroethylene (Teflon) in liquid or supercritical carbon dioxide, which would eliminate the use of chlorofluorocarbon (CFC) solvents; and synthesis of adipic acid from cyclohexene using greener oxidizing agents and water solvent rather than the traditional nitric acid oxidation of cyclohenanol/one, which produces nitrous oxide as a waste product.

At the conference Joseph Miller, who headed DuPont's R&D department for thirty-five years before retiring recently, stated that "our approach to dealing with sustainability—green chemistry, green engineering—must be multifaceted. We need to collaborate and cooperate. . . . We need to find ways to bring science, technology and policy closer together."[28]

More Testing: Benefits but Major New Costs

Progressive adoption of green chemistry will take some time, and there are arguably more urgent, current issues the industry must deal with. Probably the most important of these is the need to make rapid progress on increasing the amount of publicly available screening-level hazard information on high-production volume (HPV) chemicals. The goal is for companies to have such data available on a majority of 2,800 HPV chemicals by 2004. The HPV program was jointly developed in 1998 by the ACC, the Environmental Defense Fund, and the EPA to provide information on the 2,800 chemicals that are manufactured or imported in the United States in quantities exceeding one million pounds per year. To date, four hundred companies have agreed to generate basic health and environmental toxicity data on about 2,100 of these chemicals. Chemicals that do not have voluntary sponsors will be tested using mandatory regulations issued under the Toxic Substances Control Act (TSCA).

More recently, the European Commission issued a white paper stating that there is a general lack of publicly accessible safety information on most chemical substances, proposing new toxicity data requirements on a chemicals list about four times larger than the 2,800 HPV chemicals identified in the United States. There is also a proposal calling for company licensing to produce and handle certain dangerous substances. These testing programs will be very expensive. The Conseil Européen des Fédérations de l'Industrie Chimique (CEFIC), the main chemical industry association in Europe, estimates that each HPV assessment will cost up to 150,000 euros ($136,000). Thus, for 148 HPV chemicals targeted by the worldwide chlorine industry alone, the total cost is estimated at more than 22 million euros.

The Alliance for Chemical Awareness, consisting of a coalition of twenty-five chemical companies and several trade associations, was formed to improve the public's understanding of the huge amount of information being developed under the HPV testing program and to monitor and keep on time the efforts of chemical companies developing this information. Companies meanwhile are concerned about "public misinformation and alarm" about chemicals that do not pose a threat and about the potential of "market displacement" and "product deselection" as a result of generating hazard data.[29]

The issue of chemical safety will not go away, and it will be impossible to convince every sector of the public regarding the safety of many chemicals, even when HPV testing indicates that a preponderance are perfectly safe to use. The ACC's "Good Chemistry" campaign is largely aimed at this problem at a time when the industry continues to be on the receiving end of new regulatory initiatives.

* * *

Writing now, not long after the tragic bombing of the World Trade Center towers, the problems of the chemical industry, which have taken up most of this book, must necessarily be placed in perspective and evaluated against other serious issues. "Globalization," which is considered a major challenge and opportunity for the multinational firms, now also has a darker meaning, as demonstrators from many countries and particularly the "left-behind" parts of populations rail against American economic aggressiveness and imposition of its mercantile system and values on other cultures. Refusal of the United States to sign the Kyoto treaty, which embodies protocols that would force countries to limit emission of greenhouse gases, is resented in most other parts of the world. The type of terrorism that led to the events of 11 September 2001 in New York could easily be turned against the refining and chemical industry in the United States and in other countries, which has led to major initiatives by the ACC to help companies review their state of preparedness against such attacks. The downturn of world economics after a prolonged period of economic growth, and in particular the worsening economic situation in Japan, will have long-term effects. And while crude oil prices are now again at attractive levels, except for periodic "blips," the Middle East remains a highly volatile region, and a new oil shock is not out of the question.

The twentieth century was unique in having seen the explosive rise of an industry—the chemical industry as we know it—that became as ubiquitous and pervasive as any industry created by man. To quote the ACC: "Chemistry touches people's lives. When the science of chemistry is applied, it helps make the lives of [people] throughout the world safer, healthier and more efficient. Indeed, our food, safe water supply, clothing, shelter, health care, computer technology, transportation and every other facet of modern life, all depend upon the business of chemistry."[30] Unquestionably, the industry must continue to be monitored and must monitor itself in order to protect the health and safety of its workers and of the public, and it is right for regulatory agencies and for groups in the public sector to vigilantly and responsibly make sure this occurs. But irresponsible attacks on the industry and calls for broadscale deselection of whole classes of chemicals, such as chlorine, must be fought when they occur. Chemicals are the lifeblood of our society and will remain so as long as the world continues to function in a traditional manner.

It is difficult for managers not to feel depressed about the state of the chemical industry, at least for the next year or even beyond that. Some have said that this period is as bad or even worse than the early 1980s. The problems, strongly linked to global economic trends, began in the last quarter of 2000 and worsened considerably thereafter. Thus in the United States earnings of twenty-five chemical companies tracked by *Chemical and Engineering News* fell 44 percent (relative to the same period in 2000) over the first nine months of 2001. Struggling to reduce costs, the industry had shed 16,000 jobs by November, bringing employment to the lowest level since 1987. In Europe overall production growth for chemicals slowed from 4.6 percent in 2000 to 1.8 percent in 2001. Japan is in its fourth recession in ten years, and for the first time in ten years Japanese production of chemicals (in 2001) is down relative to the previous year.[31] One good sign is the continued high growth rate of the GNP in China, which accounts for such a large portion of world chemical exports. However, the Chinese chemical industry is expanding rapidly and will not always be such a good market for exports.

The U.S. economy has been one of the main engines for global economic growth, and there are signs that the country's economy, including the stock market, has bottomed out and that growth will resume. It is more than timely for the management of companies to recognize that the mature characteristics of the chemical industry preclude a return to business as usual. There are no proven formulas for success, although it is hoped that this book has provided some ideas for new and creative thinking, as well as to identify outdated strategies that have not worked in the past and are even less likely to work in the future. Much more consolidation will no doubt occur, as companies that depend on the investment community for raising debt and equity reach for size and scope while continuing to shed businesses that do not earn the cost of capital.

The outlook for the U.S. and European industries may be better than participants now believe. Favored by large markets, historically able to conduct breakthrough research in private laboratories and universities, and containing a number of forward-looking managers, the chemical industry has an opportunity to at least partly reinvent itself rather than become another "smokestack industry."

Further restructuring to reduce the number of producers seems essential in creating an industry where there will be less mindless competition between too many players. This has effectively resulted in giving most of the benefits of innovation away to the industry's customers. We believe much more consolidation will happen as producers in the United States and elsewhere recognize that this will be essential for profitable survival. Further, the days when companies simply invent new products and push them into the market are over. The old way of doing business was the legacy of the traditional commodity producers, who had less interest in marketing than in bringing new capacity on stream. Recognizing now that production of many of the traditional chemicals will in the future be less profitable than before, with lower margins and less opportunity for exports, the surviving producers in the OECD countries will either learn to thrive with lower margins or concentrate more and more on value-added products—the only sensible way for the future. There will eventually be a smaller domestic industry, as the remaining commodity producers go abroad to build large, cost-effective plants alone or in partnership with local firms. This strategy is already evident when looking at recent investments by Exxon-Mobil, BP Chemicals, Chevron-Phillips, and Dow Chemical. Specialty chemical firms are also moving some of their production to Asia, particularly to China, where they will supply not only some of their traditional OECD customers that have moved their own production abroad but also indigenous firms that are springing up rapidly to supply the burgeoning Chinese market, which has over a billion customers.

In the 1990s chemical companies were emphasizing cost reduction, information technology, and portfolio rationalization, as well as considering how best to serve the growing East Asian markets. In the first decade of the twenty-first century they will continue on this course, but they will also concentrate on other issues that are now increasingly preoccupying the attention of top management.

Probably heading the list is the concern about energy and feedstock prices. In the United States, natural gas prices have spiked twice over the last two to three years to well over five dollars per million BTUs. Relatively expensive natural gas versus crude oil caused particular hardship for Gulf Coast producers, who still produce most of their ethylene from natural gas liquids. Both commodity and specialty firms are increasingly squeezed between energy and feedstock suppliers and customers who want steady reductions in price for commodity chemicals and for specialties that have become commoditized. This leads to another key issue: companies' need to develop differentiated products that can command higher margins on a sustainable basis over time. The chemical industry has a strong record of innovation, but less so recently. There does now appear to be a stronger move to look at increased research spending to develop new products that can serve the growing markets of the new century. Some of these include health care, alternative energy, food, electronics, information, and materials. As the main enabling science, chemistry holds the key to developing many new products needed in these markets. Nanotechnology, biochemistry, and material sciences are the technologies most often mentioned.

There is now also reason to believe that the petrochemical industry in North America, Europe, and Japan is finally starting to come to grips with the over-capacity and "overinvestment" problems that have plagued it with pernicious pricing cycles for thirty years. Given the buildup of large new plants in Asia, U.S. producers are less likely to add large "lumps" of petrochemical capacity, and, when they do build, they tend to embrace partnerships, as exemplified by the recently constructed BASF-FINA cracker. In this way the partners can get true economies of scale while reducing the total new capacity that would have otherwise been built by the individual firms.

The chemical industry is alive but not particularly well. It faces not only intense global competition but also a continuing credibility problem with the public, although that situation is thought to be improving. The world will continue to need chemicals and pharmaceuticals, and the industry will continue to produce and market them responsibly, recognizing that different stakeholders will have different views about where to draw an appropriate cost-benefit analysis.

Endnotes

1. Graham L. Copley and Hassan I. Ahmed, "U.S. Chemicals: The End of the 'Good Old Days,' " Bernstein Equity Research Analysis presented at the International Petrochemical Conference of the National Petroleum Refiners Association, San Antonio, 25 Mar. 2002.
2. Gary K. Adams, Chemical Market Associates, Inc., presentation, Salomon Smith Barney Chemical Conference, New York, 5 Dec. 2001.
3. Hilfra Tandy, "Overview and World Petrochemical Analysis," *Chemical Matters* [London issue], no. 379, p. 3.
4. Bill Schmitt, "Chemical Century: Formulating for the Future," *Chemical Week* 111 (22/29 Dec. 1999), 24–27.
5. Fred Aftalion, *A History of the International Chemical Industry*, 2nd ed. (Philadelphia: Chemical Heritage Press, 2001), 383.
6. John Wyatt, "Olefins—Global Ethylene Capacity Development," Tecnon-PTM Petrochemical Workshop, Vergiate, Italy, 7 June 2002.
7. Florian F. Budde et al., eds., *Value Creation: Strategies for the Chemical Industry* (Weinheim, Germany: Wiley-VCH, 2001), 97.
8. Ibid., 101–103.
9. Peter Martin, "New Logic for Chipmakers," *Financial Times*, 6 Dec. 2001, p. 14.
10. John Roberts, "And Then There Was One? Consolidation to Remain Major Industry Theme," Merrill Lynch Equity Research report, 6 Mar. 2001.
11. John Aalbregtse, personal communication.
12. Budde et al., *Value Creation* (cit. note 7), 57–58.
13. Ibid., 60–61.
14. A. R. Shapiro and John Sisto, "The Case for High Value Growth: Strategies for Value Creation," *Chemical Market Reporter* 261 (14 Jan. 2002), 13–15.
15. Anthony Durante, personal communication.
16. Michael Roberts, "Using SKU Remediation to Eliminate Low-Value Products," *Chemical Week* Special Supplement (24 Mar. 1999), 59.
17. Michael E. Porter, "Strategy and the Internet," *Harvard Business Review* 79 (Mar. 2001), 63–78.
18. Don Tapscott, "Rethinking Strategy in a Networked World (or Why Michael Porter Is Wrong about the Internet)," *Strategy and Business* (Third Quarter 2001), 34–41.
19. Fernand Kaufmann, personal communication.
20. *Technology Vision 2020: The U.S. Chemical Industry* (Washington, D.C.: American Chemical Society, Department of Government Relations and Science Policy, 1996).

21. James Mack, "Specialty Chemicals: Competitiveness in Global Markets," American Chemistry Council Leadership Conference, Houston, 29 Oct. 2001.

22. *Technology Vision 2020* (cit. note 20).

23. Mark R. Gulley, "Innovation Drives Competition," American Chemistry Council Leadership Conference, Houston, 29 Oct. 2001.

24. Leading sustainability companies set industrywide best practices in strategy, innovation, governance, shareholder relations, employee relations, and other stakeholder relations, in all cases integrating long-term economic, environmental, and social aspects into their conduct and strategies.

25. Thomas Burke, "A Sustainable Future for the Chemical Industry," American Chemistry Council Leadership Conference, Houston, 29 Oct. 2001.

26. Ibid.

27. Cynthia Challener, "Sustainable Development at a Crossroads," *Chemical Market Reporter* 260 (16 July 2001), Fr3–Fr4.

28. Stephen K. Ritter, "Green Chemistry," *Chemical and Engineering News* 79 (16 July 2001), 34.

29. Glenn Hess, "Chemical Group Launches Web Site on HPV Communications Challenge," *Chemical Market Reporter* 259 (11 June 2001), 4.

30. *Guide to the Business of Chemistry* (Arlington, Va.: American Chemistry Council, Policy, Economics and Risk Analysis, 2001), 5.

31. Jean-François Tremblay, "Asia-Pacific: Despite Areas of Growth, the Fragility of the Japanese Economy Threatens the Entire Region," *Chemical and Engineering News* 79 (17 Dec. 2001), 37–40.

Glossary

ABS resins: a group of resins made by reacting acrylonitrile, butadiene, and styrene.

acetate filter tow: an acetate fiber used to produce cigarette filters.

alcohols: organic liquids containing a hydroxyl group (-OH) attached to a carbon atom.

aliphatics: hydrocarbons that do not have a ring structure.

alkane: a straight-chain hydrocarbon without double bonds—also called *paraffin*.

alkylation: usually the reaction of a hydrocarbon, such as an alkane or aromatic, with an olefin (double-bond-containing molecule), using an acid or other catalyst.

aromatics: hydrocarbon derivatives with a ring structure.

autoclave plant: as used in this book, a polyethylene plant with stirred, high-pressure reactors.

back-integration: the linking of a chemical plant or business on a corporate basis to an *upstream* plant or business from which it receives its key raw material(s).

basic chemicals: term used to denote high-volume, commodity chemicals.

battery limits: refinery and chemical plants usually have two parts. The production part is termed *battery limits* and includes the chemical process operation; the other part is "offsites." See also *offsites*.

bridge loan: term used by banks when supplying short-term financing (usually at high interest rates to allow for greater risk) to allow a transaction (e.g., an acquisition) to proceed before it can be financed on a long-term basis.

BTX aromatics: mixtures containing at least two of the following components—benzene, toluene, xylene.

business chain profitability: the profits gained from a vertically linked chain of plants or businesses.

business process reengineering (BPR): a methodology that came into vogue in the 1980s that involved examining and then improving a company's business processes in a relatively untraditional manner.

carbonylation: reaction, usually involving synthesis gas, that attaches a carbonyl (C=O) group to another molecule.

cash cost (of production): manufacturing cost, including raw materials, utilities, labor, maintenance, and plant overhead costs but excluding depreciation and corporate overheads.

cash cost margin: the difference between the sales price realized and the cash cost of production.

catalytic reformate: the product from the reforming process, as used to produce aromatics from a naphtha fraction.

chain economics: term used when the manufacturing cost of the end product(s) from two or more vertically integrated plants (chain of plants) is based on combining the cash costs of production achieved in each of these plants.

close-coupled cracker: a cracker that is adjacent to one or more derivatives plants using products (e.g., ethylene or propylene) emanating from the cracker.

commoditization: the situation in which a specialty chemical can be produced by several competitors, therefore in a sense becoming a commodity chemical. It occurs when products are increasingly produced by more and more firms and are therefore not particularly profitable.

commodity chemical: a chemical that is produced by a number of firms, thereby creating a highly competitive situation.

cracker, or cracking: the process whereby a hydrocarbon or a complex mixture of longer-chain hydrocarbons is broken into smaller molecules through the use of heat, with or without a catalyst. In the latter cases the process is often referred to as *pyrolysis.*

derivatives: a term generally used to describe products made by reacting first-stage petrochemicals (e.g., ethylene and benzene) with other chemicals (e.g., chlorine and oxygen) to produce higher-value chemicals (e.g., vinyl chloride and ethylene oxide).

downstream: a business or plant receiving its raw material from another plant, considered *upstream.*

economic value added (EVA): the best-known measure (according to Stern Stuart, a management consulting firm) for evaluating the creation of shareholder wealth, defined as "net operating profit minus an appropriate charge for the opportunity cost of all capital invested in an enterprise."

elastomer: a synthetic polymer with rubberlike characteristics.

enterprise resource planning (ERP): a comprehensive software-based system used to monitor and control the complex transactions (for example, "order to cash") carried out by manufacturing and other firms.

exit: as used in this book, the decision by a company to terminate the production of a chemical or chemical business by selling or shutting down the business.

feedstock: raw material.

fiber intermediates: monomers used to produce synthetic fibers.

Fischer-Tropsch synthesis: a process originally developed in Germany before World War II that converts synthesis gas to a combination of hydrocarbons and oxygenates at high pressure.

flaring, or flare gas: natural gas associated with crude oil production for which there is no local use and which is therefore disposed of by burning on high flares.

forward-integration: the situation in which a chemical plant or business is linked on a corporate basis to a *downstream* plant or business to which it supplies a key raw material or other feedstock.

gas liquids: higher hydrocarbons (e.g., ethane and propane) usually contained in natural gas.

gas-phase plant: a polyolefin production plant where the reaction is carried out in the gaseous rather than the liquid phase.

gasoline fraction: the portion of hydrocarbons distilled from crude oil or recovered from other processes that boils in the gasoline (that is, naphtha) range.

hydrocarbons: chemicals containing only carbon and hydrogen atoms.

intermediates: chemicals formed as a middle step in a series of chemical reactions. They can often be converted to different end products.

laggard: as used in this book, a manufacturing plant that has comparatively unfavorable economics of production relative to other plants making the same product.

leader: as used in this book, a manufacturing plant that has the best or close to the best economics of manufacture relative to other plants making the same product(s).

legacy system: a software system in place prior to implementing a newer system, such as enterprise resource planning (ERP) software.

merchant market: the term used when a chemical producer sells a product to other companies (the "merchant" market) to differentiate that part of its production that transfers the product to its own downstream operation (the "captive" market).

metallocene: a type of catalyst used in more recent polyethylene plants to achieve a higher-valued polymer.

metathesis: a process that reacts one or more olefins to produce a mixture of other olefins with different distributions of chain lengths.

monomer: a reactive chemical of relatively low molecular weight that may be reacted with itself or other reactive chemicals to form various types and lengths of molecular chains called polymers or copolymers.

MRO: maintenance, repair, and operating articles. This term is used with respect to the purchase of items other than raw materials, catalysts, and chemicals.

natural gas: found either separately or in association with crude oil. This is primarily methane, with some ethane, propane, and higher hydrocarbons. Natural gas often contains hydrogen sulfide, carbon dioxide, and other nonhydrocarbon constituents.

needle coke: a form of coke from crude oil distillation bottom residues that is used to produce graphite electrodes for aluminum production.

netback: price received by a seller after deducting freight and other logistical costs associated with delivery to the customer; also the net proceeds from a sale after accounting for charges incurred as part of the sale (e.g., freight absorbed by the seller).

nutriceutical: a food-oriented synthetic chemical, such as a synthetic sweetener.

offsites: parts of a plant not used for direct manufacture (e.g., buildings, tankage, and laboratories).

olefins: hydrocarbons containing one double bond in their structure (double bond–containing hydrocarbons are much more reactive than single bond–containing hydrocarbons that are more prevalent in fractions of crude oil and natural gas). Ethylene is the simplest olefin, with two carbon atoms. Then comes propylene (three carbon atoms, one double bond), butylene (four carbon atoms, one double bond), and so forth. Compounds with double (or triple) bonds are also called "unsaturates."

operating rate: usually defined as the total industry product demand divided by the effective product supply, as represented by all producers in a country or region.

organics: term used to include compounds of carbon, which may contain hydrogen, oxygen, nitrogen, sulfur, and other heteroatoms.

oxidation: the reaction of oxygen with a hydrocarbon molecule to produce an oxygenate or, if carried further, to produce carbon monoxide, carbon dioxide, and water.

oxygenates: chemicals containing carbon, hydrogen, and oxygen in their structure (e.g., alcohols, aldehydes, and ketones).

performance chemicals: specialty chemicals used in small quantities that are sold for their ability to produce certain effects in their application.

petrochemical cycle: a period (usually seven to ten years) during which operating rates and pricing go from very high to very low and back again.

petroleum fractions: crude oil is a mixture of light and heavy hydrocarbons separated into various fractions in a refinery distillation column (pipe still) and then further processed or treated and sold as products.

phosgenation: a reaction involving the use of phosgene.

plasticizer: material used to soften a hard polymer, such as polyvinyl chloride (PVC).

polymer: a high-molecular-weight material obtained by joining together many simple molecules (monomers) linked end to end or cross-linked.

polymerization: the formation of very large molecules from small molecules, a catalyst being normally used for this purpose. The resulting material is a polymer.

polyolefin: polymer made from ethylene, propylene, or other olefins.

pyrolysis gasoline: high boiling fraction recovered from distillation of cracker products.

reforming: process that uses heat and usually a catalyst to transform hydrocarbons into other hydrocarbons (e.g., in naphtha reforming used to make aromatics from paraffins and naphthenes) or mixtures of hydrocarbons and oxides of carbon into hydrogen, with air or steam taking part in the reaction. Hydrocarbon reforming is used to improve the quality (octane number) of gasoline.

resins: usually synthetic or natural high boiling polymers.

return on sales (ROS): the percentage of sales revenues representing profit.

selectivity: the percentage of desired product in the stream leaving a reaction.

"semiworks" plant: a very small manufacturing plant, generally for a new process, that is built to demonstrate the viability of the process and to provide samples.

SG&A: sales, general, and administrative costs—a term that covers many company overhead expenses.

single-site catalyst: a catalyst (for example, in olefin polymerization) that produces a polymer with a narrow weight distribution (see also *metallocene*).

single-train, or single-line, plant: term used to describe a relatively large plant that does not use two or more reaction trains or systems in parallel. *Single-line* is a term used in the polyethylene and other industries.

slurry reactor: a reactor in which a catalyst mixes with reactants that are in the liquid phase.

specialty chemicals: the term used to differentiate relatively small-volume chemicals (versus commodity chemicals) that are used for certain characteristics.

specification products: chemicals of known molecular composition as shown by analysis.

spot market: the market that exists for chemicals not sold on a contract basis. These tend to have wider price swings than industrywide contract prices.

spot price: the price of a chemical sold in the spot market.

steam cracker, or steam cracking: the high-temperature cracking of hydrocarbons in the presence of steam to give high yields of olefins (e.g., ethylene and propylene).

"stranded" gas: natural gas found in remote locations that has no local use.

strategic business unit (SBU): a group of businesses within a company for which a relatively common business strategy can be employed.

synthesis gas (syngas): a mixture of carbon monoxide and hydrogen.

"take-or-pay" contract: an agreement between a buyer and seller, whereby the buyer agrees to purchase a certain quantity (usually a maximum and a minimum amount) at an agreed-on (e.g., formula) price and incurs a financial penalty if a decision is made to take less product.

tar sands: hydrocarbon-containing deposits, such as those found in Alberta, Canada, that can be mined and processed into a form of synthetic crude oil.

thermoplastics: polymers that soften without chemical change when heated or cooled.

toll arrangement, or tolling: a business arrangement common in the chemical industry where, instead of buying a product, the "buyer" supplies raw material(s) to the seller who in turn produces the desired product at the producer's manufacturing cost plus a "tolling" fee.

tubular plant: a polyethylene plant that uses reactors in the shape of long tubes to carry out high-pressure synthesis.

ultraviolet absorber: an additive to a polymer that prevents or slows degradation caused by ultraviolet light.

upstream: a business or plant sending its product(s) to another plant, considered *downstream*.

vacuum gas oil: the heaviest distilled fraction from crude oil distillation.

value-based management (VBM): the approach that ensures that corporations are run consistently to achieve corporate values as embodied in mission, strategy, performance, and reward measures and corporate culture (see also *economic value added*).

value chain: the series of production steps (e.g., feedstock to first-stage petrochemical, to monomer, to polymer, to compound, to extruded form, to finished article) that add value to each step.

variable margin: the difference between selling price and raw materials cost.

vertical integration: the term used to describe a situation where a company operates several plants that send material from one to the next, sequentially producing chemical(s) of higher value at each step.

world-sized, or world-scale: refers to a manufacturing plant of a large size, as currently built or planned to be built in new installations around the globe.

zeolite catalysts: catalysts in the form of microporous crystalline solids with well-defined structures, generally containing silicon, aluminum, and oxygen in their framework.

Bibliography

Chapter 1

"A Craving for Chemicals Changes to Aversion." *Chemical Week* 130 (21 Apr. 1982), 32–33.

Fred Aftalion. *A History of the International Chemical Industry.* Philadelphia: University of Pennsylvania Press/Chemical Heritage Foundation, 1991.

John Akitt. "The U.S. Chemical Industry—New Directions." Society of Chemical Industry conference, Florence, 10 October 1983.

Ashish Arora; Ralph Landau; Nathan Rosenberg, editors. *Chemicals and Long-Term Economic Growth.* New York: Wiley Interscience, 1998.

Joseph L. Bower. *When Markets Quake.* Boston: Harvard Business School Press, 1986.

Joseph. L. Bower et al. "Reinventing an Italian Chemical Company: Montedison." Harvard Business School case no. 9-392-049. Boston: Harvard Business School Press, 20 December 1991.

Langdon Brockinton. "Quantum Unveils Recap Plan: Emery for Sale." *Chemical Week* 144 (4/11 Jan. 1989), 9–10.

William A. Brophy. "New Routes to Chemicals from Synthesis Gas." Chemical Marketing Research Association conference, San Antonio, 15–18 March 1982.

Bryan Burroughs; John Helyar. *Barbarians at the Gate.* New York: Harper & Row, 1990.

Gordon Cain. *Everybody Wins! A Life in Free Enterprise.* Second edition. Philadelphia: Chemical Heritage Press, 2001.

"Capital Spending: First Drop Since 1968." *Chemical Week* 132 (12 Jan. 1983), 41.

Keith Chapman. *The International Petrochemical Industry: Evolution and Location.* Oxford: Basil Blackwell, 1991.

Chem Systems U.S. Annual Report. "Price Forecasts." October 1999.

Chemical and Engineering News 58 (9 June 1980), 44, and 68 (18 June 1990), 46.

Chemical Data Inc. "Monthly Petrochemical and Plastics Analysis." Feb. 1999.

"Coping with the Cost-Price Squeeze." *Chemical Week* 130 (20 Jan. 1982), 40.

"Costlier Crude Oil: No Snag for Synthetics." *Chemical Week* 124 (28 Mar. 1979), 19–20.

Jack DeWitt. "The Needs and Progress in U.S. Chemical Rationalization." European Chemical Marketing Research Association conference, Venice, 10 October 1983.

Charles M. Doscher. "Restructuring: What Still Needs to Be Done." European Chemical Marketing Research Association conference, Antwerp, 1986.

"Facts and Figures for the Chemical Industry." *Chemical and Engineering News* 68 (18 June 1990), 46–47.

"For CPI Leaders: Pre-dawn Darkness." *Chemical Week* 132 (2 Mar. 1983), 66.

Marshall Frank. "Petrochemical Industry Profitability Cycle: Where Are We Now? Where Are We Going?" Chem Systems Annual Chemical Conference, Houston, January 1995.

R. L. Grandy. "The Promise of the 1980s." Chemical Marketing Research Association conference, New York, 5–7 May 1980.

David Hunter. "Overcapacity Is the Big Concern at ECMRA in Paris." *Chemical Week* 145 (1 Nov. 1989), 14–15.

"It Was a Grim First Quarter." *Chemical Week* 130 (28 Apr. 1982), 10.

H. Kamata; M. Hongoh. "Reconstructing the Japanese Petrochemical Industry—The Japanese Petrochemical Plan." European Chemical Marketing Research Association conference, Venice, 10 October 1983.

Lewis Koflowitz. "Set for a Sparking Second Half." *Chemical Business* 10 (June 1988), 43–49.

"Making Money at Lower Operating Rates." *Chemical Week* 132 (13 Apr. 1983), 28–31.

Gunther Metz. "Solving Structural Problems of the European Chemical Industry." *Chemistry and Industry* no. 24 (17 Dec. 1984), 871–877.

"New Directions for the Chemical Industry." Papers presented at the Society of Chemical Industry conference, Florence, 10 October 1983.

Chemistry and Industry no. 24 (19 Dec. 1983), 911–926.

J. A. Philpot. "Climbing the Barriers." European Chemical Marketing Research Association conference, Amsterdam, October 1984.

"The Reworking of Monsanto." *Chemical Week* 132 (12 Jan. 1983), 42–47.

"Rising Oil Prices Spur Growth of New Technology." *European Chemical News, European Review '79*, supplement to 23 July 1979 issue of *Chemscope*, 18–30.

Peter Savage. "Cain Chemical Builds a Name in Basic Chemicals." *Chemical Week* 142 (23 Mar. 1988), 22–26.

Martin B. Sherwin. "Chemicals from Methanol." *Hydrocarbon Processing* (Mar. 1981), 79–84.

Harold A. Sorgenti. "Restructuring towards a New Age of Vitality." Society of Chemical Industry conference, Madrid, 6 October 1986.

Peter H. Spitz. *Petrochemicals: The Rise of an Industry.* New York: John Wiley & Sons, 1988.

L. E. St. Pierre; G. R. Brown. "Future Sources of Raw Materials." World Conference of Future Sources of Organic Raw Materials, Toronto, 10–13 July 1978.

William J. Storck. "C&EN's Top Fifty Chemical Products and Producers." *Chemical and Engineering News* 58 (5 May 1980), 33–40.

———. "Foreign Investment in U.S. Chemicals Picking Up Again." *Chemical and Engineering News* 63 (5 Aug. 1985), 9–11.

———. "Union Carbide Is a Weaker, Smaller Company following GAF Fight." *Chemical and Engineering News* 64 (27 Jan. 1986), 11–12.

Sumio Takeichi. "The Nature of the Chemical Industry in Japan." Chemical Management and Resources Association conference, New York, 9 May 1990.

Lynn Tattum. "Monsanto Is Putting the Accent on Europe-Africa Operations." *Chemical Week* 147 (1 Aug. 1990), 20.

Larry Terry; J. Robert Warren. "The Siege Is Over at Union Carbide." *Chemical Business* 15 (Feb. 1993), 6–9.

Philip Townsend. "Analyzing the End Use Competition between Plastics and Other Raw Materials." Chemical Marketing Research Association conference, Atlanta, February 1984.

Italo Trapasso. "Rationalization in Western Europe." European Chemical Marketing Research Association conference, Venice, 10 October 1983.

U.S. Chemical Industry Statistical Handbook. Arlington, Va.: Chemical Manufacturers Association, 1994.

Michael Valenti. "Where Did the Money Go?" *Chemical Business* 10 (June 1988).

Andrew Wood. "No Easy Balance in Petrochemicals." *Chemical Week* 146 (28 Mar. 1990), 23.

———. "Polyethylene Prices Bottom Out, but Oversupply Looms." *Chemical Week* 148 (8 May 1991), 12.

Andrew Wood; Ian Young. "Oversupply Dogs the Petrochemicals Arena." *Chemical Week* 148 (10 Apr. 1991), 24.

Chapter 2

John N. Armor. "New Catalytic Technology Commercialized in the U.S.A. during the 1980s." *Applied Catalysis* 78 (1991), 141–173.

———. "New Catalytic Technology Commercialized in the U.S.A. during the 1990s." *Applied Catalysis: A, General* 222:1 (2002), 407–426.

Jim Barber. "On a Single Site." *Asia-Pacific Chemicals* (Oct. 1999), 30–31.

"Biotech and IT Find the Right Chemistry." *European Chemical News* 69 (30 Nov. 1998), 31–39.

"Catalyst for Change." *Asia-Pacific Chemicals* (April 1999), 15–16.

Gabriele Centi; Siglinda Perathoner. "Unraveling a Catalytic Pathway: Acrylonitrile from Propane." *Chemtech* (Feb. 1998), 13–18.

"Choices Widen for Acetic Acid Producers." *European Chemical News* 65 (11 Mar. 1996), 17.

Committee on Visionary Manufacturing Challenges, National Research Council. "Visionary Manufacturing Challenges for 2020." *Chemtech* (May 1999), 49–57.

Robert H. Crabtree. "Speeding Catalyst Discovery and Optimization." *Chemtech* (April 1999), 21–30.

Norman De Lue. "Combinatorial Chemistry Moves Beyond Pharmaceuticals." *Chemical Innovation* 31 (Nov. 2001), 33–39.

James Denyer. "The Survival of the Global Chemical Industry." *Chemical Innovation* 30 (July 2000), 40–44.

Benjamin Eisenberg et al. "The Evolution of Advanced Gas-to-Liquids Technology." *Chemtech* (Oct. 1999), 32–37.

Peter Fairley. "Directed Evolution: Enzymes Enter the New Economy." *Chemical Week* 162 (5 Apr. 2000), 29–34.

"Finding the Route to Success." *ECN Chemscope* (May 1999), 23–24.

Clive M. Freeman et al. "Molecular Simulations in Heterogeneous Catalysis." *Chemtech* (Sept. 1999), 27–33.

Guide to the Business of Chemistry. Arlington, Va.: American Chemistry Council, 2001.

Joseph R. Herkert. "Sustainable Development: Ethical and Public Policy Implications." *Chemtech* (Nov. 1998), 47–53.

John Kollar. "Ethylene Glycol from Syngas." *Chemtech* (Aug. 1984), 504.

Mark Morgan. "Alkane Aims." *European Chemical News* 77 (18 Nov. 2002), 26–28.

"New Process Developed by Sabic Oxidises Ethane." *European Chemical News* 73 (20 Nov. 2000), 29.

Jeffrey S. Plotkin. "Melting Away." *European Chemical News* 77 (8 July 2002), 32–34.

Jeffrey S. Plotkin; Andrew B. Swanson. "New Technologies Key to Revamping Petrochemicals," *Oil and Gas Journal* 97 (13 Dec. 1999), 108–114.

Rob Schoevaart; Tom Kieboom. "Combined Catalytic Reactions—Nature's Way." *Chemical Innovation* 31 (Dec. 2001), 33–38.

Larry Song. "Taking the Direct Approach." *European Chemical News* 71 (27 Sept. 1999), 24–27.

Peter H. Spitz. *Petrochemicals: The Rise of an Industry.* New York: John Wiley & Sons, 1988.

William J. Storck. "Industry's Bright Outlook." *Chemical and Engineering News* 76 (12 Jan. 1998), 155–162.

Symyx Web site, www.symyx.com, 8 August 2001.

Harry H. Szmant. *Organic Building Blocks of the Chemical Industry.* New York: Wiley-Interscience, 1989.

Larry Terry. "Demand, Technology Gains Tempt New Entrants." *Chemical Week* 160 (29 July 1998), 33–34.

Steven A. Weiner; Bruce Cranford; Helena Chum. "The Chemical Industry of the Future: Two Views." *Chemtech* (April 1998), 10–15.

Klaus Weissermel; Hans-Jurgen Arpe. *Industrial Organic Chemistry.* Third edition. New York: VCH Publishers, 1997.

Harold A. Wittcoff; Bryan G. Reuben. *Industrial Organic Chemicals.* New York: John Wiley & Sons, 1996.

Chapter 3

Fred Aftalion. *A History of the International Petrochemical Industry.* Second edition. Philadelphia: Chemical Heritage Press, 2001.

David Begleiter; Patricia D. Jones. "Specialty Chemicals—Basic Ingredients: Value-Based Analysis." ABN-AMRO Equity Research, May 2001.

Andrew Boccone. "Specialty Chemicals—The Miracle Solution Revisited." Société de Chimie Industrielle—American Section meeting, New York, 13 December 2001.

James A. Cederna. "Challenges of Global Chemicals Growth." Commercial Development and Marketing Association fall meeting, Marco Island, Florida, 22–25 October 2000.

Joseph Chang. "Innovation Drives Competition in Specialty Chemicals Industry." *Chemical Market Reporter* 260 (5 Nov. 2001), 1, 19.

———. "Worst of Times Bring out Best in Spec Chems." *Chemical Market Reporter* 260 (19 Nov. 2001), 1, 23.

Chemical Week 164 (26 June 2002), 23.

Ciba Specialty Chemicals. "Solid Performance in a Challenging Environment." Salomon Smith Barney Twelfth Annual Chemical Conference, New York, 4 December 2001.

Allan Cohen; Tracy Marshbanks. "Specialty Chemicals: Outlook and Strategies for Shareholder Value." Report given at First Analysis Securities Corporation, Chicago, 16 February 2001.

Esther D'Amico. "Image Adjustment—Jumping on the Brand Wagon." *Chemical Week* 163 (14 Feb. 2001), 23–27.

Hans von Doesberg. "Embracing the Service Economy." *Chemical Specialties* 1 (Jan. 1999), 39–40.

Davis Dyer; David B. Sicilia. *Labors of a Modern Hercules.* Boston: Harvard Business School Press, 1991.

Tim Gerdeman. "The Search for Value in Global Chemicals." Salomon Smith Barney Equity Research report, July 1999.

Mark Gulley; Kevin McCarthy. "Industry Structure in Specialty Chemicals." Bank of America Securities Equity Research report, 10 January 2001.

Sheldon Hochheiser. *Rohm and Haas: History of a Chemical Company.* Philadelphia: University of Pennsylvania Press, 1986.

David Hunter et al. "Leveraging Innovation: Specialties Firms Reembrace R&D." *Chemical Week* 163 (19/26 Dec. 2001), 19–22.

Sean Milmo. "Rising Critical Mass among European Specialty Chemical Producers." *Chemical Market Reporter* 260 (11 Dec. 2000), Fr3–6.

Rick Mullin. "Hoping for a Breakthrough." *Chemical Specialties* 1:4 (Nov/Dec 1999), 20–25.

Michael E. Porter. *Competitive Strategy: Techniques for Analyzing Industries and Competitors.* New York: Free Press, 1980.

Bill Schmitt; Claudia Hume; Kerri Walsh. "Specialty Chemicals Becoming a Commodity Market?" *Chemical Week* 163 (30 May/6 June 2001), 27.

Phil Schoepke; Jim Welch. "Five Routes to Deep Collaboration in the Chemical Industry." *Chemical Innovation* 31 (Oct. 2001), 41.

Adrian J. Slywotzky et al. *Profit Patterns: 30 Ways to Anticipate and Profit from Strategic Forces Reshaping Your Business.* New York: Random House, 1999.

"Specialty Chemicals Outlook 1999–2000: On the Edge of a New Era." BT Alex Brown Equity Research report, 25 January 1999.

"Specialty Chemicals: Who Holds the Optimal Product Mix?" (special report). *Chemical Market Reporter* 259 (12 June 2000), 1–32.

Kerri Walsh. "Mill Closures, Price Cuts Put a Tear in Profits." *Chemical Week* 163 (14 Feb. 2001), 29–30.

Claire Wilson. "The Quest for Nominal Distinction." *Chemical Specialties* (May/June 2000), 43–47.

Chapter 4

Peter W. Beck. "Corporate Planning for an Uncertain Future." *Long Range Planning* 15:4 (2001), 12–21.

Florian Budde et al., editors. *Value Creation: Strategies for the Chemical Industry.* Weinheim, Germany: Wiley-VCH, 2001.

Robert D. Buzzell. "Is Vertical Integration Profitable?" *Harvard Business Review* 61 (Jan.-Feb. 1983), 92–101.

Thomas G. Cody. *Management Consulting: A Game without Chips.* Fitzwilliam, N.H.: Kennedy & Kennedy, 1986.

Cheryl Currid and Company. *The Reengineering Toolkit: 15 Tools and Technologies for Reengineering Your Organization.* Rocklin, Calif.: Prima Publishing, 1994.

Peter F. Drucker. *Managing in Turbulent Times.* New York: Harper & Row, 1980.

George Eckes. *Making Six Sigma Last: Managing the Balance between Cultural and Technical Change.* New York: John Wiley & Sons, 2001.

Richard Foster; Sarah Kaplan. *Creative Destruction: Why Companies That Are Built to Last Underperform the Market—and How to Successfully Transform Them.* New York: Doubleday, 2001.

Pankaj Ghemawat. "Building Strategy on the Experience Curve." *Harvard Business Review* 63 (Mar.-Apr. 1985), 143–149.

———. "Competition and Business History in Perspective." *Business History Review* 76 (Spring 2002), 37–74.

————. "Sustainable Advantage." *Harvard Business Review* 64 (Sept.-Oct. 1986), 53–57.

Brian J. Hall; Jeffrey B. Liebman. "Are CEOs Really Paid Like Bureaucrats?" *Quarterly Journal of Economics* 113:3 (1998), 653–691.

Michael Hammer. *Beyond Reengineering: How the Process-Centered Organization Is Changing Our Work and Our Lives.* New York: Harper Collins, 1997.

Michael Hammer; James A. Champy. *Reengineering the Corporation.* New York: HarperBusiness, 1993.

Kathryn Rudie Harrigan; Michael E. Porter. "End-Game Strategies in Declining Industries." *Harvard Business Review* 61 (July-Aug. 1983), 111–120.

Bruce Henderson. *Henderson on Corporate Strategy.* Cambridge, Mass.: Abt Books, 1979.

V. Daniel Hunt. *Process Mapping: How to Reengineer Your Business Processes.* New York: John Wiley & Sons, 1996.

D. E. Hussey. "Portfolio Analysis: Practical Experience with the Directional Policy Matrix." *Long Range Planning* 11 (Aug. 1978), 2–8.

Simon London. "The Growing Pains of Business." *Financial Times*, 5 May 2003, p. 5.

Michael E. Porter. *Competitive Advantage: Creating and Sustaining Superior Peformance.* New York: Free Press, 1985.

————. *Competitive Strategy: Techniques for Analyzing Industries and Competitors.* New York: Free Press, 1980.

————. "What Is Strategy?" *Harvard Business Review* 74 (Nov.-Dec. 1996), 61–78.

C. K. Prahalad; Gary Hamel. "The Core Competence of the Corporation." *Harvard Business Review* 68 (May-June 1990), 79–91.

Mark R. Pratt. "Experience Is a Tough Taskmaster." *Chemical Engineering Progress* 7 (July 1988), 50–55.

Joseph W. Raksis. "Growth through Focus in Specialty Chemicals." Chemical Management and Resources Association conference, Newport Beach, California, 13–16 September 1992.

Alfred Rappaport. *Creating Shareholder Value: The New Standard for Business Performance.* New York: Free Press, 1986.

————. "Linking Competitive Strategy and Shareholder Value Analysis." *Journal of Business Strategy* (1986), 58–67.

Darrell K. Rigby. *Management Tools and Techniques: An Executive's Guide.* Boston: Bain & Company, 1997.

S. J. Q. Robinson; R. E. Hitchens; D. P. Wade. "The Directional Policy Matrix: Tool for Strategic Planning." *Long Range Planning* 11 (June 1978), 8–15.

Balaji B. Singh. "Strategic Options for Commodity Chemical Producers in Transition to Specialty Markets." *Chemical Marketing and Management* (Winter 1987), 72–74.

Benson P. Shapiro; Thomas V. Bonoma. "How to Segment Industrial Markets." *Harvard Business Review* 62 (May-June 1984), 104–110.

Stuart S. P. Slatter. "Common Pitfalls in Using the BCG Product Portfolio Matrix." *London Business School Journal* (Winter 1980), 18–22.

Adrian J. Slywotzky; Richard Wise; Karl Weber. *How to Grow When Markets Don't.* New York: Warner Books, 2003.

Peter H. Spitz. *Petrochemicals: The Rise of an Industry.* New York: John Wiley & Sons, 1988.

J. Fred Weston; Eugene F. Brigham. *Managerial Finance.* Sixth edition. Hinsdale, Ill.: Dryden Press, 1978.

Chapter 5

Ira Bleskin. "Reengineering Critical to Turnaround at Geon." *Chemical Week* 153 (8 June 1994), 48.

Bruce Brocka; M. Suzanne Brocka. *Quality Management: Implementing the Best Ideas of the Masters.* Homewood, Ill.: Business One Irwin, 1992.

Robert C. Camp. "Benchmarking: The Search for Best Practices That Lead to Superior Peformance." *Quality Progress* (Jan. 1989), 61–68.

Lee Clifford. "Why You Can Safely Ignore Six Sigma: The Management Fad Gets Raves from Jack Welch, but It Hasn't Boosted the Stocks of Other Devotees." *Fortune* 143 (22 Jan. 2001), 140.

Bill Creech. *The Five Pillars of TQM: How to Make Total Quality Management Work for You.* New York: Truman Talley Books, 1994.

Thomas Davenport. "Business Process Engineering: Its Past, Present and Possible Future." Harvard Business School case no. 9-196-082, November 1995.

W. Edwards Deming. *Out of the Crisis.* Cambridge, Mass.: MIT Press, 2000.

J. Robb Dixon et al. "Business Process Reengineering: Improving in New Strategic Directions." *California Management Review* 36:4 (Summer 1994), 93–108.

———. *The New Performance Challenge: Measuring Operations for World Class Competition.* Homewood, Ill.: Business One Irwin, 1990.

"The Dow Chemical Company Receives CIO-100 Award for Excellence in Customer Service." *Business Wire* (15 Aug. 2000), 2192.

Ernest K. Drew. "Managing Assets in a Changing Environment." Chemical Marketing Research Association conference, New York, 5–7 May 1986.

Armand V. Feigenbaum. *Total Quality Control.* New York: McGraw-Hill, 1961.

Francis J. Gouillart; James N. Kelly. *Transforming the Organization.* New York: McGraw-Hill, 1995.

Gene Hall; Jim Rosenthal; Judy Wade. "How to Make Reengineering Really Work." *Harvard Business Review* 71 (Nov.-Dec. 1993), 119–131.

Michael Hammer; James A. Champy. *Reengineering the Corporation.* New York: HarperBusiness, 1993.

Michael Hammer; Steven A. Stanton. *The Reengineering Revolution.* New York: HarperBusiness, 1995.

Karen Heller, with Rick Mullin. "ISO 9000: A Framework for Continuous Improvement." *Chemical Week* 153 (22 Sept. 1993), 30–31.

J. M. Juran. *Juran on Planning for Quality.* New York: Free Press, 1988.

Allison Lucas. "Union Carbide Changes Its Mindset." *Chemical Week* 153 (24 Nov. 1993), 39.

Alex Markin. "How to Implement Competitive Cost Benchmarking." *Journal of Business Strategy* (May-June 1992), 14–20.

Rick Mullin. "CEOs on the Long March to Quality: Extracting Values along the Way." *Chemical Week* 151 (30 Sept. 1992), 58, 60.

———. "Manufacturers Determine the Scope of Change." *Chemical Week* 153 (8 June 1994), 25.

———. "Quality: Dictates from the Top, Suggestions from the Floor." *Chemical Week* 155 (21 Sept. 1994), 33–36.

———. "Quality: Evolving through the Backlash." *Chemical Week* 151 (30 Sept. 1992), 41.

Rick Mullin, with Michael Roberts. "A Focus on Customer Satisfaction." *Chemical Week* 151 (30 Sept. 1992), 41, 44.

Richard L. Nolan. "Reengineering: Competitive Advantage and Strategic Jeopardy." Harvard Business School case no. 9-196-019, 8 February 1995.

———. "Role of Management Consulting in Reengineering." Harvard Business School case no. 9-195-200, 8 February 1995.

Emily Plishner. "Third Quarter 1991: Exports Buoy Volume but Margins Shrink On." *Chemical Week* 149 (4 Dec. 1991), 32–38.

Clyde V. Prestowitz, Jr. *Trading Places: How We Allowed Japan to Take the Lead.* New York: Basic Books, 1988.

Lawrence S. Pryor. "Benchmarking: A Self-Improvement Strategy." *Journal of Business Strategy* (Nov.-Dec. 1989), 28–32.

———. "Benchmarking in the Specialty Chemicals Industry." Chemical Management and Resources Association conference, Cincinnati, January 1993.

"Quality: Back to Basics?" *Chemical Week* 151 (30 Sept. 1992), 41–62.

"Reengineering." Series of articles. *Chemical Week* 152 (24 Nov. 1993), 28–48.

"Reengineering: Manufacturers Determine the Scope of Change." *Chemical Week* 153 (8 June 1994), 24–49.

Bill Schmitt. "Moving Ahead with Six Sigma." *Chemical Week* 162 (26 April 2000), 64.

Peter H. Spitz. *Petrochemicals: The Rise of an Industry.* New York: John Wiley & Sons, 1988.

U.S. Chemical Industry Statistical Handbook. Arlington, Va.: Chemical Manufacturers Association, 1995.

"Weighing the Benefits of Quality Initiatives." *Chemical Week* 149 (25 Sept. 1991), 34–56.

Chapter 6

Doug Bartholomew. "Vinyl Victory—How One Company Found That Big Problems Require Big Solutions." *InformationWeek* (4 May 1992), 32.

E. N. Brandt. *Growth Company: Dow Chemical's First Century.* East Lansing: Michigan State University Press, 1997.

Bruce Caldwell. "Y2K under Control—By Spending Tens of Billions of Dollars, IT Organizations Think They Have the Year 2000 Issue in Hand." *InformationWeek* (10 May 1999), 52.

Cynthia Challener. "Extensions in e-Business, SCM and CRM Are Key for ERP Vendors." *Chemical Market Reporter* 258 (25 Sept. 2000), 16.

———. "Leveraging ERP Investments in the 21st Century." *Chemical Market Reporter* 257 (24 Jan. 2000), 24.

"The Challenge of e-Commerce (in the Chemical Industry)." *Specialty Chemicals* 20 (1 Oct. 2000), 319.

Susan Connor. "Data Watch—Fatal Flaws (4344)." *Data Quest* (FT Asia Intelligence Wire) (31 Jan. 1999), 1.

Kazim Isfahani. "Evaluating Y2K Expenditures." Giga Information Group—Ideabyte. www.rightnow.com/news/giga.html, 7 February 2000.

Monua Janah. "A Search for the Right Formula." *InformationWeek* (9 Sept. 1996), 92.

Cindy Jutras. *ERP Optimization: Using your Existing System to Support Profitable e-Business Initiatives.* Boca Raton, Fla.: Saint Lucie Press, 2002.

Deborah Mendez-Wilson. "Voice Portal Din Ups Intensity." *Wireless Week* (13 Nov. 2000), 30.

Roger K. Mowen, Jr. Keynote address. e-Commerce in Chemicals Conference, Amsterdam, 21 February 2000.

Tom Mulligan. "e-Commerce Comes to Chemicals." *Specialty Chemicals* 20 (1 June 2000), 190.

Rick Mullin. "IT Centers Launched by Consultants DuPont, Monsanto Are Springboards." *Chemical Week* 160 (19 Aug. 1998), s38.

———. "IT Integration: Programmed for Global Operation." *Chemical Week* 159 (12 Feb. 1997), 21.

———. "Software's New Promise: The Global Network—Decision Makers Key-in on Application Programs." *Chemical Week* 152 (19 May 1993), 24–27.

Teresa Ortega; Patricia Van Arnum. "Untangling a Web of Promise." *Chemical Market Reporter* 254 (14 Sept. 1998), Fr16.

Kevin Parker. "Batch Is Unmatched in Automation Gains." *Manufacturing Systems* 10 (Oct. 1992), 30.

Robert Preston. "Andersen Delivers for Dow." *InternetWeek* 822 (24 July 2000), 55.

Alan Radding. "Third Parties Plug in to R/3—Customer Demand Spurs a Growing Market for Add-on Applications and Tools." *InformationWeek* (9 Sept. 1996), AD01.

Christopher Reilly. "Providers Expand Services as Technology Evolves." *Purchasing* 129 (2 Nov. 2000), 48C30.

Mike Ricciuti. "Connect Manufacturing to the Enterprise (Manufacturing Software Vendors Offer Enterprise Resource Planning Packages)." *Datamation* 38 (15 Jan. 1992), 42.

Don Richards. "Chemical Industry Unvexed by Y2K Computer Problems." *Chemical Market Reporter* 257 (17 Jan. 2000), 5.

Michael Roberts. "Chemical Industry e-Commerce Has Landed: Hype Becomes Reality." *Chemical Week* (Focus 2000: Special supplement: Internet) 162 (26 July 2000), s5.

David Rotman. "New Wave of Software Entices the Industry." *Chemical Week* 148 (27 Mar. 1991), 37.

Dan Scheraga. "SAP Leads the Way as Chemical Companies Focus on ERP." *Chemical Market Reporter* 255 (25 Jan. 1999), 10–11.

Suzzane Shelley. "Launching a Battery of Dot.com Offerings." *Chemical Specialties* 2 (1 Sept. 2000), 49.

Kara Sissell. "Competing Software Muscles in on SAP: New Uses Tackle Internet Concerns." *Chemical Week* 158 (21 Aug. 1996), 26.

Howard Solomon. "SAP AG Has Leg Up on Competition: ARC." *Computing Canada* 25 (26 Nov. 1999), 4.

Jake Sorofman. "European Chemical News: m-Commerce Expected to Boom." *Chemical Business Newsbase* (online service), 4 July 2000, 14.

Martin Stone. "ERP Davids Nipping at Goliath SAP's Heels: Study." *Computing Canada* 24 (2 Nov. 1998), 23.

Thomas M. Stout; Theodore J. Williams. "Pioneering Work in the Field of Computer Process Control." *IEEE Annals on the History of Computing* 17:1 (Spring 1995), 6–18.

Hilfra Tandy. "Chemicals Industry—On-line Sales May Slash Prices and Profits." *Financial Times*, 3 July 2000, p. 4.

U.S. Department of Commerce, Economics and Statistics Administration, Office of the Chief Economist. *The Economics of Y2K and the Impact on the United States*, by William B. Brown and Laurence S. Campbell, November 1999.

Alessandro Vitelli. "What's Up with Energy e-Commerce?" *Global Energy Business* 2 (1 Aug. 2000), 45.

Chapter 7

"3M May Be the First to Get under the Bubble." *Chemical Week* 127 (3 Sept. 1980), 47.

"Air Is Charged over Air Rules Change." *Chemical Week* 127 (2 July 1980), 38.

J. Gordon Arbucke et al. *Environmental Law Handbook.* Rockville, Md.: Government Institutes, 1983.

Ashish Arora; Ralph Landau; Nathan Rosenberg. *Chemicals and Long-Term Economic Growth.* New York: John Wiley & Sons; Philadelphia: Chemical Heritage Foundation, 1998.

Arnold L. Aspelin. "Economic Aspects of Current Pesticide Regulatory Programs." Chemical Marketing Research Association conference, Houston, February 1983.

Joseph L. Badaracco; George C. Lodge. "Allied Chemical Corporation." Harvard Business School case no. 379-137, 1979.

"A Blow to OSHA's Benzene Rules." *Chemical Week* 127 (9 July 1980), 11–12.

Michael Brower; Warren Leon. *The Consumer's Guide to Effective Environmental Choices.* New York: Three Rivers Press, 1999.

Jessica Brown. "Industry Attacks White Paper." *Chemical Week* 163 (21 Feb. 2001), 9.

Frances Cairncross. *Costing the Earth: The Challenge for Government, the Opportunities for Business.* Boston: Harvard Business School Press, 1992.

Chemalliance Web site, operated by Pacific Northwest National Laboratory. www.chemalliance.org, accessed 2001.

Emma Chynoweth; Michael Roberts. "Europe Progresses Despite Financial Hardships." *Chemical Week* 153 (8 Dec. 1993), 62–65.

Craig E. Colten. "Creating a Toxic Landscape: Chemical Waste Disposal Policy and Practice (1900–1960)." *Environmental History Review* 18 (Spring 1994), 85–116.

Alex Crawford. "What's the Best Approach?" *Chemistry and Industry* no. 18 (18 Sept. 2000), 597.

Hugh D. Crone. *Chemicals and Society.* Cambridge: Cambridge University Press, 1987.

J. Clarence Davies; Barbara S. Davies. *The Politics of Pollution.* Second edition. Indianapolis, Ind.: Pegasus Press, 1970.

J. Clarence Davies; Jan Mazurek. *Pollution Control in the United States.* Washington, D.C.: Resources for the Future, 1998.

Lee Niedringhaus Davis. *Corporate Alchemists: Profit Takers and Problem Makers in the Chemical Industry.* New York: William Morrow, 1984.

"Dow Annual Report." Midland, Mich.: Dow Chemical Company, 2000.

Dow Chemical Public Report 1999. Midland, Mich.: Dow Chemical Company, 1999.

"DuPont Annual Report." Wilmington, Del.: DuPont Company, 2000.

Gail Dutton. "Green Partnerships." *Management Review* 85 (Jan. 1996), 24–25.

Lois R. Ember. "Chemical Makers Pin Hopes on Responsible Care to Improve Image." *Chemical and Engineering News* 70 (5 Oct. 1992), 13–39.

Environmental Protection Agency, Office of Air Quality Planning and Standards. "Cleaning up Air Pollution: The Programs in the 1990 Clean Air Act," 23 March 2001.

"EPA in Disarray: Can It Do the Job Industry Expects?" *Chemical Week* 129 (21 Oct. 1981), 82–85.

"Facing Down the Critics." *Chemical Week* 106 (15 Apr. 1970), 55–56.

Peter Fairley. "TRI: Growing Pains, Expansions Rile Industry." *Chemical Week* 158 (12 June 1996), 18–20.

Daniel J. Fiorino. *Making Environmental Policy.* Berkeley/Los Angeles: University of California Press, 1995.

"For Disposal Firms: Big Revenues, Big Headaches." *Chemical Week* 127 (26 Nov. 1980), 31.

G. F. Fort. "Report to CMA Board of Directors." Chemical Manufacturers Association State Affairs Committee, Arlington, Virginia, 8 September 1985.

Richard C. Fortuna; David Lennett. *Hazardous Waste Regulation: The New Era.* New York: McGraw-Hill, 1987.

Franklin Associates. "An Energy Study of Plastics and Their Alternatives in Packaging and Disposable Consumer Goods." Prairie Village, Kansas, November 1992.

Michael Fumento. *Science under Siege.* New York: William Morrow, 1993.

Joe W. Grisham, editor. *Health Aspects of the Disposal of Waste Chemicals.* New York: Pergamon Press, 1986.

Guide to the Business of Chemistry 2001. Arlington, Va.: American Chemistry Council, 2001.

David J. Hanson. "Cooperation Key to EPA's Disaster Plan." *Chemical and Engineering News* 64 (6 Jan. 1986), 20–22.

John A. Hird. *Superfund: The Political Economy of Environmental Risk.* Baltimore: Johns Hopkins University Press, 1994.

Sheldon Hochheiser. *Rohm and Haas: History of a Chemical Company.* Philadelphia: University of Pennsylvania Press, 1986.

Andrew J. Hoffman. *From Heresy to Dogma: An Institutional History of Corporate Environmentalism.* San Francisco: New Lexington Press, 1997.

Cheryl Hogue. "Chemical Producers' TRI Ranking Falls." *Chemical and Engineering News* 78 (29 May 2000), 46–47.

David Hunter. "New Global Program on Chemicals Management." *Chemical Week* 163 (22 Aug. 2001), 3.

———. "Trade Secrets" [Editorial]. *Chemical Week* 163 (4 April 2001), 5.

"It Costs More to Run a Clean Chemical Plant." *Chemical Week* 109 (8 Sept. 1971), 31.

Jeff Johnson. "Chemical Accident Debate Rolls On." *Chemical and Engineering News* 79 (9 April 2001), 22.

Kirk Johnson. "Gipper Meets 'Survivor' as G.E.'s Image Hardens." *New York Times*, 4 March 2001, Metro Section.

Andrew A. King; Michael J. Lenox. "Industry Self-Regulation without Sanctions: The Chemical Industry's Responsible Care Program." *Academy of Management Journal* 4 (Aug. 2000), 698–716.

Derek Knight. "Communicating Hazards in Europe." *Chemistry and Industry* no. 11 (4 June 2001), 337–340.

Patricia L. Layman. "Rhine Spills Force Rethinking of Potential for Chemical Pollution." *Chemical and Engineering News* 65 (23 Feb. 1987), 7–10.

"Learning to Cope with RCRA." *Chemical Week* 127 (12 Nov. 1980), 64–71.

"Learning to Live with Ecology." *Chemical Week* 108 (16 June 1971), 11–12.

Jay H. Lehr. *Rational Readings on Environmental Concerns.* New York: Van Nostrand Reinhold, 1992.

Will Lepowski. "Bhopal: Indian City Begins to Heal but Conflicts Remain." *Chemical and Engineering News* 63 (2 Dec. 1985), 18–21.

H. W. Lewis. *Technological Risk.* New York: W. W. Norton, 1990.

Bjorn Lomborg. *The Skeptical Environmentalist: Measuring the Real State of the World.* Cambridge: Cambridge University Press, 2001.

Robert T. Martinott. "How to Limit the Rising Costs of Stricter Regulation." *Chemical Week* 128 (21 Jan. 1981), 36–40.

H. Eugene McBrayer. Oral history interview by James J. Bohning, Mercer Island, Washington, 11 May 1995. Transcript with subsequent corrections and additions, Chemical Heritage Foundation, Philadelphia.

Ralph Nader; William Taylor. *The Big Boys: Power and Position in American Business.* New York: Pantheon, 1986.

"New Global Program on Chemicals Management." *Chemical Week* 163 (22 Aug. 2001), 3.

"A New Relationship? Washington Says So, Industry Is Wary." *Chemical Week* 127 (29 Oct. 1980), 70–74.

New York State Department of Environmental Conservation, Division of Hazardous Waste Remediation. "1989 Love Canal Annual Report," February 1989, and "1990 Love Canal Annual Report," May 1991.

"On Tap: More Spill Regulators." *Chemical Week* 107 (23 Sept. 1970), 42.

Mark R. Powell. *Science at the EPA: Information in the Regulatory Process.* Washington, D.C.: Resources for the Future, 1999.

Jeffrey F. Rappaport; George C. Lodge. "Responsible Care." Harvard Business School case no. 9-391-135, 18 March 1991.

J. S. Robinson. *Hazardous Chemical Spill Cleanup.* Park Ridge, N.J.: Noyes Data Corporation, 1979.

Walter A. Rosenbaum. *Environmental Politics and Policy.* Fourth edition. Washington, D.C.: CQ Press, 1998.

Bill Schmitt. "Public Disclosure: Warts and All." *Chemical Week* 162 (5/12 July 2000), 43.

———. "Responsible Care: Responding to Public Opinion Shifts." *Chemical Week* 162 (22/29 Nov. 2000), 24–30.

Irwin Schwartz. "Challenge of the '80s: Zero Pollution." *Chemical Week* 126 (5 Mar. 1980), 26–32.

———. "Environmental Control." *Chemical Week* 106 (17 June 1970), 82.

S. Fred Singer. *My Adventures in the Ozone Layer: Rational Readings on Environmental Concerns.* Edited by J. H. Lehr. New York: Van Nostrand Reinhold, 1992.

Kara Sissell. "Asbestos: The Push for Litigation Reform." *Chemical Week* 165 (5 Mar. 2003), 16–20.

Peter H. Spitz. *Petrochemicals: The Rise of an Industry.* New York: John Wiley & Sons, 1988.

"Taking Aim at the Clean Air Act." *Chemical Week* 127 (24 Dec. 1980), 32–35.

Ann M. Thayer. "Pollution Reduction." *Chemical and Engineering News* 70 (16 Nov. 1992), 22–53.

Trade Secrets: A Moyers Report. Produced by Public Affairs Television, by Sherry Jones and Bill Moyers, 1 hour and 57 min., Public Broadcasting Service, 26 March 2001.

Cathy Trost. *Elements of Risk: The Chemical Industry and Its Threat to America.* New York: Times Books, 1984.

U.S. Chemical Industry Statistical Handbook. Arlington, Va.: Chemical Manufacturers Association, 1995 and 1998.

U.S. House Committee on Interstate and Foreign Commerce, Subcommittee on Oversight and Investigations. *Waste Disposal Site Survey*, 96th Cong., 1st sess, Oct. 1979, Committee Print 96-IIC33.

"Washington Facilities, Cash Needed." *Chemical Week* 107 (17 June 1970), 98–100.

Elizabeth M. Whelan. *Toxic Terror: The Truth behind the Cancer Scares*. Amherst, N.Y.: Prometheus Books, 1993.

"Which Way Is Right?" [Editorial]. *Pittsburgh Post-Gazette*, 17 Oct. 1987.

"Will There Ever Be Another New DDT?" *Chemical Week* 108 (2 June 1971), 65–67.

Eric Zuesse. "Love Canal. The Truth Seeps Out." *Reason* (Feb. 1981), 16–33.

Chapter 8

1998 Barometer of Competitiveness. Brussels: Conseil Européen des Fédérations de l'Industrie Chimique (CEFIC), 1998.

Fred Aftalion. *A History of the International Chemical Industry*. Second edition. Philadelphia: Chemical Heritage Press, 2001.

Ashish Arora; Ralph Landau; Nathan Rosenberg. *Chemicals and Long-Term Economic Growth*. New York: John Wiley & Sons, 1998.

"Asia Pacific Petrochemicals." Special issue prepared by Asian Chemical News, Chemical Management and Resources Association, and European Chemical News. Singapore, May 2001.

Bank of America. *Guide to Petrochemicals in Asia*. Hong Kong: EFP International, 1997.

Mark A. Berggren. "Lessons Learned by Post-Crisis Asia." Chemical Market Associates, Inc., Fifteenth Annual World Petrochemical Conference, Houston, 29–30 March 2000.

E. N. Brandt. *Growth Company: Dow Chemical's First Century*. East Lansing: Michigan State University Press, 1997.

Lowell Bryan et al. "Corporate Strategy in a Globalizing World." *McKinsey Quarterly* 3 (1998), 7–19.

Florian Budde et al., editors. *Value Creation: Strategies for the Chemical Industry*. Weinheim, Germany: Wiley-VCH, 2001.

Keith Chapman. *The International Petrochemical Industry*. Oxford: Basil Blackwell, 1991.

"The Chemical Industry in Singapore" [brochure]. Singapore: Singapore Economic Development Board, 1998.

Chemical Week 158 (18/25 Dec. 1996), 21–29.

C. L. Dmytrk. "The Petrochemical Industry in Alberta." European Chemical Marketing Research Association conference, Barcelona, November 1987.

John H. Dunning. *Multinationals, Technology and Competitiveness*. London: Unwick Hyman, 1988.

Facts and Figures: The European Chemical Industry in a Worldwide Perspective. Brussels: Conseil Européen des Fédérations de l'Industrie Chimique (CEFIC), November 1999.

Peter Fairley. "Canada's Big Freeze: Can Arctic Gas Save Alberta?" *Chemical Week* 163 (1 Aug. 2001), 17–26.

———. "Canadian Chemicals: Running on Empty." *Chemical Week* 162 (19 July 2000), 20–29.

Richard D. Freeman. "The Chemical Industry: A Global Perspective." *Business Economics* 34 (Oct. 1999), 16–22.

"Getting Ahead in the Asia-Pacific Market." *Chemical Week* 148 (3 Apr. 1991), 21–38.

Guide to the Business of Chemistry 2002. Arlington, Va.: American Chemistry Council, 2002.

Guide to Petrochemicals in Asia. Hong Kong: EFP International (HK), 1997.

Ryota Hamamoto. "Outlook for Asian Petrochemical Trade." Petrochemical Industry Consulting Conference, Beijing, 24–25 May 2000.

Cheryl Hogue. "Responding to Globalization." *Chemical and Engineering News* 79 (30 Apr. 2001), 11.

"Looking Up Dow South: Chemical Investment Soars." *Chemical Week* 158 (13 Nov. 1996).

Morgan Stanley and Company. "Chemical Industry Equity Investment Research Report," 7 November 1996.

Gregory Morris. "Japan's Rush to Rationalize." *Chemical Week* 157 (29 Nov. 1995), 27–28.

Gregory D. L. Morris; Suzanne McElligott. "Traders Cozy to Suppliers: Producer Ties Eclipse Spot Trading." *Chemical Week* 158 (18/25 Dec. 1996), 25.

Gregory Morris; Robert Westerwelt; Sylvia Pfeifer. "Plastics' Clear Objectives: Regional Players Challenge Lenders." *Chemical Week* 159 (18 June 1997), 27–30.

Roy Neresian. *Ships and Shipping: A Comprehensive Guide.* Tulsa, Okla.: Pennwell, 1981.

Andrew Pettman. "The Middle East: The Waves to Come." *Chemical Management Review* (May-June 2000), 22–29.

Michael E. Porter. *The Competitive Advantage of Nations.* New York: Free Press, 1990.

———. *Competitive Strategy: Techniques for Analyzing Industries and Competitors.* New York: Free Press, 1980.

"Present Situation of Petrochemicals Markets in Asian Countries and Their Position in the World." *Chemical Economy and Engineering Review* 18 (Oct. 1986), 20–38.

"Production Trends in South Korean Petrochemicals Industry." *Chemical Economy and Engineering Review* 18 (Oct. 1986), 24–27.

Lesley C. Ravitz. "Global Competitiveness in the Chemical Industry." Morgan Stanley and Company, U.S. Investment Research, New York, 7 November 1996.

Balaji Singh; Irma Tan. "Global Watch: The Rise of Petrochemicals in China." *Chemical Market Reporter* 262 (15 July 2002), 25–27.

Kara Sissell. "Crafting a New Pemex." *Chemical Week* 163 (20 June 2001), 23–24.

———. "Private Sector Gets a Voice in Government." *Chemical Week* 163 (20 June 2001), 26–29.

———. "Streamlining Brazil's Chemicals Ownership." *Chemical Week* 163 (31 Oct. 2001), 21.

Peter Spitz. Keynote speech. Organization of Petroleum Exporting Countries (OPEC) Secretariat, Vienna, October 1978.

Robert B. Stobaugh. *Innovation and Competition: The Global Management of Petrochemical Products.* Cambridge, Mass.: Harvard Business School Press, 1988.

———. "The Neotechnology Account of International Trade: The Case of Petrochemicals." *Journal of International Business Studies* Fall (1971), 54, 57.

———. *Nine Investments Abroad and Their Impact at Home.* Cambridge, Mass.: Harvard University Press, 1976.

Andrew Swanson. "China Petrochemicals after WTO." Chem Systems Annual Conference, Houston, January 2001.

T. Kevin Swift. "The Business of Chemistry in the U.S. Performance and Outlook." *Chemical Management Review* (Jan.-Feb. 2000), 40–46.

Madoka Tashiro. "Opportunities in Asia." Society of Chemical Industry conference, Budapest, November 1994.

Jean-François Tremblay. "Change Arrives in South Korea." *Chemical and Engineering News* 79 (6 Aug. 2001), 15–19.

———. "Rapid Changes Come to China." *Chemical and Engineering News* 78 (21 Aug. 2000), 25–34.

Sergey Vasnetsov. "North American Chemical Trade: Cyclical and Secular Trends." *Chemical Market Reporter* 262 (17 Feb. 2003), 17–18.

Eduardo Eugenio Gouvea Vieira. "South America's Petrochemicals on the Eve of the 21st Century." National Petroleum Refiners Association conference, San Antonio, 24 March 1997.

Pedro Wongtschowski. *Industria quimica: Riscos e opportunidades* (The chemical industry: Risks and opportunities). São Paulo: Edgard Bluecher, 1998.

J. F. Wyatt. "Changing Trade Flows and Investment Needs." European Chemical Marketing Research Association conference, Barcelona, October 1987.

Ian Young; Natasha Alperowicz. "Mideast Builds Export Power." *Chemical Week* 159 (14 May 1997), 23–28.

Chapter 9

Richard Bernstein. "Quantitative Strategy Review." Merrill Lynch, 30 June 2001.

Richard Brealey. *An Introduction to Risk and Return of Common Stocks.* Cambridge, Mass.: MIT Press, 1987.

John M. Dalton, Esq. *How the Stock Market Works.* New York: Institute of Finance, 1988.

"Executive Compensation Review." *Barron's* (25 June 2001).

Norman G. Fosback. *Stock Market Logic.* Fort Lauderdale, Fla.: Institute for Economic Research, 1976.

Benjamin Graham. *The Intelligent Investor.* New York: Harper and Row, 1973.

Sumner N. Levine. *The Financial Analyst's Handbook.* Homewood, Ill.: Dow Jones-Irwin, 1988.

Burton G. Malkiel. *A Random Walk down Wall Street.* New York: W. W. Norton, 1990.

Karol Nielson. "The Last of the Big Deals? M&A Values Begin to Slide." *Chemical Week* 162 (13 Sept. 2000), 31.

John Roberts. "And Then There Was One? Consolidation to Remain Major Industry Theme." Merrill Lynch Equity Research report, 6 March 2001.

———. "The Prisoner's Dilemma." Merrill Lynch, 23 September 1998.

———. "Quarterly Segment Review." Merrill Lynch, 23 August 2001.

———. "To Be or Not to Be-Integrated." Merrill Lynch, 13 October 1997.

———. "What a CROC!" Merrill Lynch, 1 May 1998.

———. "When Margins May Not Matter." Merrill Lynch, 28 January 1997.

Howard M. Schilt. *Financial Shenanigans.* New York: McGraw-Hill, 1993.

William F. Sharpe. *Investments.* Englewood Cliffs, N.J.: Prentice Hall, 1985.

Jeremy J. Siegel. *Stocks for the Long Run.* Burr Ridge, Ill.: Irwin Press, 1994.

Wall Street Journal, 30 June 2001, p. 18.

John J. Wild; Leopold A. Bernstein; K. R. Subramanyam. *Financial Statement Analysis.* Seventh edition. Boston: McGraw-Hill, 2001.

Chapter 10

Gary K. Adams. Chemical Market Associates, Inc., presentation, Salomon Smith Barney Chemical Conference, New York, 5 December 2001.

———. "Outlook for Chemicals and Plastics 2001: A Profit Odyssey." *Chemical Management Review* (May-June 2000), 10–20.

Fred Aftalion. *A History of the International Chemical Industry.* Second edition. Philadelphia: Chemical Heritage Press, 2001.

"Billion Dollar Club." *Chemical Week* 164 (4 Dec. 2002), 25–36.

Florian F. Budde et al., editors. *Value Creation: Strategies for the Chemical Industry.* Wein-heim, Germany: Wiley-VCH, 2001.

Thomas Burke. "A Sustainable Future for the Chemical Industry." American Chemistry Council Leadership Conference, Houston, 29 October 2001.

Cynthia Challener. "Sustainable Development at a Crossroads." *Chemical Market Reporter* 260 (16 July 2001), Fr3–Fr4.

"China: On the Brink of Change." *Chemical Week* 164 (28 Aug./4 Sept. 2002), 23–28.

Graham L. Copley; Hassan I. Ahmed. "U.S. Chemicals: The End of the 'Good Old Days.' " Bernstein Equity Research Analysis presented at the International Petrochemical Conference of the National Petroleum Refiners Association, San Antonio, 25 March 2002.

Nigel Davis. "The Chemical Industry in 2010—Learning from the Past?" Scenario Project Chemical Industry 2010, European Chemical Marketing and Strategy Association (London)–Commercial Development and Marketing Association (Philadelphia), December 2002.

Pankaj Ghemawat; Fabiborz Ghadar. "The Dubious Logic of Global Megamergers." *Harvard Business Review* 78 (July-Aug. 2000), 64–75.

"Global Top 50." *Chemical and Engineering News* 79 (23 July 2001), 24.

Guide to the Business of Chemistry. Arlington, Va.: American Chemistry Council, Policy, Economics and Risk Analysis, 2001.

Mark R. Gulley. "Innovation Drives Competition." American Chemistry Council Leadership Conference, Houston, 29 October 2001.

Chad O. Halliday. "The Challenge of Sustainable Growth." *Chemical Business* (May 2001), 19–23.

Glenn Hess. "Chemical Group Launches Web Site on HPV Communications Challenge." *Chemical Market Reporter* 259 (11 June 2001), 4.

David Hurwitz; Gordon Nechvatal. "Industry Outlook Report: The Chemical Industry in 2010." *Prism* (First Quarter 1999), 83–96.

Jeff Johnson. "Getting a Grip on Wasted Energy." *Chemical and Engineering News* 78 (24 July 2000), 31–33.

Jim Langabeer; Amol Joshi; with Demantra Inc. "Margin and Performance Improvement: Using Supply Chain Excellence to Improve Competitiveness." *Chemical Market Reporter* 260 (3 Dec. 2001), 29–31.

Erik Leon; Peter H. Spitz. "e-Business: Taking the Chemical Industry into the 21st Century." *Chemical Week* (special supplement) (24 Mar. 1999).

James Mack. "Specialty Chemicals: Competitiveness in Global Markets." American Chemistry Council Leadership Conference, Houston, 29 October 2001.

Peter Martin. "New Logic for Chipmakers." *Financial Times,* 6 Dec. 2001, p. 14.

"Movers and Shakers." *Chemical and Engineering News* 79 (7 May 2001), 28.

"Nanomaterials: A Big Market Potential." *Chemical Week* 164 (16 Oct. 2002), 17–21.

Thomas M. Parris. "Transition toward Sustainability." *Environment* 45 (Jan./Feb. 2003), 15–20.

Michael E. Porter. "Strategy and the Internet." *Harvard Business Review* 79 (Mar. 2001), 63–78.

Tom Reilly. "Sustainable Development." *Chemistry and Industry* no. 21 (4 Nov. 2002), 16–17.

Marc S. Reisch. "Patience and Persistence: Dow's Stavropoulos, Palladium Medal Winner, Recounts His Career and Goals for the Future." *Chemical and Engineering News* 79 (14 May 2001), 22–25.

Stephen K. Ritter. "Green Chemistry." *Chemical and Engineering News* 79 (16 July 2001), 27–34.

John Roberts. "And Then There Was One? Consolidation to Remain Major Industry Theme." Merrill Lynch Equity Research report, 6 March 2001.

Michael Roberts. "Using SKU Remediation to Eliminate Low-Value Products." *Chemical Week* (special supplement) (24 Mar. 1999), 59.

Pamela Sauer. "Chemical Industry Management under the Magnifying Glass." *Chemical Market Reporter* 259 (3 June 2000), Fr3–11.

Bill Schmitt. "Chemical Century: Formulating for the Future." *Chemical Week* 111 (22/29 Dec. 1999), 24–27.

A. R. Shapiro; John Sisto. "The Case for High Value Growth: Strategies for Value Creation." *Chemical Market Reporter* 261 (14 Jan. 2002), 13–15.

Peter H. Spitz, with *Chemical Week* staff writers. "Petrochemicals: From Dyestuffs to Hydrocarbons and NPRA." *Chemical Week* (special supplement, "NPRA: A Century of Achievement and Excellence") 164 (18 Sept. 2002), 62–73.

Peter H. Spitz; Andrew B. Swanson. "Evolving Strategies in the Petrochemical Landscape." World Chemical Congress, Newport Beach, California, 14-17 September 1997.

T. Kevin Swift. "Where Is the Chemical Industry Going?" *Chemistry Business* (Dec. 1999–Jan. 2000), 8–11.

Hilfra Tandy. "Mergers and Acquisitions: The Survival of the Fattest." *Chemical Matters* no. 379, 1–8.

———. "Overview and World Petrochemical Analysis." *Chemical Matters* [London issue], no. 379, 3.

Don Tapscott. "Rethinking Strategy in a Networked World (or Why Michael Porter Is Wrong about the Internet)." *Strategy and Business* (Third Quarter 2001), 34–41.

Technology Vision 2020: The U.S. Chemical Industry. Washington, D.C.: American Chemical Society, Department of Government Relations and Science Policy, 1996.

Ann M. Thayer. "Chemical e-Business: Are We There Yet?" *Chemical and Engineering News* 81 (10 Feb. 2003), 13–17.

Jean-François Tremblay. "Asia-Pacific: Despite Areas of Growth, the Fragility of the Japanese Economy Threatens the Entire Region." *Chemical and Engineering News* 79 (17 Dec. 2001), 37–40.

John Wyatt. "Olefins—Global Ethylene Capacity Development." Tecnon-PTM Petrochemical Workshop, Vergiate, Italy, 7 June 2002.

Contributors

Peter H. Spitz, Editor
Peter Spitz was founder and managing director at Chem Systems, a highly respected global chemical consulting firm, before it was acquired by IBM in 1998. At Chem Systems he concentrated on corporate and divisional strategy engagements and also founded the firm's financial practice. Before Chem Systems he held management positions at Scientific Design Company, a petrochemical technology development and engineering firm, and earlier at Esso Research and Engineering. He is recognized as an authority in his field and has addressed a number of groups with presentations on trends in the chemical industry. His earlier book, *Petrochemicals: The Rise of an Industry* was published by John Wiley & Sons in 1988.

Andrew Boccone
Andrew Boccone joined Kline & Company, a leading global business research consulting firm, in 1974. He became the firm's president in 1990 and served in that capacity until his retirement in 2001. Before joining Kline & Company, he held sales and business development positions at American Cyanamid. He has addressed numerous symposiums and conferences on business characteristics, trends, and strategies in specialty chemicals. He currently serves on the board of directors of United-Guardian, Inc., and on the advisory board of InMat Corporation.

David A. Crow
Dave Crow is the managing partner for Accenture's Global Chemicals Practice. In this role he leads an international group of partners and has responsibility for a broad portfolio of leading global companies. Prior to this he served as the managing partner for the Americas Chemicals Portfolio and earlier for the Global Process/Energy Practice. His experience spans a broad range of industries, including energy, chemicals, and pharmaceuticals, and he has worked closely with many industry-leading companies. He is a frequent speaker on topics of globalization, forging effective alliances, and innovative approaches to enhance workforce effectiveness, and has contributed to and been quoted in a number of industry and business publications.

Michael Eckstut
Michael Eckstut headed the chemical industry practices at Booz-Allen & Hamilton and was a member of the firm's board of directors. Later, he headed the chemicals industry practice A. T. Kearney. Mr. Eckstut was also a senior executive at ChemConnect, an e-commerce leader in helping companies opti-

mize their purchasing and sales processes for chemical feedstock, chemicals, plastics, and related products. He has a broad range of business, technical, marketing, organizational, and operational experience across all segments of the chemical industry as well as extensive experience in operational business transformation and the role of information technology in driving business success. He has been a frequent contributor to industry forums on such topics as corporate-business strategy, globalization, and management of innovation.

Jeffrey S. Plotkin

Jeffrey Plotkin is the director of Nexant/Chem Systems' Process Evaluation/ Research Planning (PERP) program. Managing this activity involves working closely with technology developers, including operating companies and engineering contractors. His interests are in all phases of process R&D, with special emphasis on gas-to-chemicals processes, selective oxidations, alkane activation, and biocatalyzed routes to chemicals. His prior work experience with ISP and Exxon Chemicals provided an excellent background for his interests in commodity and specialty chemicals from both a technology and a marketing perspective. He holds over thirty U.S. patents and has coauthored twenty-eight peer-reviewed publications. He was a postdoctoral research fellow at Ohio State University.

John Roberts

John Roberts is a senior vice president at the Buckingham Research Group, responsible for the chemical and materials markets. This includes traditional commodity and specialty chemical businesses, as well as such technology materials producers as Monsanto, 3M, and Cabot Microelectronics. Mr. Roberts held a similar position at Merrill Lynch for eleven years, and at Kidder, Peabody for two years before that. Relevant earlier experience includes five years at Booz Allen as a strategy consultant to chemical companies and five years as a chemical process design engineer. He holds a master's degree in management from the MIT Sloan School. He also serves on the editorial board of *Chemical Management Review* magazine and the board of directors for the U.S. Section of the Societé de Chimie Industrielle.

Index

The notation *t* following a page number denotes a table. The notation *f* indicates an illustration (figure).